T0319342

Graph Theoretic Methods in Multiagent Networks

Graph Theoretic Methods in Multiagent Networks

Mehran Mesbahi and Magnus Egerstedt

PRINCETON UNIVERSITY PRESS

PRINCETON AND OXFORD

Published by Princeton University Press,

41 William Street, Princeton, New Jersey 08540

In the United Kingdom: Princeton University Press,

6 Oxford Street, Woodstock, Oxfordshire OX20 1TW

press.princeton.edu

Library of Congress Cataloging-in-Publication Data

Mesbahi, Mehran.
   Graph theoretic methods in multiagent networks / Mehran Mesbahi and Magnus
Egerstedt.
   p. cm.
Includes bibliographical references and index.
ISBN 978-0-691-14061-2 (hardcover : alk. paper) 1. Network analysis (Planning)–
Graphic methods. 2. Multiagent systems–Mathematical models. I. Egerstedt, Mag-
nus. II. Title.
T57.85.M43 2010
006.3–dc22

                    2010012844

British Library Cataloging-in-Publication Data is available

The publisher would like to acknowledge the authors of this volume
for providing the camera-ready copy from which this book was printed

10 9 8 7 6 5 4 3 2 1

To our very own multiagent systems

Maana, Milad, and Kathy (M.M.)
Annika, Olivia, and Danielle (M.E.)

Contents

Preface

"I don't want to achieve immortality
through my work ... I want to achieve
it through not dying." — Woody Allen

The emergence of (relatively) cheap sensing and actuation nodes, capable of
short-range communications and local decision-making, has raised a number of new system-level questions concerning how such systems should be
coordinated and controlled. Arguably, the biggest challenge facing this new
field of research is means by which local interaction rules lead to desired
global properties, that is, given that the networked system is to accomplish
a particular task, how should the interaction and control protocols be struc-
tured to ensure that the task is in fact achieved?

This newly defined area of networked systems theory has attracted wide
interest during the last decade. A number of sessions are devoted to this
problem at the major conferences and targeted conferences have emerged.
Moreover, graduate-level courses are beginning to be taught in this general
area, and major funding institutions are pursuing networked systems as integral to their missions due to the many applications where network-level
questions must be addressed. These applications include sensor networks,
multiagent robotics, and mobile ad hoc communication nets, in addition to
such areas as social networks and quantum networks.

The particular focus of this book is on *graph theoretic methods* for the
analysis and synthesis of networked dynamic systems. By abstracting away
the complex interaction geometries associated with the sensing and communication footprints of the individual agents, and instead identifying agents
with nodes in a graph and encoding the existence of an interaction between
nodes as an edge, a powerful new formalism and set of tools for networked
systems have become available. For instance, the graph theoretic framework
described in this book provides means to examine how the structure of the
underlying interaction topology among the agents leads to distinct global
behavior of the system. This graph theoretic outlook also allows for examining the correspondence between system theoretic features of networked
systems on one hand, and the combinatorial and algebraic attributes of the

underlying network on the other. By doing this, one can, for example, address questions related to the robustness of networked systems in terms of the variation of the network topology, as well as the network synthesis problem in the context of embedded networked systems.

This book builds on the foundation of graph theory and gradually paves the way toward examining graph theoretic constructs in the context of networked systems. This target is laid out in the first part of the book, which focuses on the interplay between the agreement protocol (also known as the *consensus algorithm*) and graph theory. Specifically, in Chapter 3, the correspondence between the network structure and the convergence properties of the agreement protocol is shown for both undirected and directed networks using the spectral properties of the graph. This is followed by establishing an explicit correspondence between the agreement protocol and the general area of Markov chains. The latter chapters in Part I delve into the extension of the basic setup in Chapter 3 to consider the effect of randomness, noise, and nonlinearities on the behavior of the consensus coordination protocols. This is accomplished by introducing the powerful machinery of Lyapunov theory in deterministic (Chapter 4) and stochastic (Chapter 5) settings, which provides the flexibility for analyzing various extensions of the basic agreement protocol.

In Part II, we provide various dynamical, system theoretic, and applied facets of dynamic systems operating over networks. These include formation control (Chapter 6), mobile robot networks (Chapter 7), distributed estimation (Chapter 8), and social networks, epidemics, and games (Chapter 9).

Part III provides an introduction to a perspective of viewing networks as *dynamic systems*. In Chapter 10, we discuss the controllability and observability of agreement protocols equipped with input and output nodes. In particular, this chapter is devoted to the study of how control theoretic properties of the system are dictated by the algebraic and combinatorial structure of the network. This is followed by the problem of synthesizing networks (Chapter 11), with particular attention to the dynamic, graph theoretic, and game theoretic aspects that such an endeavor entails. Another novel ramification of the graph theoretic outlook on multiagent systems is in the context of graph processes, where the network topology itself is given a dynamic role that lends itself to analysis via system theoretic methods (Chapter 12). Higher-order interconnections conclude the book (Chapter 13), demonstrating how the graph theoretic machinery can be extended to simplicial complexes, for example, in order to address sensor-coverage problems.

Pictorially, one can view the chapters in this book as nodes in a directed graph, shown below, whose edges suggest dependencies between the various chapters.

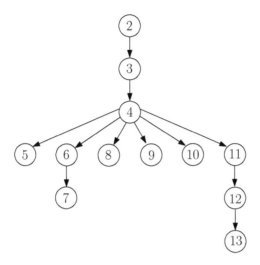

One can also think of these edges as suggesting possible routes through the book. For example, the $2 \rightarrow 3 \rightarrow 4 \rightarrow 6 \rightarrow 7$ branch constitutes a natural graduate level course on multiagent robotics, while the branch $2 \rightarrow 3 \rightarrow 4 \rightarrow 11 \rightarrow 12 \rightarrow 13$ provides a more mathematical treatment of the underlying theme of the book. Ideas for teaching from the book, additional examples and exercises, as well as other comments will be posted on the book website at *https://sites.google.com/site/mesbahiegerstedt*.

The book is suitable for graduate students and researchers in systems, controls, and robotics across various engineering departments, as well as those in applied mathematics and statistics whose work is *network-centric*. Part of this book is also suitable for senior undergraduate students in engineering and computer science programs. As such, we hope that it fills a niche by providing a timely exposition of networked dynamic systems with emphasis on their graph theoretic underpinnings. We enjoyed countless hours discussing and thinking about the topics that have found their way into the book; along the way, we have been humbled by gaining a better glimpse of the research creativity that has been expressed through scholarly works by many researchers in the general area of networks and system theory. Since our goal was expanded at some point to cover a rather broad set of topics related to graph theoretic methods for analysis and synthesis of dynamical systems operating over networks, we had to make a few compromises on the style. As such, we decided to offer proofs for most of the presented results, yet only state the results that we felt played a supportive role in each chapter. In most cases, results that are stated without a proof are discussed in the exercises and can be found in the references discussed at the end of each chapter. Our hope is that researchers and students who are new to this field will find in this book a welcoming and readable account of an active area of research- and for the experts–to stumble across new insights that further complements their research horizons.

Throughout the development of this book, we have been fortunate to be supported by a number of funding agencies, including NSF, ONR, ARO, AFOSR, Boeing, NASA/JPL, and Rockwell Collins. Their support is gratefully acknowledged.

On a final note, this book would not have been possible without help, support, and suggestions from a number of colleagues. In particular, Randy Beard, Richard Murray, and Panagiotis Tsiotras provided feedback on the book that certainly helped make it stronger. Parts of this book are based on results obtained by our current and former students and post-docs, some of whom graciously helped us proofread parts of the book. Special thanks go to Dan Zelazo, Amir Rahmani, Airlie Chapman, Marzieh Nabi-Abdolyousefi, Meng Ji, Musad Haque, Brian Smith, Patrick Martin, Philip Twu, Arindam Das, Yoonsoo Kim, Peter Kingston, Simone Martini, Mauro Franceschelli, and Abubakr Muhammad. We would like to thank Vickie Kearn at Princeton University Press for shepherding this book, from its initial conception to the final product. And finally, we are forever grateful to our parents for cultivating in us a sense of appreciation for what is beautiful yet–at times– enigmatic.

M.M. (Seattle), M.E. (Atlanta)
April 2010

Notation

Graph Theory

$\mathcal{G}$: undirected graph; also referred to as graph

$\mathcal{D}$: directed graph; also referred to as digraph

$\widetilde{\mathcal{D}}$: graph obtained after removing the orientation of the directed edges of $\mathcal{D}$; also referred to as disoriented digraph

$\overline{\mathcal{G}}$: complement of undirected graph $\mathcal{G}$

$\mathcal{G}^o$: oriented version of graph $\mathcal{G}$

V: vertex set; when necessary, also denoted by $V(\mathcal{G})$ or $V(\mathcal{D})$

∂S: boundary of vertex set S (with respect to an underlying graph)

cl S: closure of vertex set S (with respect to an underlying graph)

v_i, $i = 1, \ldots, n$: vertex i; also used for denoting the ith entry of vector v

E: edge set; when necessary, also denoted by $E(\mathcal{G})$ or $E(\mathcal{D})$

$e_{ij} = \{v_i, v_j\}$: edge in a graph; also denoted by $v_i v_j$ or ij

$i \sim j$: edge $\{v_i, v_j\}$ is present in the graph

dist(i, j): the length of the shortest path between vertices v_i and v_j

$e_{ij} = (v_i, v_j)$: edge in a digraph

$\mathcal{G} \backslash e$: graph $\mathcal{G}$ with edge e removed

$\mathcal{G} + e$: graph $\mathcal{G}$ with edge e added

$N(i)$: set of agents adjacent to i

$N(i, t)$: set of agents adjacent to i at time t

$d(v)$: degree of vertex v

$d_{in}(v)$: in-degree of vertex v

$d_{min}(\mathcal{G})$: minimum vertex degree in $\mathcal{G}$

$d_{max}(\mathcal{G})$: maximum vertex degree in $\mathcal{G}$

$\bar{d}_{in}(\mathcal{D})$: maximum (weighted) in-degree in $\mathcal{D}$

$\mathrm{diam}(\mathcal{G})$: diameter of $\mathcal{G}$

$A(\mathcal{G})$: adjacency matrix of $\mathcal{G}$

$A(\mathcal{D})$: in-degree adjacency matrix of $\mathcal{D}$

$\Delta(\mathcal{G})$: degree matrix of $\mathcal{G}$

$\Delta(\mathcal{D})$: in-degree matrix of $\mathcal{D}$

$L(\mathcal{G})$: graph Laplacian of $\mathcal{G}$

$L_e(\mathcal{G})$: edge Laplacian of $\mathcal{G}$

$L(\mathcal{D})$: in-degree Laplacian of $\mathcal{D}$

$L_o(\mathcal{D})$: out-degree Laplacian of $\mathcal{D}$

$\mathbf{L}(\mathcal{G})$: line graph of $\mathcal{G}$

$D(\mathcal{D})$: incidence matrix of $\mathcal{D}$

C_n: cycle graph on n vertices

P_n: path graph on n vertices

K_n: complete graph on n vertices

S_n: star graph on n vertices

$\mathbf{G}(n,p)$: set of random graphs on n vertices, with edge probability p

$\mathbf{G}(n,r)$: set of random geometric graphs on n vertices, with edge threshold distance r

$\mathcal{G}_1 \square \mathcal{G}_2$: Cartesian product of two graphs $\mathcal{G}_1$ and $\mathcal{G}_2$

Linear Algebra

$\mathbf{R}^n$: Euclidean space of dimension n

$\mathbf{R}^n_+$: nonnegative orthant in $\mathbf{R}^n$

$\mathbf{R}^{m\times n}$: space of $m \times n$ real matrices

$\mathcal{S}^n$: space of $n \times n$ symmetric matrices over reals

$\mathcal{S}^n_+$: space of $n \times n$ (symmetric) positive semidefinite matrices

I_n: $n \times n$ identity matrix; also denoted as I if the dimension is clear from the context

$0_{m\times n}$: $m \times n$ zero matrix; also denoted as 0 if the dimension is clear from the context

$M^{-1}, M^\dagger$: respectively, inverse and pseudo-inverse of M

M^T, M^{-T}: respectively, transpose and inverse transpose of M

$\mathcal{N}(M)$: null space of M

$\mathcal{R}(M)$: range space of M

$[A]_{ij}$: entry of matrix A on ith row and jth column

$\det(M)$: determinant of (square) matrix M

rank M: rank of M

trace M: trace of M

e^M: matrix exponential of square matrix M

$M_1 \otimes M_2$: Kronecker product of two matrices M_1 and M_2

$L_{[i,j]}$: matrix obtained from L by removing its ith row and jth column

diag(M): vector comprised of the diagonal elements of M

Diag(v): diagonal matrix with the vector v on its diagonal

Diag$(v_k), k = 1, 2, \cdots, n$: **Diag**$([v_1, \cdots, v_n]^T)$

$M > 0$ (M a symmetric matrix): M is positive definite

$M \geq 0$ (M a symmetric matrix): M is positive semidefinite

$\lambda_i(M)$: ith eigenvalue of M; M is symmetric and its eigenvalues are ordered from least to greatest value

v_i: ith entry of the vector v; also used for denoting vertex i in a graph

$\rho(M)$: spectral radius of M, that is, the maximum eigenvalue of M in magnitude

span$\{x\}$: span of vector x, that is, the subspace generated by scalar multiples of x

$\langle x, y\rangle$: inner product between two vectors x and y; real part of the inner product x^*y if x and y are complex-valued

1: vector of all ones

$\mathbf{1}_n$: $n \times 1$ vector of all ones

$\mathbf{1}^{\perp}$: subspace orthogonal to **span**$\{1\}$

$\|x\|$: 2-norm of vector x; $\|x\| = (x^Tx)^{1/2}$ unless indicated otherwise

Other

dist: distance function

j: $\sqrt{-1}$

$|z|$: modulus of complex number $z = \alpha + j\beta$, that is, 2-norm of vector $[\alpha, \beta]^T$

$V \backslash W$: elements in set V that are not in set W

$\prod_i \alpha_i$: product of α_is

$\sum_i \alpha_i$: sum of α_is

$\approx$: approximately equal to

$\ll$: much less than

x^*: complex conjugate transpose for complex-valued vector x

$x_i(t) \in \mathbf{R}^p$: state of agent i at time t

$\mathcal{A}$: agreement set, equal to **span**$\{1\}$

$\mathbf{E}\{x\}$: expected value of random variable x

var$\{x\}$: variance of random variable x

$\widehat{x}$: estimate of random variable (vector) x

$[n]$ (n a positive integer): set $\{1, 2, \ldots, n\}$

mod p: $a = b \pmod{p}$ if $a - b$ is an integer multiple of p

2^V (V a finite set): the power set of V, that is, the set of its subsets

$\binom{n}{m}$: number of ways to choose m-element subsets of $[n]$, that is, $n!/(m!(n - m)!)$

$\text{card}(A)$: cardinality of set A

$\arg \min f$: argument of the function f that minimizes it over its domain or constraint set

$\arg \max f$: argument of the function f that maximizes it over its domain or constraint set

$\mathbf{R}[x_1, \ldots, x_n]$: set of polynomials over the reals with indeterminants $x_1, \ldots, x_n$

$\mathcal{O}(f(n))$: $g(n) = \mathcal{O}(f(n))$ if $g(n)$ is bounded from above by some constant multiple of $f(n)$ for large enough n

$\Omega(f(n))$: $g(n) = \Omega(f(n))$ if $g(n)$ is bounded from below by some constant multiple of $f(n)$ for large enough n

PART 1
FOUNDATIONS

Chapter One

Introduction

> "If a man writes a book,
> let him set down only what he knows.
> I have guesses enough of my own." — Goethe

In this introductory chapter, we provide a brief discussion of networked multiagent systems and their importance in a number of scientific and engineering disciplines. We particularly focus on some of the theoretical challenges for designing, analyzing, and controlling multiagent robotic systems by focusing on the constraints induced by the geometric and combinatorial characters of the information-exchange mechanism.

1.1 HELLO, NETWORKED WORLD

Network science has emerged as a powerful conceptual paradigm in science and engineering. Constructs and phenomena such as interconnected networks, random and small-world networks, and phase transition nowadays appear in a wide variety of research literature, ranging across social networks, statistical physics, sensor networks, economics, and of course multiagent coordination and control. The reason for this unprecedented attention to network science is twofold. On the one hand, in a number of disciplines– particularly in biological and material sciences–it has become vital to gain a deeper understanding of the role that inter-elemental interactions play in the collective functionality of multilayered systems. On the other hand, technological advances have facilitated an ability to synthesize networked engineering systems–such as those found in multivehicle systems, sensor networks, and nanostructures–that resemble, sometimes remotely, their natural counterparts in terms of their functional and operational complexity.

A basic premise in network science is that the structure and attributes of the network influence the dynamical properties exhibited at the system level. The implications and utility of adopting such a perspective for engineering networked systems, and specifically the system theoretic consequences of such a point of view, formed the impetus for much of this book.[1]

---

[1]One needs to add, however, that–judging by the vast apparatus of social networking, e.g.,

1.2 MULTIAGENT SYSTEMS

Engineered, distributed multiagent networks, such as distributed robots and mobile sensor networks, have posed a number of challenges in terms of their system theoretic analysis and synthesis. Agents in such networks are required to operate in concert with each other in order to achieve system-level objectives, while having access to limited computational resources and local communications and sensing capabilities. In this introductory chapter, we first discuss a few examples of such distributed and networked systems, such as multiple aerospace vehicles, sensor networks, and nanosystems. We then proceed to outline some of the insights that a *graph theoretic* approach to multiagent networks is expected to provide, before offering a preview of the book's content.

1.2.1 Boids Model

The Reynolds boids model, originally proposed in the context of computer graphics and animation, illustrates the basic premise behind a number of multiagent problems, in which a collection of mobile agents are to collectively solve a global task using local interaction rules. This model attempts to capture the way social animals and birds align themselves in swarms, schools, flocks, and herds. In the boids flocking model, each "agent," in this case a computer animated construct, is designed to react to its neighboring flockmates, following an ad hoc protocol consisting of three rules operating at different spatial scales. These rules are *separation* (avoid colliding with neighbors), *alignment* (align velocity with neighbors' velocities), and *cohesion* (avoid becoming isolated from neighbors). A special case of the boids model is one in which all agents move at the same constant speed and update their headings according to a nearest neighbor rule for group level alignment and cohesion. It turns out that based on such local interaction rules alone, velocity alignment and other types of flocking behaviors can be obtained. An example of the resulting behavior is shown in Figure 1.1.

1.2.2 Formation Flight

Distributed aerospace systems, such as multiple spacecraft, fleets of autonomous rovers, and formations of unmanned aerial vehicles, have been identified as a new paradigm for a wide array of applications. It is envisioned that distributed aerospace technologies will enable the implementation of a spatially distributed network of vehicles that collaborate toward

---

email, facebook, twitter, and a multitude of networked, coordinated, and harmonic behavior in nature and the arts—our fascination with multiagent networks is more intrinsic.

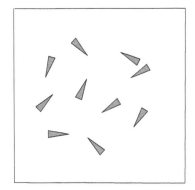

Figure 1.1: A Reynolds boids model in action. Ten agents, each with an arbitrary initial heading (given by the orientation of the triangles) and spacing, are considered (left); after a while they are aligned, moving in the same general direction at regular interagent distances (right). When this is the case, we say that *flocking* has been achieved.

a single collective scientific, military, or civilian goal. These systems are of great interest since their distributed architecture promises a significant cost reduction in their design, manufacturing, and operation. Moreover, distributed aerospace systems lead to higher degrees of scalability and adaptability in response to changes in the mission goals and system capabilities.

An example of a multiple platform aerospace system is space-borne optical interferometry. Space interferometers are distinguished by their composition and operational environment. They are composed of separated optical instruments, leading to a so-called sparse aperture. Although optical interferometers can, in principle, function on the earth's surface, there are many advantages in operating them in space. Space-borne interferometers have greater optical sensitivity and resolution, wider field of view, and greater detection capability. The resolution of these interferometers, as compared with space telescopes (e.g., Hubble), is dictated by the *separation* between the light collecting elements (called the baseline) rather than their *size*. Consequently, as the achievable imaging resolution of a space telescope is dictated by advanced manufacturing techniques, the size of the launch vehicle, and the complex deployment mechanism, the capability of a space-borne optical interferometer is limited by how accurately the operation of separated optical elements can be coordinated. These space-borne optical interferometers can be mounted on a single large space structure, composed of rigid or semirigid trusses or even inflatable membranes. In this case, the structural dynamics of the spacecraft plays a major role in the operation and the

Figure 1.2: Terrestrial Planet Finder, courtesy of JPL/NASA

success of the mission. An alternate approach is to fly the interferometer on multiple physically separated spacecraft, that is, a distributed space system. An example of such a mission is the Terrestrial Planet Finder (TPF) shown in Figure 1.2.

Another important set of applications of networked aerospace systems is found in the area of unmanned aerial vehicles of various scales and capabilities. These vehicle systems provide unique capabilities for a number of mission objectives, including surveillance, synthetic aperture imaging, mapping, target detection, and environmental monitoring.

1.2.3 Sensor Networks

A wireless sensor network consists of spatially distributed autonomous devices that cooperatively monitor physical or environmental conditions, such as temperature, sound, vibration, or pressure. Each node in a sensor network is equipped with a wireless communication device as well as an energy source–such as a battery–that needs to be efficiently utilized. The size, cost, and fidelity of a single sensor node can vary greatly, often in direct correspondence with its energy use, computational speed, and the ease by which it can be integrated within the network. Each sensor exchanges information on its local measurements with other nodes in the network in order to reach an accurate estimate of the physical or environmental variable of interest. We note that the efficiency requirement on the utilization of the energy source for each sensor often dictates a geometry on the internode communication for the sensor network.

1.2.4 Nanosystems

Recently, there has been a surge of interest by material scientists in organic compounds that are interconvertible via chemical reactions; this process is often referred to as *tautomerization*. These chemical reactions can be used for constructing molecular switches, where a molecule is steered between two or more stable states in a controlled fashion. Other electronic components such as diodes and transistors can be made that rely on similar induced transitions between structural isomers. Such molecular devices can then be put together, leading to the possibility of designing molecular circuits, networks, and more generally, *molecular dynamic systems*. An example of a molecular switch is a hydrogen tautomerization employed to manipulate and probe a naphthalocyanine molecule via low-temperature scanning tunneling microscopy. The properties and functionality of the corresponding molecular machines and networks are highly dependent on the inter-molecular bonds that can generally be manipulated by techniques such as electron beam lithography and molecular beam epitaxy.

1.2.5 Social Networks

Social networks are comprised of social entities, such as individuals and organizations, with a given set of interdependencies. The interaction between these entities can assume a multitude of relations, such as financial, social, and informational. Such networks are of great interest in a variety of fields, including theoretical sociology, organizational studies, and sociolinguistics. In fact, the *structure* of social networks has always been of fundamental importance for understanding these networks. More recently, the notion of manipulating the network structure has been contemplated as a viable means of altering the network behavior. For example, the concept of a *change agent* refers to a network entity that intentionally or indirectly causes or accelerates social, cultural, or behavioral change in the network.

1.2.6 Energy Networks

Complex, large-scale energy systems, delivering electrical and mechanical energy from generators to loads via an intricate distribution network, are among the most useful engineered networked dynamic systems. These systems often consist of a heterogeneous set of dynamic systems, such as power electronics and switching logics, that evolve over multiple timescales. Dynamics, stability, and control of individual power system elements (e.g., synchronous machines) or their interconnections (e.g., multi-machine models) have extensively been examined in the literature. However, as the need for more efficient generation and utilization of energy has become prevalent, distributed and network architectures such as the "smart grid" have gained particular prominence.

1.2.7 The Common Thread

The examples above, sampled from distinct disciplines, share a set of fundamental system theoretic attributes with a host of other networked multiagent systems. In a nutshell, such systems consist of (1) dynamic units, potentially with a decision making capability and means by which they can receive and transmit information among themselves, and (2) a signal exchange network, which can be realized via wired or wireless protocols in engineering, biochemical reactions in biological systems, and psychological and sociological interactions in the context of social networks.

The fundamental feature of networked systems, distinguishing them from systems that have traditionally been considered in system theory, is the presence of the network and its influence on the behavior of the overall system. Consequently, a successful "system theory for networked systems" has to blend the mathematics of information networks with paradigms that are at the core of dynamic system theory (stability, controllability, optimality, etc.). One of the challenging aspects facing such an interdisciplinary marriage in the context of system theory is that many network properties, for example, the network geometry, have a logical or combinatorial character.

1.3 INFORMATION EXCHANGE VIA LOCAL INTERACTIONS

In order to have a concrete model of "local interactions," in this section, we delineate the local nature of information exchange mechanisms for robotic networks.

1.3.1 Locality in Communication

One way in which agents can share information with their surroundings is through communication channels. But transmitting and receiving information requires energy, which is typically a sparse commodity in many networked applications, such as sensor networks and mobile ad hoc communication networks. Hence, only agents within a limited communication range can exchange information directly, forcing information to propagate through the network over intermediary nodes. Another communication constraint pertains to the available bandwidth. If a large collection of agents simultaneously broadcast large amounts of data, the communication channels saturate and lead to sharp deterioration of the communication system. Thus, in large networks, the information exchange should be maintained and kept parsimonious in order to satisfy the bandwidth limitations.

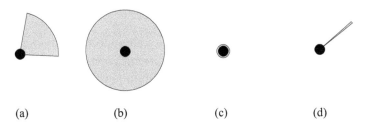

(a)   (b)   (c)   (d)

Figure 1.3: Various sensing geometries: (a) a vision-sensor with a wedge-shaped effective geometry; (b) an omnidirectional range sensor with a limited sensor range; (c) a tactile sensor provides information about the immediate surroundings; and (d) a single ray range sensor.

1.3.2 Locality in Sensing

Direct communications aside, agents can also infer information about each other and their environment through sensors. But every sensor has its own limitations in terms of range and resolution. Some of the most common sensors and their corresponding constraints are:

- **Vision-based sensors:** Cameras typically have long effective range (at least monocular vision), but they cover only a particular wedge-shaped region, as seen in Figure 1.3(a).

- **Range sensors:** The most common range sensors include sonars, infrared sensors, and laser scanners. These range sensors have very different sensing resolutions, ranging from very short range (e.g., low-cost infrared sensors) to covering hundreds of meters (e.g., high-quality laser-scanners). These sensors emit rays along a single direction but are typically ring-mounted to provide omnidirectional sensing capabilities (e.g., sonar and infrared rings) or have moving mirrors to provide scans across a larger area (e.g., laser scanners). This is shown in Figure 1.3(b).

A number of other sensing modalities are also widely used, with their own geometric constraints, as seen in Figure 1.3(c - d). However, as will be discussed throughout this book, this geometry will be subsumed by a graph theoretic interpretation of interactions as edges in the so-called *proximity graphs*, in which the existence of an edge indicates that neighboring nodes are within sensing range of each other.

1.4 GRAPH-BASED INTERACTION MODELS

The interaction geometry will indeed play an important role in the analysis and synthesis of networked multiagent systems regardless of whether the information exchange takes place over a communication network or through active sensing, or for that matter whether it assumes a wireless, chemical, physical, or sociological character. It turns out, however, that making the interaction protocol and its geometry explicit in the system-level analysis and control synthesis is far from trivial. In this direction, it becomes judicious to treat interactions as essentially combinatorial–at least initially–to codify whether an interaction exists and to what degree. An example of this abstraction is seen in Figure 1.4, in which the interaction geometry is defined by omnidirectional range sensors. As we will see throughout this book, such an abstraction, which cuts through the particular realization of the interaction, allows us to highlight the role of the interconnection topology, not only in the analysis of these systems but also in their synthesis.

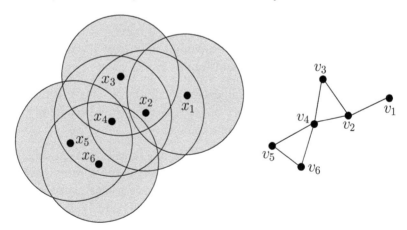

Figure 1.4: A network of agents equipped with omnidirectional range sensors can be viewed as a graph, with nodes corresponding to the agents and edges to the interactions.

1.4.1 Static, Dynamic, and Random Networks

If the edges in graphs are to be interpreted as enabling information to flow between the vertices on the corresponding edge, these flows can be directed as well as undirected. In other words, it is possible that the information will flow only in one direction. This would, for example, be the case if the vertices correspond to sensor agents, and agent i can sense agent j, while agent j can not sense agent i, for instance, due to different sensing modalities. In that case, the edge would be directed, with v_j as its "tail" and v_i as its "head." We will pictorially depict this as an arrow originating from v_j and ending at v_i. If the edge is undirected, we will simply drop the arrow and draw the edge as a line between the vertices.

However, directionality is not the only aspect of the edges that we will consider. We will also investigate different forms of temporal persistence, that is, situations in which the edges may disappear and reappear. In particular, we will group graphs into three classes:

- **Static Networks:** In these networks, the edges are static, that is, the edge set will not be time varying. This is, for example, the situation when a static communication network has been established, through which the information is flowing.

- **Dynamic, State-dependent Networks:** Here the edge set is time varying in that edges may disappear and reappear as functions of the underlying state of the network agents. For example, if the vertices in the graph correspond to mobile robots equipped with range sensors, edges will appear as agents get within the sensory range of each other, and be lost as agents get out of the sensory range.

- **Random Networks:** These networks constitute a special class of dynamic networks in that the existence of a particular edge is given by a probability distribution rather than some deterministic, geometric sensing condition. Such networks arise, for example, in the communications setting when the quality of the communication channels can be modeled as being probabilistic in nature.

It should be noted already at this point that these three types of networks will require different tools for their analysis. For static networks, we will rely heavily on the theory of linear, time-invariant systems. When the networks are dynamic, we have to move into the domain of hybrid systems, which will inevitably lead down the path of employing Lyapunov-based machinery for switched and hybrid systems. The random networks will in turn rely on a mix of Lyapunov theory and notions from stochastic stability.

1.5 LOOKING AHEAD

Graphs are inherently combinatorial objects, with the beauty but also limitations that come with such objects. Even though we will repeatedly connect with combinatorics, a host of issues pertaining to multiagent networks do not fruitfully lend themselves to a (pure) graph theoretic paradigm–at least not yet! Examples of such application domains include coverage control in sensor networks, which involves explicit partitioning of the environment and feedback control over a lossy and delayed network, where issues of delays, packet loss, and asynchronous operation, even for a pair of agents, are dominant. Moreover, the perspective adopted in this book does not include a detailed analysis of the underlying communication protocols, but instead employs a rather idealized model of information sharing, such as broadcast or single- and multi-hop strategies, and it is assumed that we can transmit and receive real numbers rather than quantized, finite bandwidth packets.

Another broad approach that we have adopted in this book is to work for the most part with simplified dynamics for the agents, that is, those with single and double integrators, linear time-invariant models, and unicycle models. In contrast, real-world networked systems are often comprised of agents with nontrivial dynamic input-output characteristics, interacting with each other via an elaborate set of interaction protocols. In this case, the behavior of the overall system depends not only on the interconnection topology and its detailed attributes, but also on how the interconnection protocol combines with the nonlinear and hybrid nature of the agents' dynamics.

Examples of topics that will be examined in this book include local interaction protocols for

- **Consensus:** having agents come to a global agreement on a state value;
- **Formations:** making the agents move to a desired geometric shape;
- **Assignments:** deciding a fair assignment of tasks among multiple agents;
- **Coverage:** producing maximally spread networks without making them disconnected or exhibit "holes" in their coverage;
- **Flocking/Swarming:** making the agents exhibit behaviors observed in nature, such as flocking birds, schooling fish, or swarming social insects;
- **Social Networks and Games:** analyzing how the outcomes of games and social interactions are influenced by the underlying interaction topology; and
- **Distributed Estimation:** organizing a group of sensors to collectively estimate a random phenomena of interest.

In later parts, we will also look at system theoretic models of controlled networks, capturing to what extend the behavior of networks can be influenced by exogenous inputs. We will examine dynamic notions of graph processes, thus allowing the graph structure itself be subject to control and time evolution. We conclude the book by providing an account of a framework for analyzing higher-dimensional interaction models via simplicial complexes.

NOTES AND REFERENCES

The boids model is due to Reynolds, who was motivated by animating movements of animal flocking [205]; this model was later employed by Vicsek, Czirók, Ben-Jacob, Cohen, and Shochet [238] for constant speed particles, mainly as a way to reason about self-organizing behaviors among large numbers of self-driven agents. This so-called Vicsek model, in turn, has provided an impetus for system theoretic analysis, such as the work of Jadbabaie, Lin, and Morse [124], which is also related to works on parallel and distributed computation [22] that in turn were inspired by works in distributed decision making examined by statisticians and economists [13], [198],[213].

Space-borne optical interferometry is an active area of research for a number of future scientific missions by NASA, such as the Terrestrial Planet Finder [3] and by the European Space Agency, such as the Darwin Mission [1]. Interferometry is one of the cornerstones of applied optics [32]; for the spaceborne application of interferometry, see [224]. Molecular switch and tautometers are of great interest in nanotechnology, examples of which can be found in [146],[172],[206]. Social networks is an active area of research in sociology, statistics, and economics; see for example, Wasserman and Faust [241]; for a more network-centric treatment, see the books by Goyal [105] and Jackson [122].

For complementary references related to this book, with somewhat different emphasis and outlook, see the books by Ren and Beard [204], and Bullo, Cortés, and Martínez [41].

Chapter Two

Graph Theory

> "The origins of graph theory are humble,
> even frivolous."
> — N. Biggs, E. K. Lloyd, and R. J. Wilson

As seen in the introductory chapter, graphs provide natural abstractions for
how information is shared between agents in a network. In this chapter, we
introduce elements of graph theory and provide the basic tools for reasoning about such abstractions. In particular, we will give an introduction to
the basic definitions and operations on graphs. We will also introduce the
algebraic theory of graphs, with particular emphasis on the matrix objects
associated with graphs, such as the adjacency and Laplacian matrices.

Graph-based abstractions of networked systems contain virtually no information about what exactly is shared by the agents, through what protocol
the exchange takes place, or what is subsequently done with the received
information. Instead, the graph-based abstraction contains high-level de-
scriptions of the network topology in terms of objects referred to as vertices
and edges. In this chapter, we provide a brief overview of graph theory. Of
particular focus will be the area of algebraic graph theory, which will provide the tools needed in later chapters for tying together inherently dynamic
objects (such as multi-agent robotic systems) with combinatorial characterization of networks (graphs).

2.1 GRAPHS

A *finite, undirected, simple graph*–or a *graph* for short–is built upon a finite
set, that is, a set that has a finite number of elements. We refer to this set as
the *vertex set* and denote it by V; each element of V is then a *vertex* of the
graph. When the vertex set V has n elements, it is represented as

$$V = \{v_1, v_2, \ldots, v_n\}.$$

Now consider the set of 2-element subsets of V, denoted by $[V]^2$. This set consists of elements of the form $\{v_i, v_j\}$ such that $i, j = 1, 2, \ldots, n$ and $i \neq j$. The finite graph $\mathcal{G}$ is formally defined as the pair $\mathcal{G} = (V, E)$, where V is a finite set of vertices and E is a particular subset of $[V]^2$; we refer to E as the set of *edges* of $\mathcal{G}$. We occasionally refer to vertices and edges of $\mathcal{G}$ as $V(\mathcal{G})$ and $E(\mathcal{G})$, respectively, and simplify our notation for an edge $\{v_i, v_j\}$ by sometimes denoting it as $v_i v_j$ or even ij.

A graph is inherently a set theoretic object; however, it can conveniently be represented graphically, which justifies its name. The graphical representation of $\mathcal{G}$ consists of "dots" (the vertices v_i), and "lines" between v_i and v_j when $v_i v_j \in E$. This graphical representation leads to many definitions, insights, and observations about graphs. For example, when an edge exists between vertices v_i and v_j, we call them *adjacent*, and denote this relationship by $v_i \sim v_j$. In this case, edge $v_i v_j$ is called *incident* with vertices v_i and v_j. Figure 2.1 gives an example of an undirected graph, $\mathcal{G} = (V, E)$, where $V = \{v_1, v_2, \ldots, v_5\}$ and $E = \{v_1 v_2, v_2 v_3, v_3 v_4, v_3 v_5, v_2 v_5, v_4 v_5\}$.

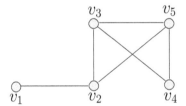

Figure 2.1: An undirected graph on 5 vertices

Analogously, the *neighborhood* $N(i) \subseteq V$ of the vertex v_i will be understood as the set $\{v_j \in V \mid v_i v_j \in E\}$, that is, the set of all vertices that are adjacent to v_i. If $v_j \in N(i)$, it follows that $v_i \in N(j)$, since the edge set in a (undirected) graph consists of unordered vertex pairs. The notion of adjacency in the graph can be used to "move" around along the edges of the graph. Thus, a *path* of length m in $\mathcal{G}$ is given by a sequence of distinct vertices

$$v_{i_0}, v_{i_1}, \ldots, v_{i_m}, \tag{2.1}$$

such that for $k = 0, 1, \ldots, m - 1$, the vertices v_{i_k} and $v_{i_{k+1}}$ are adjacent. In this case, v_{i_0} and v_{i_m} are called the end vertices of the path; the vertices $v_{i_1}, \ldots, v_{i_{m-1}}$ are the inner vertices. When the vertices of the path are distinct except for its end vertices, the path is called a cycle. A graph without cycles is called a forest.

We call the graph $\mathcal{G}$ *connected* if, for every pair of vertices in $V(\mathcal{G})$, there is a path that has them as its end vertices. If this is not the case, the graph is called *disconnected*. For example, the graph in Figure 2.1 is connected. We refer to a connected graph as having one connected component–a component in short. A component is thus a subset of the graph, associated with a minimal partitioning of the vertex set, such that each partition is connected. Hence, a disconnected graph has more than one component. A forest with one component is–naturally–called a tree.

The graphical representation of graphs allows us to consider graphs as *logical* constructions without the explicit identification of a vertex with an element of a vertex set V. This is achieved by deleting the "labels" on the dots representing the vertices of the graph; in this case, the graph is called unlabeled. An unlabeled graph thus encodes the qualitative features of the incident relation between a finite set of an otherwise unidentified objects. When the vertices in an unlabeled graph are given back their identities, the graph is called labeled. Figure 2.5 depicts two unlabeled graphs while the graph shown in Figure 2.1 is labeled.

Example 2.1. *Graphs can represent relations among social entities. For example, in a party of six consisting of Anna, Becky, Carolyn, David, Eaton, and Frank, the graph shown in Figure 2.2 depicts a scenario where all males in the group are each others' friends, all females in the group are each others' friends, and Anna and David are the only cross-gender friends in the group.*

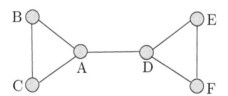

Figure 2.2: Boys and girls

Example 2.2. *Geographical locations, interconnected via roads, bridges, bike routes, and so on, can naturally be represented by graphs. For example, the graph shown in Figure 2.3 abstracts how the different land-masses of the city of Königsberg in eighteenth-century East Prussia were connected by bridges over rivers that passed through the city.*

Example 2.3. *Graphs can effectively express combinatorial relations between finite sets. Let $[n] = \{1, \ldots, n\}$ and for $n > k > m$, consider the k-element subsets of $[n]$ as vertices of a graph. Then let two vertices*

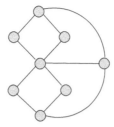

Figure 2.3: The graph abstracting the bridges of Königsberg

be adjacent when the corresponding sets intersect at m elements. The re-
sulting graphs, for various values of n, k, and m, are called the Johnson
graphs J(n, k, m). The Johnson graph J(5, 2, 0), also known as the Peter-
son graph, is shown in Figure 2.4.

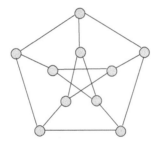

Figure 2.4: The Peterson graph $J(5, 2, 0)$

For unlabeled graphs, such as those depicted in Figures 2.3 - 2.4, it becomes imperative to define a notion for equating one graph with another.

Definition 2.4. *Two graphs $\mathcal{G} = (V, E)$ and $\mathcal{G}' = (V', E')$ are said to*
be isomorphic if they have similar vertex and edge sets in the sense that
there exists a bijection $\beta : V \to V'$ such that $v_i v_j \in E$ if and only if
$\beta(v_i)\beta(v_j) \in E'$. If this is the case, $\mathcal{G}$ and $\mathcal{G}'$ are isomorphic, denoted as
$\mathcal{G} \simeq \mathcal{G}'$.

2.1.1 Some Standard Classes of Graphs

Our first standard graph is the *complete graph* over n vertices, K_n. This is
the graph in which every vertex is adjacent to every other vertex. An example is shown in Figure 2.5(a), where the complete graph over 4 vertices, K_4,
is depicted.

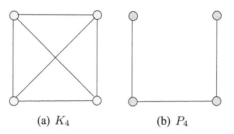

(a) K_4 (b) P_4

Figure 2.5: The complete graph and the path graph over 4 vertices

Other useful graphs include the *path graph*, the *cycle graph*, and the *star graph*. A path graph is understood to be any graph isomorphic to the graph $P_n = (\{v_1, \ldots, v_n\}, E_P)$, where $v_i v_j \in E_P$ if and only if $j = i + 1$, $i = 1, \ldots, n - 1$, as shown in Figure 2.5(b). Similarly, the n-cycle $C_n = (\{v_1, \ldots, v_n\}, E_C)$ is the graph with $v_i v_j \in E_C$ if and only if $i - j = \pm 1 \bmod n$. The star graph is given by $S_n = (\{v_1, \ldots, v_n\}, E_{\text{star}})$, with $v_i v_j \in E_{\text{star}}$ if and only if $i = 1$ or $j = 1$. These two graphs are depicted in Figure 2.6.

Two other important classes of graphs include regular and bipartite graphs. Each vertex of a k-regular graph has degree k; hence, a cycle graph is 2-regular and the complete graph on n vertices is $(n - 1)$-regular. For a bipartite graph $\mathcal{G}$, the vertex set is the union of two disjoint sets V_1 and V_2 such that $uv \in E(\mathcal{G})$ implies that either $u \in V_1$ and $v \in V_2$, or $u \in V_2$ and $v \in V_1$. If the cardinalities of the sets V_1 and V_2 are m and n, respectively, then the bipartite graph on the vertex set $V(\mathcal{G}) = V_1 \cup V_2$ with mn edges is called the complete bipartite graph $K_{m,n}$.

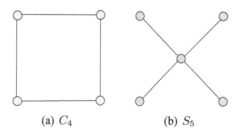

(a) C_4 (b) S_5

Figure 2.6: The cycle graph over 4 vertices and the star graph over 5 vertices

2.1.2 Subgraphs and Induced Subgraphs

Although graphs are most commonly defined as combinatorial objects, it is useful to perform set theoretic operations on graphs, such as examining their subsets and taking unions or intersections among them.

Consider a graph $\mathcal{G} = (V, E)$ and a subset of vertices $S \subseteq V$. One can let the subset of vertices "induce a subgraph" with respect to a given host graph. This induced subgraph is given by $\mathcal{G}_S = (S, E_S)$, where $E_S = \{\{v_i, v_j\} \in E \mid v_i, v_j \in S\}$. In other words, the subgraph S consists of the vertices in the subset S of $V(\mathcal{G})$ and edges in $\mathcal{G}$ that are incident to vertices in S. An example is shown in Figure 2.7, where the host graph is given in Figure 2.7(a), while the subgraph induced by the set of black vertices is given in Figure 2.7(b).

It should be noted, however, that it is not necessary to let the subgraphs be "induced." In fact, any graph $\mathcal{G}' = (V', E')$ is a subgraph of $\mathcal{G} = (V, E)$ if $V \subseteq V'$ and $E \subseteq E'$. In this case, we occasionally refer to $\mathcal{G}$ as being the "supgraph" of $\mathcal{G}'$. If $V = V'$ for a subgraph, it is referred to as a *spanning* subgraph. A spanning tree for a graph $\mathcal{G}$ is thus a subgraph of $\mathcal{G}$ that is also a tree.

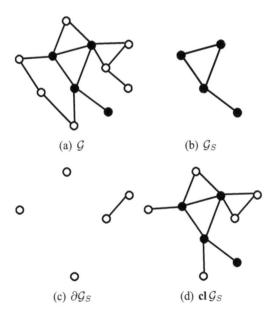

(a) $\mathcal{G}$ (b) $\mathcal{G}_S$

(c) $\partial \mathcal{G}_S$ (d) cl $\mathcal{G}_S$

Figure 2.7: A graph (a) and an induced subgraph (b) together with its boundary (c) and closure (d). The nodes in the subgraph S are shown in black while those in $V \backslash S$ are white.

2.1.3 Operations on Subgraphs

Now that we can induce subgraphs from vertex sets of graphs, we can perform a number of set theoretic operations on subgraphs as well. For example, given $S, S' \subseteq V(\mathcal{G})$, let $\mathcal{G}_S$ and $\mathcal{G}_{S'}$ be the corresponding induced subgraphs of $\mathcal{G}$. The union and intersections of these subgraphs can then be defined as the subgraphs induced by $S \cup S'$ and $S \cap S'$, respectively. In other words,

$$
\begin{aligned}
\mathcal{G}_S \cup \mathcal{G}_{S'} = \mathcal{G}_{S \cup S'} = (S \cup S', \{v_i v_j \in E \mid v_i, v_j \in S \cup S'\}), \\
\mathcal{G}_S \cap \mathcal{G}_{S'} = \mathcal{G}_{S \cap S'} = (S \cap S', \{v_i v_j \in E \mid v_i, v_j \in S \cap S'\}).
\end{aligned}
\tag{2.2}
$$

Similarly, boundaries and closures of subgraphs can be defined as

$$
\partial \mathcal{G}_S = \mathcal{G}_{\partial S} = (\partial S, \{v_i v_j \in E \mid v_i, v_j \in \partial S\}),
\tag{2.3}
$$

where $\partial S = \{v_i \in V \mid v_i \notin S \text{ and } \exists v_j \in S \text{ s.t. } v_i v_j \in E\}$. As an example, the boundary of the subgraph induced by the black vertices in Figure 2.7(a) is given in Figure 2.7(c). Following this, the closure of a subgraph $\mathcal{G}_S$ is defined as the union of the subgraph with its boundary, that is,

$$
\mathbf{cl}\, \mathcal{G}_S = \mathcal{G}_S \cup \partial \mathcal{G}_S.
\tag{2.4}
$$

2.2 VARIATIONS ON THE THEME

The notion of graphs can be generalized in various ways; in this section, we introduce two natural ones.

2.2.1 Weighted Graphs

If, together with the edge and vertex sets, a function $w : E \to \mathbf{R}$ is given that associates a value to each edge, the resulting graph $\mathcal{G} = (V, E, w)$ is a *weighted graph*. On such graphs, one can consider shortest paths, or *geodesics*, between vertices, through the notion of path *length*, defined as the sum of all the weights along the path. Specifically, by letting $\pi(v_i, v_j)$ be the set of all paths connecting v_i and v_j, a (not necessarily unique) geodesic between v_i and v_j is a minimizer to

$$
\min_{p \in \pi(v_i, v_j)} \text{length}(p).
$$

Similarly, the *diameter* of a weighted, connected graph is the length of any of its longest geodesics.

2.2.2 Digraphs

When the edges in a graph are given directions, for example as shown in Figure 2.8, the resulting interconnection is no longer considered an undirected graph. A *directed graph* (or *digraph*), denoted by $\mathcal{D} = (V, E)$, can in fact be obtained in two different ways. The first is simply to drop the requirement that the edge set E contains unordered pairs of vertices. What this means is that if the ordered pair $(v_i, v_j) \in E$, then v_i is said to be the *tail* (where the arrow starts) of the edge, while v_j is its *head*. The other manner in which a directed graph can be constructed is to associate an *orientation* o to the unordered edge set E. Such an orientation assigns a direction to edges in the sense that $o : E \rightarrow \{-1, 1\}$, with $o(v_i, v_j) = -o(v_j, v_i)$. An edge (v_i, v_j) is said to originate in v_i (tail) and terminate in v_j (head) if $o(v_i, v_j) = 1$, and vice versa if $o(v_i, v_j) = -1$.

Notions of adjacency, neighborhood, subgraphs, and connectedness can be extended in the context of digraphs. For example, a *directed path* of length m in $\mathcal{D}$ is given by a sequence of distinct vertices

$$v_{i_0}, v_{i_1}, \ldots, v_{i_m}, \tag{2.5}$$

such that for $k = 0, 1, \ldots, m - 1$, the vertices $(v_{i_k}, v_{i_{k+1}}) \in E(\mathcal{D})$. A digraph is called strongly connected if for every pair of vertices there is a directed path between them. The digraph is called weakly connected if it is connected when viewed as a graph, that is, a disoriented digraph. Analogous

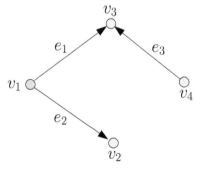

Figure 2.8: A directed graph over 4 vertices that is not strongly connected

to the case of graphs, a subgraph of a digraph $\mathcal{D} = (V, E)$, denoted by $\mathcal{D}' = (V', E')$, is such that $V' \subseteq V$ and $E' \subseteq E'$.

Figure 2.8 provides an example of a digraph. In fact, this digraph is $\mathcal{D} = (V, E)$, where $V = \{v_1, v_2, v_3, v_4\}$ while the edge set E is the set of ordered pairs $\{(v_1, v_3), (v_1, v_2), (v_4, v_3)\}$. In this figure, the edges have been given labels as well, and if we assume that such a labeling has been provided (possibly in an arbitrary fashion), the edge set can be written as $E = \{e_1, e_2, \ldots, e_m\}$, where m is the total number of edges in the graph. For example, $E = \{e_1, e_2, e_3\}$ in Figure 2.8.

2.3 GRAPHS AND MATRICES

As we have seen so far, graphs are constructs for representing relations between a finite number of objects, while admitting a straightforward graphical representation in terms of vertices and edges. Graphs also admit a representations in terms of matrices. Some of these matrices will be examined subsequently.

2.3.1 Adjacency and Degree

For an undirected graph $\mathcal{G}$, the *degree* of a given vertex, $d(v_i)$, is the cardinality of the neighborhood set $N(i)$, that is, it is equal to the number of vertices that are adjacent to vertex v_i in $\mathcal{G}$. Hence, for the graph shown in Figure 2.1, the degrees of the vertices are

$$d(v_1) = 1, d(v_2) = 3, d(v_3) = 3, d(v_4) = 2, d(v_5) = 3.$$

The degree sequence of a graph is the set of degrees of its vertices, often written in an increasing order. Based on the notions of degree and adjacency, one can associate certain matrices with graphs. The *degree matrix* of $\mathcal{G}$ is the diagonal matrix, containing the vertex-degrees of $\mathcal{G}$ on the diagonal, that is,

$$\Delta(\mathcal{G}) = \begin{pmatrix} d(v_1) & 0 & \cdots & 0 \\ 0 & d(v_2) & \cdots & 0 \\ \vdots & \vdots & \ddots & \vdots \\ 0 & 0 & \cdots & d(v_n) \end{pmatrix}, \qquad (2.6)$$

with n being the number of vertices.

The *adjacency matrix* $A(\mathcal{G})$ is the symmetric $n \times n$ matrix encoding of the adjacency relationships in the graph $\mathcal{G}$, in that

$$[A(\mathcal{G})]_{ij} = \begin{cases} 1 & \text{if } v_i v_j \in E, \\ 0 & \text{otherwise.} \end{cases} \qquad (2.7)$$

Returning to the example in Figure 2.1, the corresponding degree and adjacency matrices are

$$
\Delta(\mathcal{G}) = \begin{pmatrix} 1 & 0 & 0 & 0 & 0 \\ 0 & 3 & 0 & 0 & 0 \\ 0 & 0 & 3 & 0 & 0 \\ 0 & 0 & 0 & 2 & 0 \\ 0 & 0 & 0 & 0 & 3 \end{pmatrix} \quad \text{and} \quad A(\mathcal{G}) = \begin{pmatrix} 0 & 1 & 0 & 0 & 0 \\ 1 & 0 & 1 & 0 & 1 \\ 0 & 1 & 0 & 1 & 1 \\ 0 & 0 & 1 & 0 & 1 \\ 0 & 1 & 1 & 1 & 0 \end{pmatrix}.
$$

2.3.2 Incidence Matrix

Under the assumption that labels have been associated with the edges in a graph whose edges have been arbitrarily oriented, the $n \times m$ *incidence matrix* $D(\mathcal{G}^o)$ is defined as

$$
D(\mathcal{G}^o) = [d_{ij}], \quad \text{where } d_{ij} = \begin{cases} -1 & \text{if } v_i \text{ is the tail of } e_j, \\ 1 & \text{if } v_i \text{ is the head of } e_j, \\ 0 & \text{otherwise.} \end{cases} \tag{2.8}
$$

The interpretation here is that $D(\mathcal{G}^o)$ captures not only the adjacency relationships in the graph, but also the orientation that the graph now enjoys; the incidence matrix associated with a graph $\mathcal{G}$ that has been oriented as $\mathcal{G}^o$ shown in Figure 2.8 is

$$
D(\mathcal{G}^o) = \begin{pmatrix} -1 & -1 & 0 \\ 0 & 1 & 0 \\ 1 & 0 & 1 \\ 0 & 0 & -1 \end{pmatrix}.
$$

As can be seen from this example, this incidence matrix has a column sum equal to zero, which is a fact that holds for all incidence matrices since every edge has to have exactly one tail and one head. We note that the incidence matrix for a digraph $\mathcal{D}$ can be defined analogously by skipping the preorientation that is needed for graphs. In this case, we denote the incidence matrix by $D(\mathcal{D})$.

The linear algebraic properties of the incidence matrix of graphs and digraphs provide insights into their many structural aspects. We elaborate on this connection via the notion of a *cycle space* for a weakly connected digraph $\mathcal{D}$, which is defined as the null space of the incidence matrix, that is, the set of vectors z such that $D(\mathcal{D})z = 0$.

Definition 2.5. *Given the incidence matrix $D(\mathcal{D})$, a signed path vector is a vector z corresponding to a path in $\mathcal{D}$, such that the ith index of z takes* *the value of $+1$ if the edge i is traversed positively, -1 if it is traversed negatively, and 0 if the edge is not used in the path.*[1]

The following two observations point to the convenient means of expressing graph theoretic facts using linear algebra.

Lemma 2.6. *Given a path with distinct initial and terminal vertices described by a signed path vector z in the digraph $\mathcal{D}$, the vector $y = D(\mathcal{G})z$ is such that its ith element takes the value of $+1$ if the vertex i is the initial vertex of the path, -1 if is the terminal vertex of the path, and 0 otherwise.*

Theorem 2.7. *Given a weakly connected digraph $\mathcal{D}$, the null space of $D(\mathcal{D})$ is spanned by all linearly independent signed path vectors corresponding to the cycles of $\mathcal{D}$.*

It is thus natural to refer to the null space of $D(\mathcal{D})$ as the cycle space of the digraph. The orthogonal complement of the cycle space, on the other hand, is called the *cut space* of $\mathcal{D}$, which is characterized by the range space of $D(\mathcal{D})^T$.

2.3.3 The Graph Laplacian

Another matrix representation of a graph $\mathcal{G}$, which plays an important role in this book, is the *graph Laplacian*, $L(\mathcal{G})$. This matrix can be defined in different ways, resulting in the same object. The most straightforward definition of the graph Laplacian associated with an undirected graph $\mathcal{G}$ is

$$L(\mathcal{G}) = \Delta(\mathcal{G}) - A(\mathcal{G}), \tag{2.9}$$

where $\Delta(\mathcal{G})$ is the degree matrix of $\mathcal{G}$ and $A(\mathcal{G})$ is its adjacency matrix. From this definition, it follows that for all graphs the rows of the Laplacian sum to zero. For example, the graph Laplacian associated with the graph in Figure 2.1 is

$$L(\mathcal{G}) = \begin{pmatrix} 1 & -1 & 0 & 0 & 0 \\ -1 & 3 & -1 & 0 & -1 \\ 0 & -1 & 3 & -1 & -1 \\ 0 & 0 & -1 & 2 & -1 \\ 0 & -1 & -1 & -1 & 3 \end{pmatrix}.$$

Alternatively, given an (arbitrary) orientation to the edge set $E(\mathcal{G})$, the graph Laplacian of $\mathcal{G}$ can be defined as

$$L(\mathcal{G}) = D(\mathcal{G}^o)D(\mathcal{G}^o)^T, \tag{2.10}$$

---

[1] An edge is traversed positively in the path if the orientation of the edge conforms with how the path is traversed.

where $D(\mathcal{G}^o)$ is the corresponding incidence matrix for the oriented graph $\mathcal{G}$. This definition directly reveals that the graph Laplacian is in fact a symmetric and positive semidefinite matrix.

It should be noted that since the two definitions (2.9) and (2.10) are equivalent, and since no notion of orientation is needed in (2.9), the graph Laplacian is orientation independent. We will therefore adopt the convention of using $D(\mathcal{G})$ for the incidence matrix of the graph when the orientation of $\mathcal{G}$ is arbitrary. Regardless of this fact, sometimes it proves useful to use one of these two definitions for the graph Laplacian. As an example, one can form the *weighted* graph Laplacian associated with the weighted graph $\mathcal{G} = (V, E, w)$ as

$$L_w(\mathcal{G}) = D(\mathcal{G})WD(\mathcal{G})^T, \tag{2.11}$$

where W is an $m \times m$ diagonal matrix, with $w(e_i)$, $i = 1, \ldots, m$, on the diagonal. Note here that a labeling has been assumed over the edge set, which is also needed in order to define the incidence matrix $D(\mathcal{G})$.

2.3.4 Edge Laplacian

The edge Laplacian for an arbitrary oriented graph $\mathcal{G}$ is defined as

$$L_e(\mathcal{G}) = D(\mathcal{G})^T D(\mathcal{G}). \tag{2.12}$$

Two key linear algebraic properties of $L_e(\mathcal{G})$ are as follows: (1) the set of nonzero eigenvalues of $L_e(\mathcal{G})$ is equal to the set of nonzero eigenvalues of $L(\mathcal{G})$, and (2) the nonzero eigenvalues of $L_e(\mathcal{G})$ and $L(\mathcal{G})$ are equal to the square of the nonzero singular values of $D(\mathcal{G})$. Moreover, consider the graph $\mathcal{G}$ with p connected components $\mathcal{G}_i$ and associated incidence matrices $D(\mathcal{G}_i)$, and let

$$D(\mathcal{G}) = [D(\mathcal{G}_1) \ \cdots \ D(\mathcal{G}_p)].$$

Then the edge Laplacian of $\mathcal{G}$ has the block diagonal form

$$L_e(\mathcal{G}) = \begin{bmatrix} D(\mathcal{G}_1)^T D(\mathcal{G}_1) & & 0 \\ & \ddots & \\ 0 & & D(\mathcal{G}_p)^T D(\mathcal{G}_p) \end{bmatrix}. \tag{2.13}$$

The edge Laplacian can thus be thought of as an "edge adjacency matrix" in that edges that do not share a common vertex are considered nonadjacent and the corresponding value in $L_e(\mathcal{G})$ becomes zero. On the other hand, edges that do share a vertex are considered adjacent, and the sign of the corresponding entry in $L_e(\mathcal{G})$ gives information on the direction of both edges relative to the vertex they share. Finally, each edge is always considered adjacent to itself; the number of common vertices between the edge and itself is thereby two. Hence, all diagonal entries of the edge Laplacian $L_e(\mathcal{G})$ have the value 2.

2.3.5 Laplacian for Digraphs

We first define the notions of *adjacency* and *degree* matrices for directed weighted graphs. Let $\mathcal{D}$ denote the underlying digraph; for the adjacency matrix, we let

$$[A(\mathcal{D})]_{ij} = \begin{cases} w_{ij} & \text{if } (v_j, v_i) \in E(\mathcal{D}), \\ 0 & \text{otherwise,} \end{cases} \tag{2.14}$$

and for the diagonal degree matrix $\Delta(\mathcal{D})$ we set

$$[\Delta(\mathcal{D})]_{ii} = d_{in}(v_i) \quad \text{for all } i, \tag{2.15}$$

where $d_{in}(v)$ is the *weighted in-degree* of vertex v, that is,

$$d_{in}(v_i) = \sum_{\{j \mid (v_j, v_i) \in E(\mathcal{D})\}} w_{ij}.$$

We note that

$$\Delta(\mathcal{D}) = \mathbf{Diag}\,(A(\mathcal{D})\mathbf{1}).$$

The corresponding (in-degree) weighted Laplacian is now defined by

$$L(\mathcal{D}) = \Delta(\mathcal{D}) - A(\mathcal{D}).$$

Note that by construction, for every digraph $\mathcal{D}$, one has

$$\mathbf{1} \in \mathcal{N}\,(L(\mathcal{D})),$$

that is, the vector of all ones is the eigenvector associated with the zero eigenvalue of $L(\mathcal{D})$. Our choice of "in-degree" as opposed to "out-degree" to define the adjacency and Laplacian matrices for digraphs is primarily motivated by how they will be used in the context of networked systems. Essentially, the "in-degree" versions of these matrices capture more directly how the dynamics of an agent is influenced by others.[2]

---

[2]In the same vein, the out-degree Laplacian captures how each node in the network influences other nodes.

2.4 ALGEBRAIC AND SPECTRAL GRAPH THEORY

Algebraic graph theory associates algebraic objects, such as matrices and polynomials, to graphs, and by doing so makes available a range of algebraic techniques for their study. Examples of objects that can represent graphs and be algebraically manipulated include matrices and their eigenvalues. As such, the degree, adjacency, incidence, and Laplacian matrices associated with a graph are examples of objects in algebraic graph theory. In fact, the study of the eigenvalues associated with these matrices belong to its own subdiscipline of graph theory, namely *spectral graph theory*.

As an example of what can be accomplished by associating matrices with graphs, consider the graph Laplacian $L(\mathcal{G})$. This matrix is known to be symmetric and positive semidefinite; hence its real eigenvalues can be ordered as

$$\lambda_1(\mathcal{G}) \leq \lambda_2(\mathcal{G}) \leq \cdots \leq \lambda_n(\mathcal{G}),$$

with $\lambda_1(\mathcal{G}) = 0$.

Theorem 2.8. *The graph $\mathcal{G}$ is connected if and only if $\lambda_2(\mathcal{G}) > 0$.*

Proof. Since the null spaces of $D(\mathcal{G})^T$ and $L(\mathcal{G})$ are the same, it suffices to show that the null space of $D(\mathcal{G})^T$ has dimension one when the graph $\mathcal{G}$ is connected. Suppose that there exists a vector $z \notin \mathbf{span}\{\mathbf{1}\}$, with $\mathbf{1}$ being the vector with 1s in all its entries, such that

$$z^T D(\mathcal{G}) = 0,$$

that is, when $uv \in E$ then $z_v - z_u = 0$. However, since $\mathcal{G}$ is connected, this implies that $z_v = z_u$ for all $u, v \in V$ and $z \in \mathbf{span}\{\mathbf{1}\}$. Thus, the dimension of the null space of $D(\mathcal{G})^T$ is one if and only if the geometric, and hence algebraic, multiplicity of the zero eigenvalue of the Laplacian, namely $\lambda_1(\mathcal{G})$, is one.   ∎

Another classic result in algebraic graph theory is the matrix-tree theorem. We state it in two pieces without proof; see notes and references. First, let L_v denote the matrix obtained after removing the row and column that index the vertex v from $L(\mathcal{G})$.

Proposition 2.9. *Consider the graph $\mathcal{G}$ on n vertices with $n-1$ edges. Then* $\det L_v = 1$ *if and only if $\mathcal{G}$ is a spanning tree.*

Theorem 2.10. *Let $t(\mathcal{G})$ be the number of spanning trees in $\mathcal{G}$. Then*

$$t(G) = \det L_v$$

for any $v \in \mathcal{G}$.

A generalization of the matrix-tree theorem for weighted digraphs is as follows; first a definition.

Definition 2.11. *A digraph $\mathcal{D}$ is a* rooted out-branching *if (1) it does not contain a directed cycle and (2) it has a vertex v_r (the root) such that for every other vertex $v \in \mathcal{D}$ there is a directed path from v_r to v. In this case, we refer to the out-branching as diverging from v_r, or in short, a v_r out-branching.*

An out-branching in $\mathcal{D}$ is spanning if its vertex set coincides with the vertex set of $\mathcal{D}$.

Theorem 2.12. *Let v be an arbitrary vertex of a weighted digraph $\mathcal{D}$. Then*

$$\det L_v(\mathcal{D}) = \sum_{T \in \mathcal{T}_v} \prod_{e \in T} w(e),$$

where $\mathcal{T}_v$ is the set of spanning v out-branchings in $\mathcal{D}$, $\prod_{e \in T} w(e)$ is the product of weights on the edges of out-branching T, and $L_v(\mathcal{D})$ is the matrix obtained from $L(\mathcal{D})$ by deleting the row and column that index the vertex v.

2.4.1 Laplacian Spectra for Specific Graphs

Although in general finding the Laplacian spectrum of arbitrary graphs is far from trivial, there are certain classes of graphs whose spectrum, as well as the associated eigenvectors, can be precisely characterized. In this section, we present a few such examples.

Example 2.13. *The Laplacian spectrum of the complete graph K_n:* As $L(K_n) = -\mathbf{1}\mathbf{1}^T + nI$, the spectrum of $L(K_n)$ is that of $-\mathbf{1}\mathbf{1}^T$ shifted by n. Since the spectrum of the rank one matrix $\mathbf{1}\mathbf{1}^T$ is $\{0, 0, \ldots, 0, n\}$, the Laplacian spectrum of K_n is $\{0, n, \ldots, n, n\}$.

A general technique that is often very useful for finding the spectrum of the Laplacian, as well as the spectrum of the adjacency matrix of a graph, is to interpret the definition of the eigenvalues and eigenvectors of a matrix in terms of the means by which each node in the graph is assigned an eigenvector entry. For example, from the equation $L(K_n)x = \lambda x$, defining the eigenvalue λ corresponding to the eigenvector x, it follows that

$$\sum_{j \neq i} (x_i - x_j) = \lambda x_i,$$

where x_i is, without loss of generality, nonzero. However, since

$$\sum_{j\neq i}(x_i - x_j) = -\left(\sum_{j=1}^{n} x_j\right) + nx_i,$$

it follows that for all $x \perp 1$, $nx_i = \lambda x_i$, implying that $\lambda_i = n$.

Example 2.14. *Spectrum of the n-cycle: since the n-cycle C_n is 2-regular, it is sufficient to find the spectrum of the adjacency matrix of the n-cycle. Let ω be the nth root of unity,*

$$e^{j\frac{2\pi}{n}} = \cos\frac{2\pi}{n} + j\sin\frac{2\pi}{n},$$

for a positive integer n; see Figure 2.9. Now, let $x = [1, \nu, \nu^2, \ldots \nu^{n-1}]^T$, where $\nu \in \{1, \omega, \omega^2, \ldots, \omega^{n-1}\}$, and consider

$$A(\mathcal{G})x = \lambda x. \tag{2.16}$$

The eigenvalue equation (2.16) implies that for $i = 1, 2, \ldots, n$,

$$x_{i+1} + x_{i-1} = \nu^i + \nu^{i-2} = (\nu + \bar{\nu})\nu^{i-1} = (\nu + \bar{\nu})x_i,$$

where $\bar{\nu}$ denotes the complex conjugate of ν, the cycle graph has been labeled in an increasing manner, and the arithmetic for the indices above is mod n.

This implies that $\nu + \bar{\nu}$ is an eigenvalue of the adjacency matrix of C_n. As we have n candidates for ν, we conclude that the eigenvalues of $A(C_n)$ are

$$2, 2\cos\frac{2\pi}{n}, \ldots, 2\cos\frac{2(n-1)\pi}{n}.$$

As C_n is 2-regular, it follows that the Laplacian spectrum of C_n is

$$0, 2 - 2\cos\frac{2\pi}{n}, \ldots, 2 - 2\cos\frac{2(n-1)\pi}{n}.$$

Example 2.15. *A graph on n vertices is* circulant *if the ith vertex is adjacent to $(i + j)$th and $(i - j)$th vertex (mod n) for each j in a particular list l. Thus when $l = \{1\}$, the circulant graph is precisely the n-cycle. Moreover, when l is $\{1, 2, 3\}$ the circulant graph on seven nodes, is K_7. It follows that*

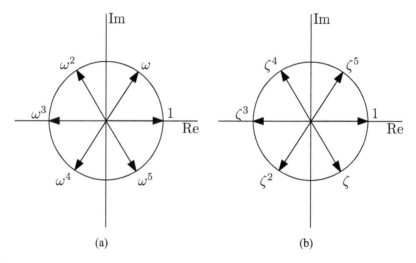

(a)   (b)

Figure 2.9: The nth roots of unity for $n = 6$ in the complex plane: (a) powers proceed counterclockwise, (b) powers proceed clockwise; note that $\zeta = \bar{\omega}$.

the Laplacian of a circulant graph is itself a circulant matrix, which is of the general form

$$
\begin{bmatrix}
c_0 & c_{n-1} & \cdots & c_1 \\
c_1 & c_0 & \cdots & c_2 \\
c_2 & c_1 & \cdots & c_3 \\
\vdots & \vdots & \cdots & \vdots \\
c_{n-1} & c_{n-2} & \cdots & c_0
\end{bmatrix}
\qquad (2.17)
$$

for some $c_0, c_1, c_2, \ldots, c_{n-1}$. The matrix of eigenvectors of the circulant matrix is the Fourier matrix

$$
\begin{bmatrix}
1 & 1 & \cdots & 1 & 1 \\
1 & \zeta & \zeta^2 & \cdots & \zeta^{n-1} \\
1 & \zeta^2 & \zeta^4 & \cdots & \zeta^{n-2} \\
\vdots & \vdots & \vdots & \vdots & \vdots \\
1 & \zeta^{n-1} & \zeta^{n-2} & \cdots & \zeta
\end{bmatrix},
$$

that is, the $(i+1)(j+1)$th entry of the Fourier matrix is ζ^{ij}, where $0 \leq i, j \leq n-1$ and $\zeta = \bar{\omega}$, with ω being a root of unity. The Laplacian spectrum of the circulant graph, on the other hand, is specified by the values

$$
p(1), p(\zeta), \ldots, p(\zeta^{n-1}),
$$

where $p(x) = c_o + c_1 x + \cdots + c_{n-1} x^{n-1}$ when the Laplacian is put in the form (2.17).

2.4.2 Eigenvalue Bounds

As we have seen above, the spectrum of the Laplacian[3] contains information about the structural properties of the graph, including its connectivity
(expressed in terms of an inequality) and the number of spanning trees (expressed in terms of an identity). In this subsection, we point out another set
of relations, expressed in terms of eigenvalue bounds for a few other features of graphs. We will also discuss a useful machinery for insights into
how eigenvalues of the Laplacian change as the graph undergoes structural
surgery, such as edge addition or removal; see for example Exercise 2.15.

Let us first start by enhancing the definition of graph connectivity.

Definition 2.16. *A vertex cut-set for $\mathcal{G} = (V, E)$ is a subset of V whose removal results in a disconnected graph. The vertex connectivity of the graph
$\mathcal{G}$, denoted by $\kappa_o(\mathcal{G})$, is the minimum number of vertices in any of its vertex
cut-sets.*

It is only natural to consider the analogous notion of connectivity pertaining to the edges of the graph as well.

Definition 2.17. *An edge cut-set in $\mathcal{G}$ is the set of edges whose deletion
increases the number of connected components of $\mathcal{G}$. The edge connectivity
of the graph $\mathcal{G}$, denoted by $\kappa_1(\mathcal{G})$, is the minimum number of edges in any
of its edge cut-sets.*

The variational characterization of eigenvalues of symmetric matrices
turns out to provide a convenient machinery for generating a host of inequalities between graph parameters, such as vertex and edge cut-sets, and
Laplacian eigenvalues. This characterization asserts, for example, that

$$\lambda_2(\mathcal{G}) = \min_{x \perp 1, \|x\|=1} x^T L(\mathcal{G}) x$$

and

$$\lambda_n(\mathcal{G}) = \max_{\|x\|=1} x^T L(\mathcal{G}) x.$$

One important ramification of such a variational characterization of eigenvalues is that

---

[3]Our emphasis on the spectrum of the Laplacian is mainly motivated by our applications
in the subsequent sections, rather than a lack of appreciation for the well-developed theory
of spectral graph theory via the adjacency matrix of the graph.

$$\lambda_2(\mathcal{G}) \leq \kappa_o(\mathcal{G}) \leq \kappa_1(\mathcal{G}) \leq d_{\min}(\mathcal{G}),$$

provided that $\mathcal{G}$ is not the complete graph, and where $d_{\min}$ is the minimum degree of the vertices in $\mathcal{G}$.

We conclude this subsection by mentioning yet another useful way that $\lambda_2(\mathcal{G})$ for a connected $\mathcal{G}$, shows up, namely, in the context of Cheeger's inequality. The setup is as follows. If we consider a subset of vertices $S \subset V$ together with its complement $S^c = V \backslash S$, we can ask how many edges need to be cut in order to completely separate S from S^c, that is, to quantify

$$\varepsilon(S, S^c) = \text{card}(\{v_i v_j \in E \mid (v_i \in S, v_j \in S^c) \text{ or } (v_j \in S, v_i \in S^c)\}).$$

As an example, consider the graph in Figure 2.10 where the number of edges that must be cut to separate S from S^c is 5, that is, $\varepsilon(S, S^c) = 5$.

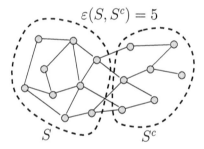

Figure 2.10: The number of edges that must be cut to separate S from its complement is 5.

Now, assume that the nodes in the graph belong to a network, and if two subsets get disconnected from each other, the agents that get separated are essentially lost. Since there really are two sets that are lost from each other, we consider the smaller to be the one that is actually lost. As such, we can define the *ratio* of the cut to be

$$\phi(S) = \frac{\varepsilon(S, S^c)}{\min\{\text{card}(S), \text{card}(S^c)\}}.$$

If we return to the example in Figure 2.10, we see that in this case $\text{card}(S) = 9 > 7 = \text{card}(S^c)$, and hence $\phi(S) = 5/7$.

The worst one can do in terms of losing vertices as compared to how many edges need to be cut can thus be thought of as a measure of robustness in the graph; this quality is known as the *isoperimetric number* of the graph

$$\phi(\mathcal{G}) = \min_{S \in 2^V} \phi(S). \tag{2.18}$$

Cheeger's inequality states that

$$\phi(\mathcal{G}) \geq \lambda_2(\mathcal{G}) \geq \frac{\phi(\mathcal{G})^2}{2 d_{\max}(\mathcal{G})}, \tag{2.19}$$

where $d_{\max}(\mathcal{G})$ is the maximum degree achieved by any vertex in $\mathcal{G}$.

In the following chapters, we will see that $\lambda_2(\mathcal{G})$ is important not only as a measure of the robustness (or level of connectedness) of the graph, but also for the convergence properties of a collection of distributed coordination algorithms.

2.5 GRAPH SYMMETRIES

Graph theory has a number of intriguing connections with other areas of discrete mathematics and in particular with abstract algebra. In this section, we give an introduction to two important constructs associated with graphs that are distinctively algebraic, namely, the symmetry structure in the graph and its equitable partitions.

Definition 2.18. *An automorphism of the graph $\mathcal{G} = (V, E)$ is a permutation ψ of its vertex set such that*

$$\psi(i)\psi(j) \in E \iff ij \in E.$$

The set of all automorphisms of $\mathcal{G}$, equipped with the composition operator, constitutes the automorphism group of $\mathcal{G}$; note that this is a "finite" group.[4] It is clear that the degree of a node remains unchanged under the action of the automorphism group, that is, if ψ is an automorphism of $\mathcal{G}$ then $d(v) = d(\psi(v))$ for all $v \in V$.

Proposition 2.19. *Let $A(\mathcal{G})$ be the adjacency matrix of the graph $\mathcal{G}$ and ψ a permutation on its vertex set V. Associate with this permutation the permutation matrix Ψ such that*

$$[\Psi]_{ij} = \begin{cases} 1 & if \quad \psi(i) = j, \\ 0 & otherwise. \end{cases}$$

---

[4]A finite group consists of a finite set of objects and a binary operation. The operation is assumed to be closed with respect to the set and admits an identity and is associative; moreover each element has an inverse with respect to this operation.

Then ψ is an automorphism of $\mathcal{G}$ if and only if

$$\Psi A(\mathcal{G}) = A(\mathcal{G}) \Psi.$$

In this case, the least positive integer z for which $\Psi^z = I$ is called the order of the automorphism.

Of course, we cannot avoid mentioning a beautiful connection between the graph automorphism and eigenvalues of the adjacency matrix.

Theorem 2.20. *If all eigenvalues of the adjacency matrix for the graph are simple, then every non-identity automorphism of $\mathcal{G}$ has order two.*

We will not provide the proof of Theorem 2.20; however, we will see an analogous statement and proof for the graph Laplacian in Chapter 10.

2.5.1 Interlacing and Equitable Partitions

A cell C is a subset of the vertex set $V = [n]$. A *partition* of the graph is then a grouping of its node set into different cells.

Definition 2.21. *An r-partition π of V, with cells $C_1, \ldots, C_r$, is said to be* equitable *if each node in C_j has the same number of neighbors in C_i, for all i, j. We denote the cardinality of the partition π by $r = |\pi|$. Let b_{ij} be the number of neighbors in C_j of a node in C_i. The directed graph, potentially containing self-loops, with the cells of an equitable r-partition π as its nodes and b_{ij} edges from the ith to the jth cells of π, is called the* quotient *of $\mathcal{G}$ over π, and is denoted by $\mathcal{G}/\pi$. An obvious trivial partition is the n-partition, $\pi = \{\{1\}, \{2\}, \ldots, \{n\}\}$. If an equitable partition contains at least one cell with more than one node, we call it a nontrivial equitable partition (NEP), and the adjacency matrix of the quotient is specified by*

$$[A(\mathcal{G}/\pi)]_{ij} = b_{ij}.$$

Equitable partitions of a graph can be obtained from its automorphisms. For example, in the Peterson graph shown in Figure 2.11(a), one equitable partition π_1 (Figure 2.11(b)) is given by two orbits of the automorphism group, namely the 5 inner vertices and the 5 outer vertices. The adjacency matrix of the quotient is then given by

$$A(\mathcal{G}/\pi_1) = \begin{bmatrix} 2 & 1 \\ 1 & 2 \end{bmatrix}.$$

The equitable partition can also be introduced by the equal distance partition. Let $C_1 \subset V$ be a given cell, and let $C_i \subset V$ be the set of vertices at distance $i-1$ from C_1. Then C_1 is said to be *completely regular* if its distance partition is equitable. For instance, every node in the Peterson graph is completely regular and introduces the partition π_2 as shown in Figure 2.11(c). The adjacency matrix of this quotient is then given by

$$A(\mathcal{G}/\pi_2) = \begin{bmatrix} 0 & 3 & 0 \\ 1 & 0 & 2 \\ 0 & 1 & 2 \end{bmatrix}.$$

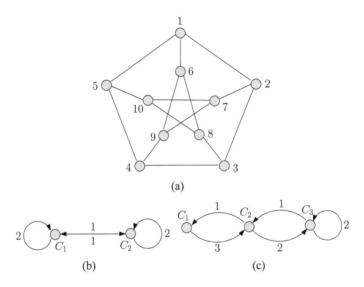

(a)

(b)   (c)

Figure 2.11: Equitable partitions on (a) the Peterson graph $\mathcal{G} = J(5,2,0)$ and the quotients, (b) the NEP introduced by the automorphism is $\pi_1 = \{C_1, C_2\}, C_1 = \{1,2,3,4,5\}, C_2 = \{6,7,8,9,10\}$, and (c) the NEP introduced by equal-distance partition is $\pi_2 = \{C_1, C_2, C_3\}, C_1 = \{1\}, C_2 = \{2,5,6\}, C_3 = \{3,4,7,8,9,10\}$.

The adjacency matrix of the original graph and its quotient are closely related through the interlacing theorem. First, let us introduce the following definition.

Definition 2.22. *A* characteristic vector $p_i \in \mathbf{R}^n$ *of a nontrivial cell* C_i *has* 1s *in components associated with* C_i *and* 0s *elsewhere.*[5] *A* characteristic matrix $P \in \mathbf{R}^{n \times r}$ *of a partition* π *of* V *is a matrix with characteristic vectors of the cells as its columns.*

---

[5] A nontrivial cell is a cell containing more than one vertex.

For example, the characteristic matrix of the equitable partition of the graph in Figure 2.12(a) is given by

$$P = \begin{bmatrix} 1 & 0 & 0 & 0 \\ 0 & 1 & 0 & 0 \\ 0 & 1 & 0 & 0 \\ 0 & 0 & 1 & 0 \\ 0 & 0 & 0 & 1 \end{bmatrix}, \tag{2.20}$$

with the corresponding quotient in Figure 2.12(b).

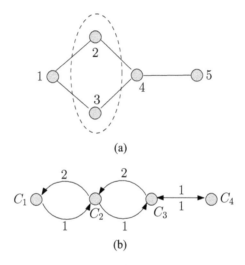

(a)

(b)

Figure 2.12: (a) Equitable partition and (b) quotient of a graph

Lemma 2.23. *Let P be the characteristic matrix of an equitable partition π of the graph $\mathcal{G}$, and let $\hat{A} = A(\mathcal{G}/\pi)$. Then $A(\mathcal{G})P = P\hat{A}$ and $\hat{A} = P^{\dagger}A(\mathcal{G})P$, where $P^{\dagger} = (P^{T}P)^{-1}P^{T}$ is the pseudo-inverse of P.*

As an example, the graph in Figure 2.12(a) has a nontrivial cell $\{2,3\}$. The adjacency matrix of original graph is

$$A(\mathcal{G}) = \begin{bmatrix} 0 & 1 & 1 & 0 & 0 \\ 1 & 0 & 0 & 1 & 0 \\ 1 & 0 & 0 & 1 & 0 \\ 0 & 1 & 1 & 0 & 1 \\ 0 & 0 & 0 & 1 & 0 \end{bmatrix},$$

while the adjacency matrix of the quotient is

$$\hat{A} = P^\dagger A(\mathcal{G})P = \begin{bmatrix} 0 & 2 & 0 & 0 \\ 1 & 0 & 1 & 0 \\ 0 & 2 & 0 & 1 \\ 0 & 0 & 1 & 0 \end{bmatrix}.$$

Lemma 2.24. *Let $\mathcal{G} = (V, E)$ be a graph with adjacency matrix $A(\mathcal{G})$, and let π be a partition of V with characteristic matrix P. Then π is equitable if and only if the column space of P is $A(\mathcal{G})$-invariant, that is, $A(\mathcal{G})\mathcal{R}(P) \subseteq \mathcal{R}(P)$.*

SUMMARY

In this chapter, we provided an introduction to graph theory at the level of providing the basic tools for reasoning about and analyzing networked systems as they appear in this book. Specifically, we provided an overview of the basic constructs in graph theory, for example, vertices and edges, graphs, subgraphs, and digraphs. We then explored connections between graphs and their algebraic representation in terms of adjacency, Laplacians, and edge Laplacian matrices, as well as the applied aspects of the spectrum of the graph Laplacian. We concluded with some of the algebraic properties of graphs, namely, their automorphism group and equitable partitions.

NOTES AND REFERENCES

Graph theory is a rich area in discrete mathematics, often considered the "other-half" of the general discipline of combinatorics. It is rather surprising that the simple structure of graphs, conveniently represented by dots and lines, lends itself to a rich area of mathematical inquiry with many applications in science and engineering. In fact, as many engineering disciplines move toward being more "networked," it is not surprising that graph theory has found itself at the heart of many networked sciences of current interest.

The origins of graph theory go back to Euler, who stated the first "theorem" in graph theory, namely, that given a graph, one can start from an arbitrary vertex, transverse every edge exactly once, and come back to the original vertex, if and if only every vertex has an even degree. The corresponding path in the graph, when one exists, is referred to as the *Eulerian cycle*. A glimpse in the beautiful historical book by Biggs, Lloyd, and Wilson [23], for example, reveals that the main thrusts in graph theory research for the first two hundred years since its inception pertained to electrical circuits, chemistry, polyhedra theory, planarity, and of course, coloring. Some of the more recent applications of graph theory are in information networks, sensor networks, social networks, large-scale networks, and network models such as those characterized by random, random geometric, and scale-free networks.

Graph theory, like other branches of mathematics, has many subareas. A few of them are extremal graph theory [29], topological graph theory [103], algorithmic graph theory [98],[233], and network optimization [5]. Extensions of basic graph theory that we believe will play important roles in networked systems research are the theory of hypergraphs [19], matroids [188], and connections with algebraic and combinatorial topology [195].

Most of the material in this chapter is standard and can be found in books on graph theory such as [71],[101], which is why many of the proofs have been omitted from this chapter. Example 2.2 refers to Euler's theorem related to the existence of an Eulerian cycle in a graph. The edge Laplacian was formally named and analyzed in [255], although other researchers have used the same construct without naming it. The statement and proof of the matrix-tree theorem (Theorems 2.9 and 2.10) is the celebrated result of Kirchhoff, who was motivated by his studies of electrical networks. The generalization of the matrix-tree theorem stated as Theorem 2.12 for weighted digraphs is due to Tutte [236].

Other names for out-branching often used by researchers are arborescence and directed rooted spanning tree.[6] Example 2.14 can be found in [148]. Example 2.15 pertains to discrete Fourier transforms and can be looked up in Meyer [159]. Eigenvalue bounds can be found in [101]; we also recommend the lecture notes by Spielman [222]. The Cheeger's inequality hints to a deep connection between differential geometry and graph theory–see the manuscript by Chung [50]. A nice treatment on graph automorphisms and equitable partitions of § 2.5 can be found in [101].

SUGGESTED READING

The suggested reading for this chapter are the books by Wilson [247], West [243], Diestel [71], and Godsil and Royle [101], the latter devoted to algebraic methods in graph theory. We also recommend the books by Bollobás [28] and Bondy and Murty [30] for a comprehensive introduction to graph theory. For a more problem-oriented approach to graph theory–and combinatorics in general–we highly recommend Lovász [148].

---

[6]We found arborescence to be a bit cumbersome to spell and directed rooted tree to be a little vague, as it does not hint that the spanning tree should be directed in the "right way."

EXERCISES

Exercise 2.1. Show that the number of edges in any graph is half the sum of the degrees of its nodes. Conclude that the trace of $L(\mathcal{G})$ is always an even number and that the number of odd degree nodes in any graph has to be even.

Exercise 2.2. The degree sequence for a graph is a listing of the degrees of its nodes; thus K_3 has the degree sequence 2, 2, 2. Is there a graph with the degree sequence 3, 3, 3, 3, 5, 6, 6, 6, 6, 6, 6? How about with the degree sequence 1, 1, 3, 3, 3, 3, 5, 6, 8, 9?

Exercise 2.3. Alkanes are chemical compounds that consist of carbon (C) and hydrogen (H) atoms, where each carbon atom has four bonds and each hydrogen atom only one. The graph of the alkane is obtained by denoting each atom by a vertex and drawing an edge between a pair of vertices if there is a bond between the corresponding atoms. Show that an alkane with n carbon atoms assumes the chemical formula $C_n H_{2n+2}$, indicating that for any alkane with n carbon atoms there are $2n + 2$ hydrogen atoms. Show that the graph of an alkane is a tree. Draw two realizations of $C_4 H_{10}$.

Exercise 2.4. A graph is k-regular if the degree of every vertex is k; thus K_3 is 2-regular. What is the relationship between k in a k-regular graph and the number of nodes in the graph other than $k \leq n - 1$?

Exercise 2.5. Let $\mathcal{G}$ be a graph on n vertices with c connected components. Show that **rank** $L(\mathcal{G}) = n - c$.

Exercise 2.6. Show that any graph on n vertices with more than $(n-1)(n-2)/2$ edges is connected.

Exercise 2.7. The complement of graph $\mathcal{G} = (V, E)$, denoted by $\overline{\mathcal{G}}$, is a graph $(V, \overline{E})$, where $uv \in \overline{E}$ if and only if $uv \notin E$. Show that

$$L(\mathcal{G}) + L(\overline{\mathcal{G}}) = nI - \mathbf{11}^T.$$

Conclude that for $2 \leq j \leq n$,

$$\lambda_j(\overline{\mathcal{G}}) = n - \lambda_{n+2-j}(\mathcal{G}).$$

Exercise 2.8. The list adjacency of a graph is an array, each row of which is initiated by a vertex in the graph and lists all vertices adjacent to it. Given the list adjacency of a graph, write an algorithm (in your favorite language) that checks whether the graph is connected.

Exercise 2.9. Recall that Cheeger's inequality states that

$$\phi(\mathcal{G}) \geq \lambda_2(\mathcal{G}) \geq \frac{\phi(\mathcal{G})^2}{2d_{\max}(\mathcal{G})},$$

where $\phi(\mathcal{G})$ is the isoperimetric number of $\mathcal{G}$ that can be used as a robustness measure of $\mathcal{G}$ to edge deletions. Construct a maximally robust graph consisting of n vertices and $n-1$ edges. Explain how would you do this and, in particular, give the value of $\phi(\mathcal{G})$ for this maximally robust graph.

Exercise 2.10. The line graph of $\mathcal{G}$ is a graph whose vertex set is the set of edges of $\mathcal{G}$, and there is an edge between these vertices if the corresponding edges in $\mathcal{G}$ are incident on a common vertex. What is the relationship between the automorphism groups of a graph and its complement and its line graphs?

Exercise 2.11. Show that any graph on n vertices that has more than $n-1$ edges contains a cycle.

Exercise 2.12. Show that the graph and its complement cannot both be disconnected.

Exercise 2.13. Show that for a graph $\mathcal{G}$, $D(\mathcal{G})D(\mathcal{G})^T = \Delta(\mathcal{G}) - A(\mathcal{G})$. Conclude that the graph Laplacian $D(\mathcal{G})D(\mathcal{G})^T$ is independent of the orientation given to $\mathcal{G}$ for constructing $D(\mathcal{G})$. Is the edge Laplacian $D(\mathcal{G})^T D(\mathcal{G})$ independent of the orientation given to $\mathcal{G}$ for constructing $D(\mathcal{G})$?

Exercise 2.14. Show that for any graph $\mathcal{G}$, $\lambda_n(\mathcal{G}) \geq d_{\max}(\mathcal{G})$.

Exercise 2.15. Let $\mathcal{G} = (V, E)$ be a graph, and let $uv \notin E$ for some $u, v \in V$. Show that

$$\lambda_2(\mathcal{G}) \leq \lambda_2(\mathcal{G} + e) \leq \lambda_2(\mathcal{G}) + 2,$$

where $\mathcal{G} + e$ is the graph $(V, E \cup \{e\})$.

Exercise 2.16. What are the eigenvalues and eigenvectors of the Laplacian matrix for the complete graph K_n, the path graph P_n, and the complete bipartite graph $K_{n,n}$?

Exercise 2.17. What is the automorphism group of the Peterson graph?

Exercise 2.18. A nontrivial equitable partition π of a graph is said to be maximal if any other nontrivial, equitable partition of the graph contains no fewer cells than π. Find the maximal, nontrivial equitable partition for the graph below.

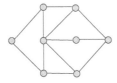

Exercise 2.19. Prove Theorem 2.7.

Chapter Three

The Agreement Protocol: Part I–The Static Case

"Whenever people agree with me
I always feel I must be wrong."
— Oscar Wilde

Agreement is one of the fundamental problems in multiagent coordination, where a collection of agents are to agree on a joint state value. In this chapter, we consider the dynamics of the so-called agreement protocol over undirected and directed static networks. Our primary goal is to highlight the intricate relationship between the convergence properties of this protocol on one hand, and the structure of the underlying interconnection on the other. We also explore connections between the agreement protocol and the theory of Markov chains in addition to a decomposition framework for the protocol's dynamics.

Consider a situation where a group of sensors are to measure the temperature of a given area. Although the temperature measured by each sensor will vary according to its location, it is required that the sensor group–using an information sharing network–agree on a single value which represents the temperature of the area. For this, the sensor group needs a *protocol* over the network, allowing it to reach consensus on what the common sensor measurement value should be.

In this first chapter devoted to the *agreement* –or the *consensus*–protocol over static networks, we explore the interdependency between the convergence properties of such a protocol and the structural attributes of the underlying network. The significance of the agreement protocol is twofold. On one hand, agreement has a close relation to a host of multiagent problems such as flocking, rendezvous, swarming, attitude alignment, and distributed estimation. On the other hand, this protocol provides a concise formalism for examining means by which the network topology dictates properties of the dynamic process evolving over it.

The agreement protocol involves n dynamic units, labeled $1, 2, \ldots, n$, interconnected via relative information-exchange links. The rate of change of each unit's state is assumed to be governed by the sum of its relative states with respect to a subset of other (neighboring) units. An example of the agreement protocol with three first-order dynamic units is shown in Figure 3.1.

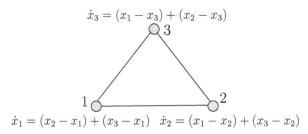

$$\dot{x}_3 = (x_1 - x_3) + (x_2 - x_3)$$

$$\dot{x}_1 = (x_2 - x_1) + (x_3 - x_1) \quad \dot{x}_2 = (x_1 - x_2) + (x_3 - x_2)$$

Figure 3.1: Agreement protocol over a triangle

Denoting the scalar state of unit i as $x_i \in \mathbf{R}$, one then has

$$\dot{x}_i(t) = \sum_{j \in N(i)} (x_j(t) - x_i(t)), \quad i = 1, \ldots, n, \tag{3.1}$$

where $N(i)$ is the set of units "adjacent to," or neighboring, unit i in the network. When the adopted notion of adjacency is symmetric, the overall system can be represented by

$$\dot{x}(t) = -L(\mathcal{G}) x(t), \tag{3.2}$$

where the positive semidefinite matrix $L(\mathcal{G})$ is the Laplacian of the agents' interaction network $\mathcal{G}$ and $x(t) = (x_1(t), \ldots, x_n(t))^T \in \mathbf{R}^n$. We refer to (3.2) as the agreement dynamics.[1]

Example 3.1. *(Symmetric Adjacency Relation) Consider the resistor-capacitor circuit shown in Figure 3.2. Letting the values of all resistances and capacitances to be 1 ohm and 1 farad, respectively, Kirchhoff's current and voltage laws lead to*

$$\dot{v}_i(t) = \sum_{j \in N(i)} (v_j(t) - v_i(t)),$$

---

[1]If $x_i \in \mathbf{R}^s$, $s > 1$, one can still obtain a compact description of (3.1), which is left to the reader in Exercise 3.4.

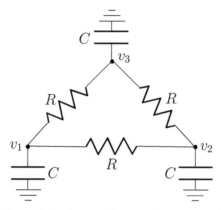

Figure 3.2: A capacitor-resistor network

describing the dynamics of the capacitors' voltages, where $N(i)$ represents the set of nodes in the circuit that are connected to node i via a resistor. This dynamics can compactly be represented in the form (3.2), where $\mathcal{G}$ is the triangle graph of Figure 3.1.

Example 3.2. *(Asymmetric Adjacency Relation) Consider a group of three robots coordinating their respective speeds according to the following chain of command: the rate of change of the second robot's speed is dictated by its speed difference with respect to the first one; the rate of change of the third robot's speed is adjusted analogously with respect to the second one. Finally, the first robot adjusts its speed by the taking the average of its speed differences with respect to the second and third robots.*

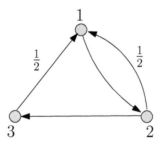

Figure 3.3: A robotic chain of command represented by the directed graph $\mathcal{D}$

Denoting the speed of robot i by s_i, the dynamics of the resulting system can be written as

$$\dot{s}_1(t) = \frac{1}{2}\left((s_3(t) - s_1(t)) + (s_2(t) - s_1(t))\right),$$
$$\dot{s}_2(t) = s_1(t) - s_2(t),$$
$$\dot{s}_3(t) = s_2(t) - s_3(t),$$

which assumes the form

$$\dot{s}(t) = \begin{bmatrix} -1 & \frac{1}{2} & \frac{1}{2} \\ 1 & -1 & 0 \\ 0 & 1 & -1 \end{bmatrix} s(t), \tag{3.3}$$

where $s(t) = [\,s_1(t)\, s_2(t)\, s_3(t)\,]^T$. We note that the matrix (3.3) corresponds to the negative of the in-degree Laplacian of the network shown in Figure 3.3; thus

$$\dot{s}(t) = -L(\mathcal{D})s(t), \tag{3.4}$$

where $\mathcal{D}$ is the underlying directed interconnection, that is, the weighted digraph of the network.

We note that in the above examples, the dynamics of each vertex in the network is "pulled" toward the states of the neighboring vertices. It is tempting then to conjecture that asymptotically, all vertices will reach some weighted average of their initial states, which also corresponds to the fixed point of their collective dynamics. As such a state of *agreement* is of great interest to us, we are obliged to formally define it.

Definition 3.3. *The agreement set $\mathcal{A} \subseteq \mathbf{R}^n$ is the subspace **span**$\{1\}$, that is,*

$$\mathcal{A} = \{x \in \mathbf{R}^n \mid x_i = x_j,\ \textit{for all } i, j\}. \tag{3.5}$$

Our first goal in this chapter is to expand upon the mechanism by which the dynamics (3.2) over an undirected graph guides the vertices of the network to their agreement state, or the consensus value. We will then revisit the agreement protocol over directed networks, for example, those that can be represented as in (3.4).

3.1 REACHING AGREEMENT: UNDIRECTED NETWORKS

Recall from Chapter 2 that the spectrum of the Laplacian for a connected undirected graph assumes the form

$$0 = \lambda_1(\mathcal{G}) < \lambda_2(\mathcal{G}) \leq \cdots \leq \lambda_n(\mathcal{G}), \tag{3.6}$$

with $\mathbf{1}$, the vector of all ones, as the eigenvector corresponding to the zero eigenvalue $\lambda_1(\mathcal{G})$. We note that $L(\mathcal{G})$ is symmetric and $L(\mathcal{G})\mathbf{1} = 0$ for an arbitrary undirected $\mathcal{G}$. Let $U = [\, u_1 \, u_2 \, \cdots \, u_n \,]$ be the matrix consisting of normalized and mutually orthogonal eigenvectors of $L(\mathcal{G})$, corresponding to its ordered eigenvalues (3.6). Furthermore, set

$$\Lambda(\mathcal{G}) = \mathbf{Diag}\left(\, [\, \lambda_1(\mathcal{G}), \dots, \lambda_n(\mathcal{G})\,]^T \,\right).$$

Using the spectral factorization of the Laplacian, one has

$$e^{-L(\mathcal{G})t} = e^{-(U\Lambda(\mathcal{G})U^T)\,t} = U\,e^{-\Lambda(\mathcal{G})t}\,U^T$$
$$= e^{-\lambda_1(\mathcal{G})t}\,u_1 u_1^T + e^{-\lambda_2(\mathcal{G})t}\,u_2 u_2^T + \cdots + e^{-\lambda_n(\mathcal{G})t}\,u_n u_n^T.$$

Hence the solution of (3.2), initialized from $x(0) = x_0$, is

$$x(t) = e^{-L(\mathcal{G})t}x_0,$$

which can be decomposed along each eigen-axis as

$$x(t) = e^{-\lambda_1(\mathcal{G})t}(u_1^T x_0)\, u_1 + e^{-\lambda_2(\mathcal{G})t}(u_2^T x_0)\, u_2$$
$$+ \cdots + e^{-\lambda_n(\mathcal{G})t}(u_n^T x_0)\, u_n. \tag{3.7}$$

Theorem 3.4. *Let $\mathcal{G}$ be a connected graph. Then the (undirected) agreement protocol (3.2) converges to the agreement set (3.5) with a rate of convergence that is dictated by $\lambda_2(\mathcal{G})$.*

Proof. The proof follows directly from (3.7) by observing that for a connected graph $\lambda_i(\mathcal{G}) > 0$ for $i \geq 2$; as always, $\lambda_1(\mathcal{G}) = 0$. Thus

$$x(t) \rightarrow (u_1^T x_0)u_1 = \frac{\mathbf{1}^T x_0}{n}\mathbf{1} \quad \text{as } t \rightarrow \infty, \tag{3.8}$$

and hence $x(t) \rightarrow \mathcal{A}$;[2] see Figure 3.4. As $\lambda_2(\mathcal{G})$ is the smallest positive eigenvalue of the graph Laplacian, it dictates the slowest mode of convergence in (3.8). ∎

---

[2] See Appendix A.1 for a definition of convergence to a set.

We note that as the states of the vertices evolve toward the agreement set, one has

$$\frac{d}{dt}\left(\mathbf{1}^T x(t)\right) = \mathbf{1}^T(-L(\mathcal{G})x(t)) = -x(t)^T L(\mathcal{G})\mathbf{1} = 0.$$

As such, the quantity $\mathbf{1}^T x(t) = \sum_i x_i(t)$, that is, the centroid of the network states, evaluated for any $t \geq 0$, is a *constant of motion* for the agreement dynamics (3.2).[3] Furthermore, the proof of Theorem 3.4 indicates that the state trajectory generated by the agreement protocol converges to the projection of its initial state, in the Euclidean norm, onto the agreement subspace, since

$$\arg\min_{x \in \mathcal{A}} \|x - x_0\| = \frac{\mathbf{1}^T x_0}{\mathbf{1}^T \mathbf{1}}\mathbf{1} = \frac{\mathbf{1}^T x_0}{n}\mathbf{1}. \qquad (3.9)$$

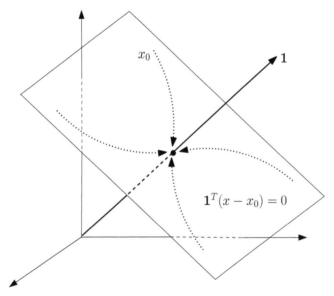

Figure 3.4: Trajectory of the agreement protocol retains the centroid of the nodes' states as its constant of motion.

The general form of the solution to the agreement dynamics, represented in (3.7), indicates that in order to have convergence to the agreement subspace from an arbitrary initial condition, it is necessary and sufficient to have $\lambda_2(\mathcal{G}) > 0$. As positivity of $\lambda_2(\mathcal{G})$ corresponds to the connectivity of $\mathcal{G}$ (see Chapter 2), one concludes that the minimum order structure needed

---

[3] In reference to quantities such as energy in conservative dynamical systems.

for asymptotic convergence to agreement is an interconnected network containing a spanning tree; see Figure 3.5.

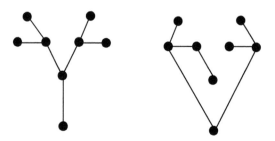

Figure 3.5: Two examples of trees on eight vertices

Proposition 3.5. *A necessary and sufficient condition for the agreement protocol (3.2) to converge to the agreement subspace (3.3) from an arbitrary initial condition is that the underlying graph contains a spanning tree.*

Example 3.6. *As an example of the agreement protocol in action, consider the so-called rendezvous problem, in which a collection of mobile agents–with single integrator dynamics–are to meet at a single location. This location is not given in advance and the agents do not have access to their global positions. All they can measure is their relative displacements with respect to their neighbors. By executing the agreement protocol*

$$\dot{x}_i(t) = - \sum_{j \in N(i)} (x_i(t) - x_j(t)),$$

where the position of agent i is given by $x_i \in \mathbf{R}^p$ (with $p = 2$ for planar robots, and so on) one obtains the response shown in Figure 3.6. The evolution of individual trajectories is shown in Figure 3.7.

3.2 REACHING AGREEMENT: DIRECTED NETWORKS

We now generalize the convergence analysis for the agreement protocol over (undirected, unweighted) graphs to those over weighted directed networks, that is, digraphs. In direct analogy with the agreement in the undirected case, let us consider the weighted digraph shown in Figure 3.8, which corresponds to the first-order dynamics

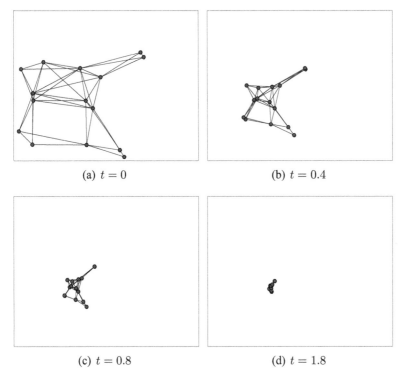

(a) $t = 0$ (b) $t = 0.4$

(c) $t = 0.8$ (d) $t = 1.8$

Figure 3.6: A progression of the group's planar configuration as 15 agents execute the agreement protocol in order to solve the rendezvous problem.

$$\dot{x}_1(t) = 0, \qquad\qquad\qquad\qquad\qquad\qquad\qquad\qquad (3.10)$$
$$\dot{x}_2(t) = w_{21}(x_1(t) - x_2(t)), \qquad\qquad\qquad\qquad (3.11)$$
$$\dot{x}_3(t) = w_{32}(x_2(t) - x_3(t)) + w_{34}(x_4(t) - x_3(t)), \qquad (3.12)$$
$$\dot{x}_4(t) = w_{42}(x_2(t) - x_4(t)) + w_{43}(x_3(t) - x_4(t)). \qquad (3.13)$$

Assuming the states to be scalar, the system of differential equations (3.10) - (3.13) can be represented as

$$\dot{x}(t) = \begin{bmatrix} 0 & 0 & 0 & 0 \\ -w_{21} & w_{21} & 0 & 0 \\ 0 & -w_{32} & w_{32} + w_{34} & -w_{34} \\ 0 & -w_{42} & -w_{43} & w_{42} + w_{43} \end{bmatrix} x(t),$$

where, as before, $x(t) = [\, x_1(t) \; x_2(t) \; x_3(t) \; x_4(t) \,]^T$.

Recalling the definition of the (in-degree) graph Laplacian for directed graphs discussed in Chapter 2, we observe that the dynamics of networked

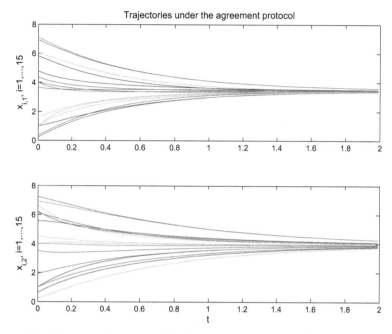

Figure 3.7: Trajectories when 15 planar agents execute the agreement protocol

systems such as (3.10) - (3.13) can be represented by

$$\dot{x}(t) = -L(\mathcal{D})\, x(t), \tag{3.14}$$

where $\mathcal{D}$ is the underlying directed interconnection between the vertices.

Our goal in the rest of this section is to identify necessary and sufficient conditions on the interconnection $\mathcal{D}$ that lead to the convergence of systems of the form (3.14) to the agreement subspace. A moment's reflection on the mechanism by which an analogous objective was achieved for the undirected network reveals the critical role played by the rank of the Laplacian, or equivalently, the multiplicity of its zero eigenvalue, and how this algebraic condition relates to the structure of the graph. We start by restating a construction for digraphs discussed in Chapter 2 that parallels the notion of spanning trees for undirected graphs.

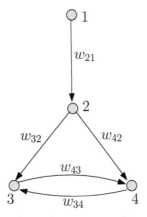

Figure 3.8: Directed graph corresponding to (3.10) - (3.13)

Definition 3.7. *A digraph $\mathcal{D}$ is a rooted out-branching if (1) it does not contain a directed cycle and (2) it has a vertex v_r (root) such that for every other vertex $v \in \mathcal{D}$ there is directed path from v_r to v; see Figure 3.9.*

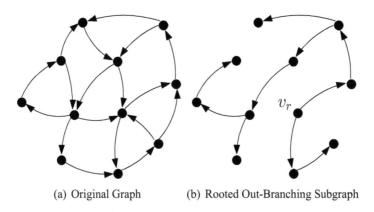

(a) Original Graph   (b) Rooted Out-Branching Subgraph

Figure 3.9: The left figure is an example of a digraph that contains a rooted out-branching as a subgraph. The corresponding rooted out-branching is given in the right figure, together with the root vertex v_r.

Proposition 3.8. *A digraph $\mathcal{D}$ on n vertices contains a rooted out-branching as a subgraph if and only if* **rank** $L(\mathcal{D}) = n - 1$. *In that case, $\mathcal{N}(L(\mathcal{D}))$ is spanned by the vector of all ones.*

Proof. There are several proofs for this statement; our favorite one is due to Tutte. The statement of the proposition is equivalent to showing that zero, as the root of the characteristic polynomial of $L(\mathcal{D})$, has algebraic multiplicity one. Let us denote this characteristic polynomial as

$$p_{\mathcal{D}}(\lambda) = \lambda^n + \alpha_{n-1}\lambda^{n-1} + \cdots + \alpha_1\lambda + \alpha_0,$$

noting that $\alpha_0 = 0$ since zero is an eigenvalue of $L(\mathcal{D})$. Thus

$$\mathbf{rank}\ L(\mathcal{D}) = n - 1$$

if and only if α_1 is nonzero. In the meantime,

$$\alpha_1 = \sum_v \det L_v(\mathcal{D}),$$

where $L_v(\mathcal{D})$ is the matrix obtained by deleting the vth row and the vth column of $L(\mathcal{D})$. Concurrently, by the matrix-tree theorem (Theorem 2.12), one has

$$\det L_v(\mathcal{D}) \neq 0$$

if and only if there is a rooted out-branching in $\mathcal{D}$ that is rooted at v. Hence, α_1 is nonzero if and only if there is a rooted out-branching rooted at some $v \in \mathcal{D}$. Then, the fact that $\mathcal{N}(L(\mathcal{D})) = \mathbf{span}\{1\}$ follows directly from the fact that $L(\mathcal{D})\mathbf{1} = 0$ and $\mathbf{rank}\ L(\mathcal{D}) = n - 1$. ∎

Since an eigenvalue with algebraic multiplicity of one also has geometric multiplicity of one (see Appendix A.2), for a digraph $\mathcal{D}$ that contains a rooted out-branching

$$L(\mathcal{D})\,p = 0 \quad \text{implies that} \quad p \in \mathbf{span}\{1\}.$$

It is instructive to examine the locations of other eigenvalues of $L(\mathcal{D})$ besides its zero eigenvalue. Let us first recall the celebrated Geršgorin disk theorem.

Theorem 3.9. *Let $M = [m_{ij}]$ be an $n \times n$ real matrix. Then all eigenvalues of M are located in*

$$\bigcup_i \left\{ z \in \mathbf{C} \,\middle|\, |z - m_{ii}| \leq \sum_{j=1,\dots,n; j \neq i} |m_{ij}| \right\}.$$

Proposition 3.10. *Let $\mathcal{D}$ be a weighted digraph on n vertices. Then the spectrum of $L(\mathcal{D})$ lies in the region*

$$\{ z \in \mathbf{C} \,|\, |z - \bar{d}_{in}(\mathcal{D})| \leq \bar{d}_{in}(\mathcal{D}) \},$$

where $\bar{d}_{in}$ denotes the maximum (weighted) in-degree in $\mathcal{D}$. In other words, for every digraph $\mathcal{D}$, the eigenvalues of $L(\mathcal{D})$ have non-negative real parts.

Proof. Viewing the spectrum of $L(\mathcal{D})$ in light of Theorem 3.9, we conclude that the eigenvalues of $L(\mathcal{D})$ lie in the region

$$\bigcup_i \{z \in \mathbf{C} \mid |z - d_{in}(v_i)| \le d_{in}(v_i)\}.$$

The statement of the proposition now follows, as illustrated in Figure 3.10. ∎

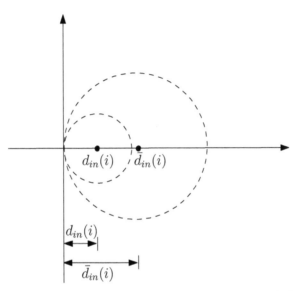

Figure 3.10: Geršgorin's regions for the eigenvalues of $L(\mathcal{D})$: the eigenvalue of $L(\mathcal{D})$ are contained in a disk of radius $\bar{d}_{in}$ centered at $\bar{d}_{in}$.

Localizing the spectrum of $L(\mathcal{D})$ has ramifications for the convergence properties of the system governed by (3.14).

Proposition 3.11. *Let $L(\mathcal{D}) = PJ(\Lambda)P^{-1}$ be the Jordan decomposition of the (in-degree) Laplacian for the digraph $\mathcal{D}$. When $\mathcal{D}$ contains a rooted out-branching, the nonsingular matrix P can be chosen such that*

$$J(\Lambda) = \begin{bmatrix} 0 & 0 & \cdots & 0 \\ 0 & J(\lambda_2) & \cdots & 0 \\ 0 & 0 & \cdots & 0 \\ \vdots & \vdots & \vdots & \vdots \\ 0 & \cdots & 0 & J(\lambda_n) \end{bmatrix},$$

where the $\lambda_i s$ (i = 2, ..., n) have positive real parts, and $J(\lambda_i)$ is the Jordan block associated with eigenvalue λ_i.[4] *Consequently,*

$$\lim_{t \to \infty} e^{-J(\Lambda)t} = \begin{bmatrix} 1 & 0 & \cdots & 0 \\ 0 & 0 & \cdots & 0 \\ 0 & 0 & \cdots & 0 \\ \vdots & \vdots & \vdots & \vdots \\ 0 & \cdots & 0 & 0 \end{bmatrix} \tag{3.15}$$

and

$$\lim_{t \to \infty} e^{-L(\mathcal{D})t} = p_1 q_1^T, \tag{3.16}$$

where p_1 and q_1^T are, respectively, the first column of P and the first row of P^{-1}, that is, where $p_1^T q_1 = 1$.

Proof. Consider the Jordan decomposition of $L(\mathcal{D})$; let the nonsingular matrix P be such that

$$P^{-1}L(\mathcal{D})P = J(\Lambda) = \begin{bmatrix} J(0) & 0 & \cdots & 0 \\ 0 & J(\lambda_2) & \cdots & 0 \\ \vdots & \vdots & \vdots & \vdots \\ 0 & 0 & 0 & J(\lambda_n) \end{bmatrix},$$

where the $\lambda_i s$ are the eigenvalues of $L(\mathcal{D})$. Since the digraph contains a rooted out-branching, by Propositions 3.8 and 3.10, $J_1(0) = 0$, and all other eigenvalues of $L(\mathcal{D})$ have positive real parts.

Now, note that

$$L(\mathcal{D})P = PJ(\Lambda),$$

which implies that $L(\mathcal{D})p_1 = 0$; as a result, p_1 belongs to **span**$\{1\}$. Similarly, the relation

$$P^{-1}L(\mathcal{D}) = J(\Lambda)P^{-1}$$

implies that the first row of P^{-1}, q_1, is the left eigenvector of $L(\mathcal{D})$ associated with its zero eigenvalue. Since $PP^{-1} = I$, it follows that $p_1^T q_1 = 1$. Putting these observations together, we conclude that

$$e^{-L(\mathcal{D})t} = P \begin{bmatrix} e^0 & 0 & \cdots & 0 \\ 0 & e^{J(-\lambda_2)t} & \cdots & 0 \\ \vdots & \vdots & \vdots & \vdots \\ 0 & 0 & 0 & e^{J(-\lambda_n)t} \end{bmatrix} P^{-1}.$$

---

[4]Note that the number of Jordan blocks is not necessary the number of vertices in the graph.

Since all nonzero eigenvalues of $L(\mathcal{D})$ have positive real parts, for all $i > 1$,

$$\lim_{t \to \infty} e^{-J(\lambda_i)t} = 0,$$

and (3.15) - (3.16) follow.   ∎

Armed with these results, we are finally ready to state the main theorem about the agreement protocol for directed, weighted networks.

Theorem 3.12. *For a digraph $\mathcal{D}$ containing a rooted out-branching, the state trajectory generated by (3.14), initialized from x_0, satisfies*

$$\lim_{t \to \infty} x(t) = (p_1 q_1^T) x_0,$$

where p_1 and q_1, are, respectively, the right and left eigenvectors associated with the zero eigenvalue of $L(\mathcal{D})$, normalized such that $p_1^T q_1 = 1$. As a result, one has $x(t) \to \mathcal{A}$ for all initial conditions if and only if $\mathcal{D}$ contains a rooted out-branching.

Proof. Choosing $p_1 = \mathbf{1}$ in Proposition 3.11, by (3.16), one has

$$\lim_{t \to \infty} x(t) = (q_1^T x_0)\mathbf{1},$$

with $q_1^T \mathbf{1} = 1$.   ∎

Recall that the constant of motion for the agreement protocol over an undirected graph is the sum of the node states at any given time. Analogously, we can identify the conserved quantity for the agreement protocol evolving over digraphs as follows.

Proposition 3.13. *Let q be the left eigenvector of the digraph in-degree Laplacian associated with its zero eigenvalue. Then the quantity $q^T x(t)$ remains invariant under (3.14).*

Proof. Since $q^T L(\mathcal{D}) = 0$, one has

$$\frac{d}{dt}\{q^T x(t)\} = -q^T\{L(\mathcal{D})x(t)\} = 0.$$

                                                                        ∎

3.2.1 Balanced Graphs

One of the consequences of Proposition 3.11 is that the agreement protocol over a digraph containing a rooted out-branching reaches the average
value of the initial states of the vertices if the left eigenvector of $L(\mathcal{D})$ corresponding to its zero eigenvalue is a scalar multiple of the vector of all
ones. In that case, $p_1 q_1^T$ in Theorem 3.12 reduces to the matrix $(1/n)\mathbf{11}^T$.
This observation leads us to the notion of balanced digraphs.

Definition 3.14. *A digraph is called* balanced *if, for every vertex, the in-
degree and out-degree are equal.*

When the digraph is balanced, in addition to having $L(\mathcal{D})\mathbf{1} = 0$, one has

$$\mathbf{1}^T L(\mathcal{D}) = 0.$$

Thus, if the digraph contains a rooted out-branching and is balanced, then
the common value reached by the agreement protocol is the average value
of the initial nodes, that is, the average consensus, since

$$\lim_{t \to \infty} x(t) = \frac{1}{n}\mathbf{11}^T x_0.$$

Let us strengthen the above observation by first introducing a few definitions.

Definition 3.15. *A digraph is* strongly connected *if, between every pair of
distinct vertices, there is a directed path.*

A digraph $\mathcal{D}$ is said to have been disoriented if all of its directed edges are
replaced by undirected ones.

Definition 3.16. *A digraph is* weakly connected *if its disoriented version is
connected.*

Some examples of digraphs illustrating these concepts are given in Figure
3.11.

Theorem 3.17. *The agreement protocol over a digraph reaches the average
consensus for every initial condition if and only if it is weakly connected and
balanced.*

Proof. A weakly connected balanced digraph is automatically strongly connected (see Exercise 3.13); hence, it contains a rooted out-branching. And,
by Theorem 3.12, the corresponding agreement protocol converges to the
agreement subspace. Moreover, since the digraph is balanced, the protocol's convergence is to the average consensus value.

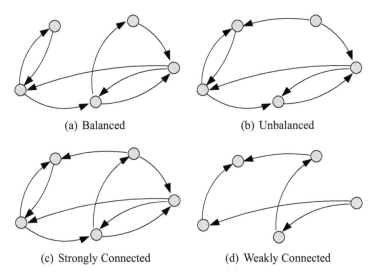

(a) Balanced   (b) Unbalanced

(c) Strongly Connected   (d) Weakly Connected

Figure 3.11: Digraphs over five nodes. The first two show balanced (in-degree = out-degree) and unbalanced graphs; the second two show strongly and weakly connected graphs.

Now, suppose that convergence to the average consensus value is in fact achieved, that is, that

$$\lim_{t \to \infty} x(t) = \lim_{t \to \infty} e^{-L(\mathcal{D})t} x_0 = \frac{1}{n} \mathbf{1}\mathbf{1}^T x_0$$

for every $x_0 \in \mathbf{R}^n$. Hence

$$\lim_{t \to \infty} e^{-L(\mathcal{D})t} = \frac{1}{n} \mathbf{1}\mathbf{1}^T, \tag{3.17}$$

where the convergence is with respect to any matrix norm. Since the left and right eigenvectors of the matrix $\lim_{t \to \infty} e^{-L(\mathcal{D})t}$, when convergent, have to be eigenvectors of $e^{-L(\mathcal{D})}$, which in turn are left and right eigenvectors of $L(\mathcal{D})$, we conclude that $\mathbf{1}$ is the left and right eigenvector of $L(\mathcal{D})$. Thus $L(\mathcal{D})\mathbf{1} = 0$ and $L(\mathcal{D})^T \mathbf{1} = \alpha \mathbf{1}$ for some α. In the meantime, $\mathbf{1}^T L(\mathcal{D})^T \mathbf{1} = (L(\mathcal{D})\mathbf{1})^T \mathbf{1} = \alpha \mathbf{1}^T \mathbf{1}$, and therefore $\alpha = 0$ and the digraph is balanced.

The asymptotic convergence (3.17) now implies that zero, as an eigenvalue of the Laplacian for the disoriented digraph $\mathcal{D}$,

$$L(\mathcal{D}) + L(\mathcal{D})^T,$$

has algebraic multiplicity of one. Therefore, $\mathcal{D}$ is weakly connected. ∎

3.3 AGREEMENT AND MARKOV CHAINS

In this section, we explore a connection between the agreement protocol and the theory of finite state, discrete time Markov chains. Markov chains are defined in the following way. We consider a stochastic process $X(k)$, $k = 0, 1, 2, \ldots$, which assumes one of the n states $x_1, \ldots, x_n$ at a given time. In addition, the chain satisfies the Markov property, that is, for all $k \geq 0$,

$$\mathbf{Pr}\{X(k+1) = x_j \,|\, X(k) = x_{i_k}, X(k-1) = x_{i_{k-1}}, \ldots, X(0) = x_{i_0}\}$$
$$= \mathbf{Pr}\{X(k+1) = x_j \,|\, X(k) = x_{i_k}\}.$$

The Markov property allows us to characterize a Markov chain by its state transition matrix P. The ijth entry of this matrix, p_{ij}, denotes the probability that the random variable X, having state i at time k, assumes state j at time $k + 1$, that is,

$$p_{ij} = \mathbf{Pr}\{X(k+1) = x_j \,|\, X(k) = x_i\}.$$

We note that since

$$\sum_{j=1}^{n} p_{ij} = 1,$$

the matrix P is a *stochastic* matrix with a unit eigenvalue. (See Appendix A.2.)

Now suppose that we define the probability distribution vector at time k, denoted by $\pi(k)$, whose ith entry encodes the probability that $X(k) = x_i$. Using the notion of a transition matrix allows us to monitor the time evolution of the distribution vector as

$$\pi(k+1)^T = \pi(k)^T P. \tag{3.18}$$

We observe that if $\pi(k)$ reaches a steady state value, say π^*, this state is characterized by the left eigenvector of P associated with its unit eigenvalue, as in this case, $\pi^* = P^T \pi^*$.

Let us now consider connections between the agreement protocol over weighted digraphs (3.14) and the discrete time evolution defined by (3.18). To this end, we monitor the progress of the agreement dynamics (3.14) at δ time intervals,

$$z(k+1) = e^{-\delta L(\mathcal{D})} z(k), \tag{3.19}$$

where $z(k) = x(\delta k)$ and $\delta > 0$. Our aim is to connect (3.19) with (3.18), and for this we proceed to gather the necessary ingredients for treating (3.19) as a Markov chain.

Proposition 3.18. *For all digraphs $\mathcal{D}$ and sampling intervals $\delta > 0$, one has*

$$e^{-\delta L(\mathcal{D})}\mathbf{1} = \mathbf{1} \quad and \quad e^{-\delta L(\mathcal{D})} \geq 0;$$

that is, for all $\mathcal{D}$ and $\delta > 0$, $e^{-\delta L(\mathcal{D})}$ is a stochastic matrix. In fact, the right and left eigenvectors of $e^{-\delta L(\mathcal{D})}$ are those of $L(\mathcal{D})$, respectively, associated with eigenvalues $e^{\delta \lambda_i}$, $i = 1, \ldots, n$.

Proof. We first observe that

$$e^{-\delta L(\mathcal{D})}\mathbf{1} = \left(\sum_{j=0}^{\infty} \frac{(-\delta)^j}{j!} L(\mathcal{D})^j \right) \mathbf{1} = \frac{(-\delta)^0}{0!} L(\mathcal{D})^0 \mathbf{1} = \mathbf{1},$$

which takes care of the first part of the proof.

Now, since $-L(\mathcal{D})$ has the property that all of its off-diagonal elements are non-negative, it is an "essentially non-negative" matrix, that is, $-L(\mathcal{D}) + sI$ is a non-negative matrix for a sufficiently large s (in this case, any $s \geq n - 1$ would suffice).

Moreover, it can be shown that for an essentially non-negative matrix C, and all $t \geq 0$, the matrix exponential e^{tC} is a non-negative matrix. Hence $e^{-\delta L(\mathcal{D})}$ is a non-negative matrix, which completes the proof. ∎

A direct consequence of Proposition 3.18 is the following fact:

Corollary 3.19. *The state of the nodes, during the evolution of the agreement protocol over a digraph $\mathcal{D}$, at any time instance, is a convex combination of the values of all nodes at the previous instance.*

Proof. By Birkhoff's theorem (see Appendix A.2), any stochastic matrix is a convex combination of permutation matrices. As the matrix $e^{-\delta L(\mathcal{D})}$ is stochastic for $\delta > 0$, the corollary follows from (3.19). ∎

Example 3.20. *Consider the agreement dynamics associated with the digraph in Figure 3.8, with weights $w_{12} = 1$, $w_{23} = 2$, $w_{43} = 4$, $w_{34} = 3$, and $w_{24} = 2$. For this digraph, we have*

$$e^{-L(\mathcal{D})} = \begin{bmatrix} 1.0000 & 0 & 0 & 0 \\ 0.6321 & 0.3679 & 0 & 0 \\ 0.3996 & 0.4651 & 0.0580 & 0.0773 \\ 0.3996 & 0.4651 & 0.0579 & 0.0774 \end{bmatrix}.$$

The corresponding Markov chain for $\delta = 1$ is shown in Figure 3.12.

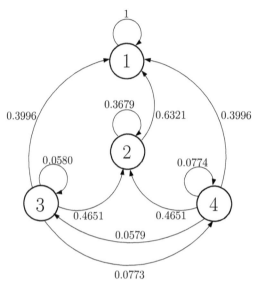

Figure 3.12: The Markov chain associated with Example 3.20 (for $\delta = 1$ after one time step)

One last piece is needed for completing the correspondence between the agreement dynamics (3.14) and Markov chains. The missing piece pertains to the non-negativity and normalization of the nodes' states. Recall that the "state" in the Markov chain governed by (3.18) is a probability distribution vector, that is, it belongs to the unit simplex. However, the state of the agreement dynamics (3.14) is an arbitrary vector in $\mathbf{R}^n$.

Proposition 3.21. *The behavior of the agreement dynamics (3.14) is characterized by its action on the unit simplex.*

Proof. Consider the normalization of the initial state of the sampled-time agreement dynamics (3.19) via

$$z(0) = \alpha \tilde{z}(0) + \beta \mathbf{1},$$

in such a way that $\tilde{z}(0)$ belongs to the unit simplex. Then

$$e^{-\delta L(\mathcal{D})} z(0) = \alpha e^{-\delta L(\mathcal{D})} \tilde{z}(0) + \beta \mathbf{1}.$$

It thus follows that the evolution of the agreement protocol can be viewed in the context of a Markov chain via an affine transformation. ∎

The close correspondence between the agreement protocol and general theory of Markov chains–as suggested by this section–provides a convenient avenue for interpreting results on the agreement protocol in terms of the well-developed theory of Markov chains.

3.4 THE FACTORIZATION LEMMA

It is interesting to see if it is possible to build complex graphs (or interaction networks) from *atomic* graphs, while at the same time being able to analyze the performance of the agreement protocol solely in terms of the individual atomic graphs. In this section we focus on this issue by investigating certain properties of the agreement protocol over Cartesian products of graphs. We will relate these properties to properties associated with the individual agreement dynamics on the corresponding atomic graphs.

As we will see, due to an intricate connection between the agreement protocol over a connected graph and its "prime factors" (a term that will be defined shortly), we will see that: (1) the trajectories generated by the agreement dynamics over the Cartesian product of a finite set of graphs is in fact the Kronecker product of the agreement trajectories over the individual graphs, and (2) the agreement dynamics over any connected graph can be factored in terms of the agreement dynamics over its prime decomposition.

The Cartesian product for a pair of graphs $\mathcal{G}_1 = (V_1, E_1)$ and $\mathcal{G}_2 = (V_2, E_2)$, denoted by

$$\mathcal{G} = \mathcal{G}_1 \square \mathcal{G}_2,$$

has its vertex set $V_1 \times V_2$, and any two vertices (v_1, v_2) and (v_1', v_2') in $V(\mathcal{G})$ are adjacent if and only if either $v_1 = v_1'$ and (v_2, v_2') is an edge in E_2, or $v_2 = v_2'$ and (v_1, v_1') is an edge in E_1. An example is given in Figure 3.13.

The Cartesian product is commutative and associative, that is, the products

$$\mathcal{G}_1 \square \mathcal{G}_2 \quad \text{and} \quad \mathcal{G}_2 \square \mathcal{G}_1$$

are isomorphic; similarly

$$(\mathcal{G}_1 \square \mathcal{G}_2) \square \mathcal{G}_3 \quad \text{and} \quad \mathcal{G}_1 \square (\mathcal{G}_2 \square \mathcal{G}_3)$$

are isomorphic.

The Cartesian product preserves connectedness properties of graphs. Thus, if both $\mathcal{G}_1$ and $\mathcal{G}_2$ are connected then $\mathcal{G} = \mathcal{G}_1 \square \mathcal{G}_2$ is connected.

One of the simplest examples of a Cartesian product is the product of two edges: it results in a cycle on four vertices. Another example is the Cartesian product of two paths that results in a rectangular grid. More elaborate

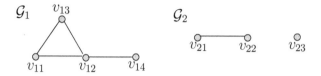

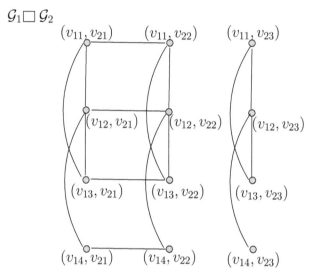

Figure 3.13: The Cartesian product of two graphs $\mathcal{G}_1$ and $\mathcal{G}_2$

examples include: (1) the product of n copies of P_2 is a hypercube Q_n,[5] (2) the product of two hypercube graphs is another hypercube, $Q_n \square Q_m = Q_{n+m}$; (3) the graph of an n-prism is the Cartesian product of an edge and the n-cycle; and (4) Rook's graph is the Cartesian product of two complete graphs.

Of fundamental importance in proving the main result of this section, the factorization lemma, is the relationship between the Laplacian of a pair of graphs and the Laplacian of their Cartesian product.[6]

Lemma 3.22. *Let $\mathcal{G}_1$ and $\mathcal{G}_2$ be a pair of graphs on n and m vertices,*

---

[5]A hypercube is a graph with vertices as the n-tuple $(b_1, b_2, \cdots, b_n)$, with $b_i \in \{0, 1\}$, and there is an edge between the vertices if the corresponding n-tuples differ at only one component.

[6]See Appendix A.2 for a review of Kronecker products.

respectively. Then

$$L(\mathcal{G}_1 \square \mathcal{G}_2) = L(\mathcal{G}_1) \otimes I_m + I_n \otimes L(\mathcal{G}_2)$$
$$= L(\mathcal{G}_1) \oplus L(\mathcal{G}_2), \tag{3.20}$$

that is, the Kronecker sum of the two graph Laplacians.

We leave the proof of this lemma as an exercise. But, as its direct consequence, using the properties of the Kronecker sum, we can conclude that the Laplacian spectrum of the Cartesian product of $\mathcal{G}_1 \square \mathcal{G}_2$, on n and m vertices, respectively, is the set

$$\{\lambda_i(\mathcal{G}_1) + \lambda_j(\mathcal{G}_2) \,|\, 1 \le i \le n, 1 \le j \le m)\}.$$

As a result, the second smallest eigenvalue is

$$\lambda_2(\mathcal{G}_1 \square \mathcal{G}_2) = \min\{\lambda_2(\mathcal{G}_1), \lambda_2(\mathcal{G}_2)\}.$$

In other words, the slowest mode of convergence in the agreement protocol over the Cartesian product of two graphs is dictated by the graph that is the least connected algebraically.

Another immediate ramification of how the Laplacian of the Cartesian product of a pair of graphs relates to the individual graph Laplacians pertains to the eigenvectors.

Lemma 3.23. *Let $\mathcal{G}_1$ and $\mathcal{G}_2$ be a pair of graphs on n and m vertices, respectively. Furthermore, assume that*

$$\lambda_1, \lambda_2, \ldots, \lambda_n \quad and \quad \mu_1, \mu_2, \ldots, \mu_m$$

are the eigenvalues of $L(\mathcal{G}_1)$ and $L(\mathcal{G}_2)$, respectively, corresponding to the eigenvectors

$$u_1, u_2, \ldots, u_n \quad and \quad v_1, v_2, \ldots, v_m.$$

Then

$$u_i \otimes v_j, \quad i = 1, 2, \ldots, n, \, j = 1, 2, \ldots, m,$$

is the eigenvector associated with the eigenvalue $\lambda_i + \mu_j$ of $L(\mathcal{G}_1 \square \mathcal{G}_2)$.

Proof. This follows directly from the properties of Kronecker sums and products (see Appendix A.2) as,

$$
\begin{aligned}
L(\mathcal{G}_1 \square \mathcal{G}_2)(u_i \otimes v_j) &= \{L(\mathcal{G}_2) \oplus L(\mathcal{G}_1)\}(u_i \otimes v_j) \\
&= (I_n \otimes L(\mathcal{G}_2))(u_i \otimes v_j) + (L(\mathcal{G}_1) \otimes I_m)(u_i \otimes v_j) \\
&= \{(I_n u_i) \otimes (L(\mathcal{G}_2) v_j)\} + \{(L(\mathcal{G}_1) u_i) \otimes (I_m v_j)\} \\
&= u_i \otimes \mu_j v_j + \lambda_i u_i \otimes v_j \\
&= (\lambda_i + \mu_j)(u_i \otimes v_j).
\end{aligned}
$$

■

3.4.1 Graph Decomposition and the Factorization Lemma

The graph Cartesian product allows us to construct large-scale graphs via a systematic procedure applied on a set of smaller-sized, *atomic* graphs. However, the application of the Cartesian product in the context of the agreement protocol over large-scale networks would be more explicit if arbitrary graphs could also be represented, or factored, in terms of the products of certain atomic graphs. Naturally, the notion of "small" (or atomic) graphs hinges upon their further factorizability.

A graph is called prime if it cannot be factored or decomposed as a product of nontrivial graphs; a graph is trivial if it consists of a single vertex. Specifically, a graph is called prime if the identity $\mathcal{G} = \mathcal{G}_1 \Box \mathcal{G}_2$ suggests that either $\mathcal{G}_1$ or $\mathcal{G}_2$ is trivial. We call a nonprime, nontrivial graph a composite graph. The importance of prime graphs in the general set of finite graphs is highlighted through the following fundamental result.

Theorem 3.24. *Every connected graph can be written as a Cartesian product of prime graphs. Moreover, such a decomposition is unique up to a reordering of factors.*

Example 3.25. *The composite graph $\mathcal{G}$ shown on the left of Figure 3.14 can be decomposed as the Cartesian product of three (prime) complete graphs on two vertices, as shown on the right.*

We now present the main result of this section, which we will refer to as the factorization lemma for the agreement dynamics.

Lemma 3.26 (Factorization Lemma). *Let $\mathcal{G}_1, \mathcal{G}_2, \ldots, \mathcal{G}_n$ be a finite set of graphs and consider $x_1(t), x_2(t), \ldots, x_n(t)$ to be states of the atomic agreement protocols,*

$$\dot{x}_1(t) = -L(\mathcal{G}_1)\,x_1(t),$$
$$\dot{x}_2(t) = -L(\mathcal{G}_2)\,x_2(t),$$
$$\vdots$$
$$\dot{x}_n(t) = -L(\mathcal{G}_n)\,x_n(t),$$

initialized from $x_1(0), \ldots, x_n(0)$. Then the state trajectory generated by the agreement protocol

$$\dot{x}(t) = L(\mathcal{G}_1 \Box \mathcal{G}_2 \Box \cdots \Box \mathcal{G}_n)\,x(t) \tag{3.21}$$

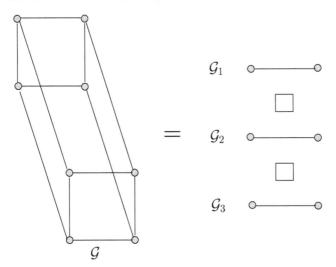

Figure 3.14: Prime factorization of a composite graph, $\mathcal{G} = \mathcal{G}_1 \square \mathcal{G}_2 \square \mathcal{G}_3$

is

$$x(t) = x_1(t) \otimes x_2(t) \otimes \cdots \otimes x_n(t), \tag{3.22}$$

when (3.21) is initialized from

$$x_1(0) \otimes x_2(0) \otimes \cdots \otimes x_n(0).$$

Proof. Due to the associativity of the Cartesian product, it suffices to prove the theorem for the case of $n = 2$. Consider graphs $\mathcal{G}_1 = (V_1, E_1)$ and $\mathcal{G}_2 = (V_2, E_2)$ with $|V_1| = n$ and $|V_2| = m$. Moreover, let (λ_j, u_j), $j = 1, 2, \ldots, n$, and (μ_i, v_i), $i = 1, 2, \ldots, m$, be the set of eigenvalues and the associated normalized, mutually orthogonal eigenvectors for the Laplacian matrices $L(\mathcal{G}_1)$ and $L(\mathcal{G}_2)$, respectively. Since

$$\dot{x}_1(t) = -L(\mathcal{G}_1)x_1(t) \quad \text{and} \quad \dot{x}_2(t) = -L(\mathcal{G}_2)x_2(t),$$

one has

$$x_1(t) = e^{-L(\mathcal{G}_1)t} x_1(0) = \sum_i e^{-\lambda_i t} u_i u_i^T x_1(0)$$

and

$$x_2(t) = e^{-L(\mathcal{G}_2)t} x_2(0) = \sum_j e^{-\mu_j t} v_j v_j^T x_2(0).$$

Thus

$$x_1(t) \otimes x_2(t)$$

$$= \sum_{i=1}^{n} (x_1(0)^T u_i) u_i e^{-\lambda_i t} \} \otimes \left\{ \sum_{j=1}^{m} (x_2(0)^T v_j) v_j e^{-\mu_j t} \right\}$$

$$= \sum_{i=1}^{n} \sum_{j=1}^{m} (x_1(0)^T u_i) u_i e^{-\lambda_i t} \otimes (x_2(0)^T v_j) v_j e^{-\mu_j t}$$

$$= \sum_{i=1}^{n} \sum_{j=1}^{m} (x_1(0)^T u_i)(x_2(0)^T v_j)(u_i \otimes v_j) e^{-(\mu_i + \lambda_j)t}$$

$$= \sum_{i=1}^{n} \sum_{j=1}^{m} \left\{ (u_i \otimes v_j)^T (x_1(0) \otimes x_2(0))(u_i \otimes v_j) e^{-(\lambda_i + \mu_j)t} \right\}.$$

Denote by $z(0) = x_1(0) \otimes x_2(0)$, $w_{ij} = u_i \otimes v_j$, and $\zeta_{ij} = \lambda_i + \mu_j$, for $i = 1, 2, \ldots, n$ and $j = 1, 2, \ldots, m$. Hence

$$x_1(t) \otimes x_2(t) = \sum_{i=1}^{n} \sum_{j=1}^{m} \{ w_{ij}^T z(0) \} w_{ij} e^{-\zeta_{ij} t},$$

which is the state trajectory generated by the agreement protocol over the product agreement dynamics

$$\dot{z}(t) = -L(\mathcal{G}_1 \square \mathcal{G}_2) z(t),$$

when initialized from $z(0)$, and the proof follows.   ∎

Using Theorem 3.24 we can also state the following corollary.

Corollary 3.27. *The agreement dynamics over a composite graph can always be represented as a Kronecker product of agreement dynamics over its prime factors. Moreover, such a factorization is unique up to a reordering.*

Example 3.28. *Consider the agreement protocol on the composite graph $\mathcal{G}$ shown in the left side of Figure 3.14. Since $\mathcal{G}$ can be decomposed as the Cartesian product of three complete graphs on two vertices, the agreement dynamics on $\mathcal{G}$ corresponds to the Kronecker product of the three appropriately initialized atomic agreement dynamics, each evolving on a complete graph on two vertices.*

SUMMARY

In this chapter we introduced the agreement protocol for static undirected and directed graphs. For undirected graphs, we saw that $\dot{x}(t) = -L(\mathcal{G})x(t)$ drives the state $x(t)$ asymptotically to the agreement set $\mathcal{A}$ as long as $\mathcal{G}$ is connected. In fact, all elements of $x(t)$ will in this case approach the initial centroid $(1/n)\mathbf{1}^T x(0)$ as $t \to \infty$. Similarly, the directed agreement protocol $\dot{x}(t) = -L(\mathcal{D})x(t)$ drives x to $\mathcal{A}$ as long as the digraph $\mathcal{D}$ contains a rooted out-branching. In both cases, the rate of convergence is dictated by the second smallest eigenvalue of the graph Laplacian. We then explored connections between the agreement protocol and Markov chains, as well as a decomposition formalism for the protocol using the Cartesian product of graphs.

NOTES AND REFERENCES

The agreement protocol (as well as its various extensions) has received considerable attention in the systems and robotic community during the past decade. The formalisms and formulations presented in this chapter owe much to a number of research papers, some more recent than others. In particular, the specific form of the protocol appeared in the work of Olfati-Saber and Murray [182], with the adjustment of using the "in-degree" Laplacian (for example, the diagonal entries of $L(\mathcal{D})$ are the in-degrees of the vertices) instead of the "out-degree" version used in [182]. The protocol can also be viewed as the discrete heat equation (without a boundary condition) on a manifold (induced by the graph), which has been studied extensively in partial differential equations and differential geometry; see [207].

The discrete-time version of the agreement protocol that leads to an iteration of the form $x(k+1) = Wx(k)$, with W a stochastic matrix, has a much longer history, for example, as studied in the theory of Markov chains. An analogous setup has also appeared in the area of chaotic iterations and asynchronous distributed multisplitting methods in numerical linear algebra, with less emphasis on the effect of the underlying information-exchange network on the convergence properties of the corresponding numerical methods; see [22],[38],[45],[75]. However, in a setting that is closer to our discussion, the discrete version of the agreement protocol was discussed in the work of DeGroot [64]; see also Chatterjee and Seneta [44]. The observation that the multiplicity of the zero eigenvalue of the Laplacian is related to the existence of an out-branching is due to Ren and Beard [203], Agaev and Chebotarev [4], and Lafferriere, Williams, Caughman, and Veerman [140].

The factorization lemma comes from Nguyen and Mesbahi [179]; however, similar points of view have been explored in the theory of random walks; see the book by Woess [248]. For a detailed convergence analysis of the agreement problem and repeated averaging, see also Olshevsky and Tsitsiklis [187]. Graph products is the subject of the book by Imrich and Klavar [121], where the Cartesian product of graphs and the corresponding factorization results are discussed. For the extension of the protocol to the case where the state of each agent is constrained to a convex set, see [173].

We also refer the reader to the notes and references for Chapter 4 for pointers to various extensions of the agreement protocol, particularly, when the underlying graph or digraph is allowed to be time-varying.

SUGGESTED READING

The suggested readings for this chapter are Ren and Beard [204] on the agreement protocol, and Chapter 8 of Meyer [159], which provides a lucid introduction to the theory of non-negative matrices and Markov chains.

EXERCISES

Exercise 3.1. Simulate the agreement protocol (3.2) for a graph on five vertices. Compare the rate of convergence of the protocol as the number of edges increases. Does the convergence of the protocol always improve when the graph contains more edges? Provide an analysis to support your observation.

Exercise 3.2. Consider the digraph $\mathcal{D}$ and the following symmetric protocol

$$\dot{x}(t) = \frac{1}{2} \left\{ L(\mathcal{D}) + L(\mathcal{D})^T \right\} x(t).$$

Does this protocol correspond to the agreement protocol on a certain graph? What are the conditions on the digraph $\mathcal{D}$ such that the resulting symmetric protocol converges to the agreement subspace?

Exercise 3.3. The reverse of $\mathcal{D}$ is a digraph where all directed edges of $\mathcal{D}$ have been reversed. A disoriented digraph is the graph obtained by replacing the directed edges of the digraph with undirected ones. Prove or disprove:

1. The digraph $\mathcal{D}$ is strongly connected if and only if its reverse is strongly connected.
2. A digraph contains a rooted out-branching if and only if its reverse digraph contains one.
3. If the disoriented graph of $\mathcal{D}$ is connected, then either the digraph or its reverse contain a rooted out-branching.
4. A digraph is balanced if and only if its reverse is balanced.

Exercise 3.4. The Kronecker product of two matrices $A = [a_{ij}] \in \mathbf{R}^{n \times m}$ and $B = [b_{ij}] \in \mathbf{R}^{p \times q}$, denoted by $A \otimes B$, is the $np \times mq$ matrix

$$
\begin{bmatrix}
a_{11}B & \cdots & a_{1m}B \\
a_{21}B & \cdots & a_{2n}B \\
a_{31}B & \cdots & a_{3n}B \\
\vdots & \vdots & \\
a_{n1}B & \cdots & a_{nm}B
\end{bmatrix}.
$$

Suppose that the state of each vertex in the agreement protocol (3.1) is a vector in $\mathbf{R}^s$, for some positive integer $s > 0$. For example, x_i might be the position of particle i along a line, that is, $s = 1$. How would the compact form of the agreement protocol (3.2) be modified for the case when $s \geq 2$? *Hint:* use Kronecker products.

Exercise 3.5. How would one modify the agreement protocol (3.1) so that the agents converge to an equilibrium $\bar{x}$, where $\bar{x} = \alpha \mathbf{1} + d$ for some given $d \in \mathbf{R}^n$ and $\alpha \in \mathbf{R}$?

Exercise 3.6. The second-order dynamics of a unit particle i in one dimension is

$$
\frac{d}{dt}
\begin{bmatrix}
p_i(t) \\
v_i(t)
\end{bmatrix}
=
\begin{bmatrix}
0 & 1 \\
0 & 0
\end{bmatrix}
\begin{bmatrix}
p_i(t) \\
v_i(t)
\end{bmatrix}
+
\begin{bmatrix}
0 \\
1
\end{bmatrix}
u_i(t),
$$

where p_i and v_i are, respectively, the position and the velocity of the particle with respect to an inertial frame, and u_i is the force and/or control term acting on the particle. Use a setup, inspired by the agreement protocol, to propose a a control law $u_i(t)$ for each vertex such that: (1) the control input for particle i relies only on the relative position and velocity information with respect to its neighbors; (2) the control input to each particle results in an asymptotically cohesive behavior for the particle group, that is, the positions of the particles remain close to each other; and (3) the control input to each particle results in having a particle group that evolves with the same velocity. Simulate your proposed control law.

Exercise 3.7. How would one extend Exercise 3.6 to n particles in three dimensions?

Exercise 3.8. Consider the uniformly delayed agreement dynamics over a weighted graph, specified as

$$\dot{x}_i(t) = \sum_{j \in N(i)} w_{ij}(x_j(t - \tau) - x_i(t - \tau)), \quad i = 1, \cdots, n,$$

for some $\tau > 0$. Show that this delayed protocol is stable if

$$\tau < \frac{\pi}{2\lambda_n(\mathcal{G})},$$

where $\lambda_n(\mathcal{G})$ is the largest eigenvalue of the corresponding weighted Laplacian. Conclude that, for the delayed agreement protocol, there is a trade-off between faster convergence rate and tolerance to uniform delays on the information-exchange links.

Exercise 3.9. A matrix M is called essentially non-negative if there exists a sufficiently large μ such that $M + \mu I$ is non-negative, that is, all its entries are non-negative. Show that e^{tM} for an essentially non-negative matrix M is non-negative when $t \geq 0$.

Exercise 3.10. An averaging protocol for n agents, with state x_i, $i = 1, 2, \ldots, n$, is the discrete-time update rule of the form

$$x(k + 1) = Wx(k), \quad k = 0, 1, 2, \ldots, \tag{3.23}$$

where $x(k) = [x_1(k), x_2(k), \ldots, x_n(k)]^T$ and W is a stochastic matrix. Derive the necessary and sufficient conditions on the spectrum of the matrix W such that the process (3.23) steers all the agents to the average value of their initial states.

Exercise 3.11. Consider vertex i in the context of the agreement protocol (3.1). Suppose that vertex i (the rebel) decides not to abide by the agreement protocol, and instead fixes its state to a constant value. Show that all vertices converge to the state of the rebel vertex when the graph is connected.

Exercise 3.12. A geometric graph on the unit square is generated by placing n points on the unit square and having $(v_i, v_j) \in E(\mathcal{G})$ when $\|x_i - x_j\| \leq \rho$, where x_i is the coordinate of vertex i and ρ is a given threshold distance for the existence of a link between a pair of vertices. Compute the Laplacian for such graphs on hundred nodes and various values of $\rho \in (0, 1)$. What is your estimate on how $\lambda_2(\mathcal{G})$ grows as a function of ρ?

Exercise 3.13. Show that a balanced digraph is weakly connected if and only if it is strongly connected.

Exercise 3.14. Show that if graphs $\mathcal{G}_1$ and $\mathcal{G}_2$ are connected, then their Cartesian product is connected.

Exercise 3.15. Prove Lemma 3.22.

Exercise 3.16. Consider a network of n processors, where each processor has been given an initial computational load to process. However, before the actual processing occurs, the processors go through an initialization phase, where they exchange certain fractions of their loads with their neighbors in the network. Specifically, during this phase, processor i adopts the *load-update* protocol

$$p_i(k+1) = p_i(k) - \sum_{j \in N(i)} w_{ij}(p_i(k) - p_j(k)), \quad k = 0,1,2\ldots, \quad (3.24)$$

that is, it sends a fraction w_{ij} of its load imbalance with its neighbors to each of them. What is the necessary and sufficient condition on the weights w_{ij} in (3.24) such that this initialization phase converges to a balanced load for all processors when the network is (1) a path graph, (2) a cycle graph, or (3) a star graph?

Exercise 3.17. Given two square matrices A and B, show that

$$e^{A \oplus B} = e^A \otimes e^B,$$

and use this to provide an alternate (shorter) proof for Lemma 3.26.

Exercise 3.18. Let the disagreement vector be

$$\delta(t) = \left(I - \frac{1}{n}\mathbf{1}\mathbf{1}^T\right)x(t).$$

Find the matrix M such that $\dot{\delta}(t) = M\delta(t)$, under the assumption that $\dot{x}(t) = -L(\mathcal{G})x(t)$ for some graph $\mathcal{G}$.

Chapter Four

The Agreement Protocol: Part II–Lyapunov and LaSalle

"Classification of mathematical problems as linear and nonlinear
is like classification of the Universe as bananas and non-bananas."
— unknown source

In this chapter, we consider variations on the basic theme of the agreement protocol. This includes viewing the protocol in the context of Lyapunov theory, which allows for a seamless generalization of its behavior when it evolves over switching networks. We also introduce an alternative representation of the agreement protocol when the dynamics of the edges of the network, as opposed to its nodes, is monitored. We then examine nonlinear extensions of the agreement problem via the passivity framework.

Lyapunov theory is an intuitive framework for the analysis of asymptotic properties of dynamical systems–one with far-reaching consequences. The power and convenience of using this framework is the relative ease by which one can analyze the stability of dynamical systems with nonlinearities, noise, and delays, and to incorporate control inputs to improve the nominal performance of the system.[1] In the first part of this chapter, we will explore the utility of the basic Lyapunov machinery in the realm of the agreement protocol.

4.1 AGREEMENT VIA LYAPUNOV FUNCTIONS

4.1.1 Agreement over Undirected Graphs

Using Lyapunov theory for analyzing the agreement protocol (3.2), at first, seems like bringing in a bulldozer for moving a piano. This is in fact the case. The key realization is that adding things on top of the piano does not prevent the machinery from going through, that is, we can expand on the

---

[1] When inspired or struck by the "right" Lyapunov function.

problem class without having to change the analysis tools. The Lyapunov approach adopted for the agreement problem (3.2) proceeds like this. Given the connected graph $\mathcal{G}$, consider

$$V(x(t)) = \frac{1}{2} x(t)^T x(t), \tag{4.1}$$

that is, half of the *sum of squares* of the vertex states. We also note that the function $V(x)$ is an affine transformation of the quadratic form

$$q(x(t)) = x(t)^T L(K_n) x(t),$$

as it, under the agreement protocol (3.2), becomes

$$\begin{aligned}
q(x(t)) &= x(t)^T (nI - \mathbf{1}\mathbf{1}^T) x(t) \\
&= n x(t)^T x(t) - (\mathbf{1}^T x(t))^2 \\
&= n x(t)^T x(t) - (\mathbf{1}^T x(0))^2,
\end{aligned}$$

since $\mathbf{1}^T x(0) = \mathbf{1}^T x(t)$ for all t; as before, K_n is the complete graph over n vertices.

Now, consider the time-evolution of the function $V(x)$ (4.1) along the trajectory generated by (3.2), which is given by

$$\dot{V}(t) = -x(t)^T L(\mathcal{G}) x(t).$$

Since $L(\mathcal{G})$ is positive semidefinite, the function V (4.1) is a weak Lyapunov function for (3.2) (see the Appendix A.3). Moreover, when $\mathcal{G}$ is connected, the largest invariant set contained in the set

$$\{x \in \mathbf{R}^n \mid \dot{V}(t) = 0\} = \mathbf{span}\ \{\mathbf{1}\}$$

is exactly the null space of $L(\mathcal{G})$. Thus, from LaSalle's invariance principle, convergence to $\mathbf{span}\{\mathbf{1}\}$ follows.

Let us see how this Lyapunov-based approach would be modified if the convergence of the agreement protocol over a strongly connected digraph is being considered. First, consider the set

$$\{x \in \mathbf{R}^n \mid \dot{V}(t) = 0\} = \{x \in \mathbf{R}^n \mid x^T\, (L(\mathcal{D}) + L(\mathcal{D})^T)\, x = 0\}. \tag{4.2}$$

As $\mathcal{D}$ is strongly connected, the largest invariant set contained in (4.2) is still the null space of $L(\mathcal{D})$, which, in turn, is parameterized as $\mathbf{span}\{\mathbf{1}\}$; hence the essential component of the approach, involving LaSalle's invariance principle for connected graphs, remains intact.

How about the more general scenario when the digraph is not necessarily strongly connected, yet contains a rooted out-branching? Of course, as we have seen in the previous chapter, the agreement protocol still converges to the agreement subspace–but we are interested in a "Lyapunov-type argument" to account for this phenomenon. It turns out that in order to adopt a Lyapunov-type argument for this case, one has to change the underlying quadratic structure of the Lyapunov function (4.1).

We will address this in the next section, but before we do, let us point out how the Lyapunov approach can be adopted for the case when the underlying connected graph, or balanced digraph, in the agreement protocol undergoes structural changes over time. When all graphs or digraphs are, respectively, connected or strongly connected in this *switched agreement protocol,* the sum of squares of the state serves as a *common weak Lyapunov function* for (3.2). (See Appendix A.3.) In this light, suppose that the digraph $\mathcal{D}$ that is undergoing structural changes can switch among a finite number of possible strongly connected digraphs

$$\{\mathcal{D}_1, \mathcal{D}_2, \ldots, \mathcal{D}_m\}.$$

Then, with respect to (4.1), one has

$$\dot{V}(t) = -x(t)^T L(\mathcal{D}_i)x(t), \tag{4.3}$$

where $i \in \{1, \ldots, m\}$; in fact, (4.3) can be written as a *differential inclusion*

$$\dot{V}(t) \in \{ x(t)^T L(\mathcal{D})x(t) \mid D \in \{\mathcal{D}_1, \ldots, \mathcal{D}_n\}\}.$$

However, as the digraphs are strongly connected, the set

$$\mathcal{F}_i = \{x \in \mathbf{R}^n \mid x^T \left(L(\mathcal{D}_i) + L(\mathcal{D}_i)^T\right) x = 0\}$$

is *independent* of the index i as it ranges over the index set $\{1, \ldots, m\}$. We have thus obtained the following fact.

LaSalle's invariance principle guarantees that the agreement protocol converges to the agreement subspace as the underlying network is switching among a set of strongly connected digraphs.

4.1.2 Agreement over Digraphs

As we pointed out in Chapter 3, when a digraph contains a rooted out-branching, the agreement dynamics over the digraph (3.4) converges to the agreement subspace. In this section, we explore how this fact can be verified via a suitable Lyapunov function. In order to do this, we will, for ease of presentation, consider the behavior of the agreement protocol at certain time intervals, namely, by letting $z(k) = x(k\delta)$ and considering

$$z(k + 1) = e^{-\delta L(\mathcal{D})}z(k), \quad k = 0, 1, \ldots \tag{4.4}$$

This sampled-data view of the protocol was used when we explored the connection between the agreement protocol and Markov chains in Chapter 3. In fact, as we noted in Chapter 3, for any $\delta > 0$, the matrix exponential $e^{-\delta L(\mathcal{D})}$ is a stochastic matrix. Moreover, the matrix exponential $e^{-\delta L(\mathcal{D})}$, viewed as a non-negative matrix, has a particular pattern for its zero entries.

Lemma 4.1. *Consider the digraph $\mathcal{D}$ and let $\delta > 0$. Then*

$$[e^{-\delta L(\mathcal{D})}]_{ij} > 0$$

if and only if $i = j$ or there is a directed path from j to i in $\mathcal{D}$.

Proof. First we notice that

$$e^{-\delta L(\mathcal{D})} = e^{-\delta \mu} e^{\delta(\mu I - L(\mathcal{D}))}$$

for any $\mu > 0$. Hence, the pattern of zeros in $e^{-L(\mathcal{D})}$ and $e^{\mu I - L(\mathcal{D})}$ are identical. Let us denote the non-negative matrix $\mu I - L(\mathcal{D})$ when

$$\mu > \max_i [L(\mathcal{D})]_{ii},$$

by L_+. The matrix L_+ is non-negative and $[L_+]_{ij} > 0$ if and only if $i = j$ or there is a directed edge from j to i in $\mathcal{D}$. We also note that $[L_+^2]_{ij} > 0$ if and only if there is a directed path of length two from j to i in $\mathcal{D}$ as

$$[L_+^2]_{ij} = \sum_k [L_+]_{ik}[L_+]_{kj}. \tag{4.5}$$

In fact, for any positive integer p, $[L_+^p]_{ij} > 0$ if there is a directed path of length p from j to i in $\mathcal{D}$. The proof of the proposition now follows by viewing the matrix exponential of L_+ in terms of its power series

$$\sum_{j=0}^{\infty} \frac{(\delta)^j}{j!} L_+^j. \tag{4.6}$$

From (4.6) we conclude that the ijth entry of e^{L_+}, and thus of $e^{-\delta L(\mathcal{D})}$, is positive if and only if a directed path–of any length–exists from j to i in $\mathcal{D}$. ∎

Corollary 4.2. *The digraph $\mathcal{D}$ contains a rooted out-branching if and only if, for any $\delta > 0$, at least of one of the columns of $e^{-\delta L(\mathcal{D})}$ is positive.*

Proof. When $\mathcal{D}$ contains a rooted out-branching, there exists a vertex in $\mathcal{D}$ that can reach any other vertex in the digraph via a directed path. Thus, by Lemma 4.1, there exists an index j such that $[e^{-\delta L(\mathcal{D})}]_{ij}$ is positive for all i. ∎

Let us now explore the ramifications of the above corollary in the context of the agreement protocol over a digraph that contains a rooted out-branching. Consider the function

$$V(z) = \max_i z_i - \min_j z_j. \tag{4.7}$$

Since the digraph contains a rooted out-branching, while $V(z) > 0$, one has

$$V(z(k+1)) < V(z(k)), \quad \text{for } k = 1, 2, \ldots$$

To see this, note that (4.4) dictates that at the end of every δ time interval, every vertex essentially updates its state by taking a convex combination of its own state and the states of other vertices in the network. Since one column of the stochastic matrix $e^{-\delta L(\mathcal{D})}$ has positive elements, the states of the nodes with the maximum and minimum entries are updated in such a way that their difference decreases as long as these states are not all equal. Hence, (4.7) is a strong (discrete time) Lyapunov function (see Appendix A.3) for the agreement protocol and (4.4) converges to the state where the Lyapunov function (4.7) vanishes.[2] In the meantime, the set where $V(z) = 0$ coincides with the agreement subspace.

4.2 AGREEMENT OVER SWITCHING DIGRAPHS

The extension of the Lyapunov argument for examining the agreement protocol over switching digraphs is now immediate. Suppose that the digraph switches, possibly at the end of every δ interval, in such a way that the union of the digraphs over some fixed interval of length $T = m\delta$, with m a positive integer, contains a rooted out-branching. It follows from Lemma 4.1 that, for some $T > 0$,

$$[e^{-TL(\cup_{k=1}^m \mathcal{D}_k)}]_{ij} > 0$$

for some j and all i. This also implies that

$$\left[\prod_{k=1}^m e^{-\delta L(\mathcal{D}_k)} \right]_{ij} > 0$$

for some j and all i. To see this, note that, for example, for an arbitrary index p,

$$[e^{-\delta L(\mathcal{D}_p)} e^{-\delta L(\mathcal{D}_{p+1})}]_{ij} = \sum_r [e^{-\delta L(\mathcal{D}_p)}]_{ir} [e^{-\delta L(\mathcal{D}_{p+1})}]_{rj},$$

---

[2]In this setting, the Lyapunov approach is invoked for a discrete-time system.

which is positive if, for some r, there exist a directed path from j to r in $\mathcal{D}_{p+1}$ and a directed path from r to i in $\mathcal{D}_p$. The aforementioned Lyapunov argument, monitoring the Lyapunov function (4.7) in relation to the sequence $z(k)$, generated by

$$z(k+1) = e^{-\delta L(\mathcal{D}_k)} z(k), \quad k = 0, 1, \ldots, \tag{4.8}$$

now implies that the agreement protocol evolving over a sequence of switching digraphs whose union over a fixed time interval T contains a rooted out-branching converges to the agreement subspace.[3]

4.3 EDGE AGREEMENT

LaSalle's invariance principle is the workhorse of convergence analysis for the agreement protocol and its various extensions. However for pedagogical and technical reasons, it is often desirable to resort to a Lyapunov-type argument, assessing the stability aspects of a dynamic system with respect to the origin. A convenient construction that allows for such an analysis is the edge Laplacian, discussed in §2.3.4, and the corresponding edge agreement.

4.3.1 From LaSalle to Lyapunov

Let us consider the system states as defined over the edges–rather than on the nodes–of the graph $\mathcal{G}$ in the agreement protocol (3.2). It is assumed that $\mathcal{G}$ has n nodes and m edges. This edge perspective is facilitated by the transformation

$$x_e(t) = D(\mathcal{G})^T x(t), \tag{4.9}$$

where, as before, $D(\mathcal{G})$ is the incidence matrix of $\mathcal{G}$ (given an arbitrary orientation) and $x_e(t) \in \mathbf{R}^m$ represents the relative internode, or edge, states. Differentiating (4.9) leads to

$$\dot{x}_e(t) = -L_e(\mathcal{G})x_e(t); \tag{4.10}$$

we refer to (4.10) as the *edge agreement protocol*. In lieu of the vertex-to-edge transformation induced by the incidence matrix of $\mathcal{G}$, it follows that "agreement" in the vertex states is equivalent to having $x_e(t) = 0$ when $\mathcal{G}$

---

[3]Note that the sampling interval δ is an arbitrary positive real number.

is connected. As a result, in the edge setting, the edge disagreement,

$$\delta_e(t) = \|x_e(t)\|,$$

rather than being the distance to a subspace, is the distance to the origin; in fact,[4]

$$\|\delta_e(t)\| = \|x_e(t)\| \leq \|D(\mathcal{G})\|\|\delta_v(t)\|, \tag{4.11}$$

where δ_v is the disagreement associated with the vertex states in a connected graph, defined as

$$\delta_v(t) = \mathbf{dist}\,(x(t), \mathcal{A}).$$

As we already know from Chapter 3, the agreement protocol over a connected graph steers the node states toward the agreement subspace. Consequently, the edge agreement protocol (4.10) over a connected graph steers the edge states to the origin. In the edge agreement the evolution of an edge state depends on its current state and the states of its adjacent edges, that is, those that share a vertex with that edge.

4.3.2 Role of Cycles in the Edge Agreement

Cycles in the graph play an important role in the agreement protocol. Recall from Chapter 2 that the null space of the edge Laplacian characterizes the cycle space of the underlying graph. In the meantime, in the agreement protocol, the agreement state is reached when the underlying state trajectory converges to the null space of $L(\mathcal{G})$ for a connected graph $\mathcal{G}$. For connected graphs, the same observation is valid when the system dynamics is specified by the edge Laplacian (4.10): when $x_e(t) \in \mathcal{N}(L_e(\mathcal{G}))$ the agreement state has been reached.

In this section our standing assumption is the connectedness of the graph under consideration. Using an appropriate permutation of the edge indices, we can partition the incidence matrix of $\mathcal{G}$ as

$$D(\mathcal{G}) = [\,D(\mathcal{G}_\tau)\ \ D(\mathcal{G}_c)\,], \tag{4.12}$$

where $\mathcal{G}_\tau$ represents a given spanning tree of $\mathcal{G}$, and $\mathcal{G}_c$ represents the remaining edges not in the tree, that is, the cycle edges; see Figure 4.1. Note that in general $\mathcal{G}_c$ does not represent a connected graph. The partitioning of the incidence matrix induces a corresponding partitioning on the graph

---

[4]Using the matrix induced 2-norm.

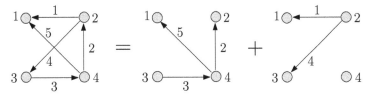

Figure 4.1: A graph can be represented (not necessarily in a unique way) as a tree and edges that complete its cycles.

Laplacian and its edge version as

$$\begin{aligned} L(\mathcal{G}) &= [D(\mathcal{G}_\tau) \ D(\mathcal{G}_c)][D(\mathcal{G}_\tau) \ D(\mathcal{G}_c)]^T \\ &= D(\mathcal{G}_\tau)D(\mathcal{G}_\tau)^T + D(\mathcal{G}_c)D(\mathcal{G}_c)^T \\ &= L(\mathcal{G}_\tau) + L(\mathcal{G}_c) \end{aligned}$$
(4.13)

and

$$\begin{aligned} L_e(\mathcal{G}) &= [D(\mathcal{G}_\tau) \ D(\mathcal{G}_c)]^T [D(\mathcal{G}_\tau) \ D(\mathcal{G}_c)] \\ &= \begin{bmatrix} D(\mathcal{G}_\tau)^T D(\mathcal{G}_\tau) & D(\mathcal{G}_\tau)^T D(\mathcal{G}_c) \\ D(\mathcal{G}_c)^T D(\mathcal{G}_\tau) & D(\mathcal{G}_c)^T D(\mathcal{G}_c) \end{bmatrix} \\ &= \begin{bmatrix} L_e(\mathcal{G}_\tau) & D(\mathcal{G}_\tau)^T D(\mathcal{G}_c) \\ D(\mathcal{G}_c)^T D(\mathcal{G}_\tau) & L_e(\mathcal{G}_c) \end{bmatrix}. \end{aligned}$$
(4.14)

This tree-cycle partitioning of the edge Laplacian as in (4.14), in turn, allows us to make the following observation. In the context of the edge agreement (4.10), the state of the edges corresponding to a spanning tree subgraph $\mathcal{G}_\tau$ evolves according to

$$\dot{x}_\tau(t) = -L_e(\mathcal{G}_\tau)x_\tau(t) - D(\mathcal{G}_\tau)^T D(\mathcal{G}_c)x_c(t),$$
(4.15)

whereas the dynamics of the cycle edges evolve according to

$$\dot{x}_c(t) = -L_c(\mathcal{G}_c)x_c(t) - D(\mathcal{G}_c)^T D(\mathcal{G}_\tau)x_\tau(t).$$
(4.16)

Thereby, the spanning tree and cycle edges act as forcing mechanisms for the mutual evolution of their respective states.

4.3.3 Minimal Edge Agreement

In the previous section we pointed out a connection between the cycles of a graph and the algebraic structure of the corresponding edge Laplacian. This observation can be used to derive a reduced order representation of the edge agreement in terms of the corresponding dynamics on the spanning tree subgraph. In this avenue, let us partition the edge state vector as in

$$x_e(t) = [\, x_\tau^T(t) \;\; x_c^T(t)\,]^T, \tag{4.17}$$

where $x_\tau(t)$ is the edge state of the spanning tree subgraph $\mathcal{G}_\tau$ and $x_c(t)$ denotes the state of the remaining edge states.

Theorem 4.3. *Consider a graph $\mathcal{G}$ with cycles, and a spanning tree subgraph $\mathcal{G}_\tau$, with the corresponding edge Laplacian partitioned as (4.14). Then there exists a matrix R such that*

$$L_e(\mathcal{G}) = R^T L_e(\mathcal{G}_\tau) R. \tag{4.18}$$

Proof. As the graph $\mathcal{G}$ has cycles, the columns of $D(\mathcal{G}_c)$ are linearly dependent on the columns of $D(\mathcal{G}_\tau)$. This can be expressed in terms of the existence of a matrix T such that

$$D(\mathcal{G}_\tau)T = D(\mathcal{G}_c). \tag{4.19}$$

Since $D(\mathcal{G}_\tau)$ has full column rank, its pseudo-inverse exists and we have

$$T = (D(\mathcal{G}_\tau)^T D(\mathcal{G}_\tau))^{-1} D(\mathcal{G}_\tau)^T D(\mathcal{G}_c). \tag{4.20}$$

Therefore, the incidence matrix of $\mathcal{G}$ can be written as

$$D(\mathcal{G}) = [\, D(\mathcal{G}_\tau) \;\; D(\mathcal{G}_\tau)T\,]. \tag{4.21}$$

We can thereby find the edge Laplacian for $\mathcal{G}$ in terms of the matrices $D(\mathcal{G}_\tau)$ and T (4.20) as

$$
L_e(\mathcal{G}) = \begin{bmatrix} L_e(\mathcal{G}_\tau) & L_e(\mathcal{G}_\tau)T \\ T^T L_e(\mathcal{G}_\tau) & T^T L_e(\mathcal{G}_\tau)T \end{bmatrix}
$$

$$
= \begin{bmatrix} I \\ T^T \end{bmatrix} L_e(\mathcal{G}_\tau) \begin{bmatrix} I & T \end{bmatrix}.
$$

The matrix R in the statement of the theorem can now be defined via

$$R = [\, I \;\; T\,]. \tag{4.22}$$

∎

Theorem 4.3 will, through the following proposition, be used to highlight the supporting role that cycles of the graph play in the convergence of the agreement protocol. In fact, all cycle states can be reconstructed from the spanning tree states through the linear relationship derived in (4.20). This is made explicit by the following proposition.

Proposition 4.4. *Consider a graph $\mathcal{G}$ with incidence matrix partitioned as (4.12). Analogously, partition the edge state vector as (4.17). Then the edge agreement (4.10) is equivalent to the descriptor system*

$$R^T \dot{x}_\tau(t) = -R^T L_e(\mathcal{G}_\tau) R R^T x_\tau(t), \tag{4.23}$$

where R is as defined in (4.22). Furthermore, the reduced order system described by

$$\dot{x}_\tau(t) = -L_e(\mathcal{G}_\tau) R R^T x_\tau(t) \tag{4.24}$$

captures the behavior of the edge agreement protocol (4.10). In fact, the cycle edge states can be reconstructed by using the matrix T (4.20) via

$$x_c(t) = T^T x_\tau(t). \tag{4.25}$$

Proof. Using (4.13) and (4.21), the agreement protocol can be written as

$$\dot{x}(t) = \left(-L(\mathcal{G}_\tau) + D(\mathcal{G}_\tau) T T^T D(\mathcal{G}_\tau)^T\right) x(t). \tag{4.26}$$

The edge agreement protocol can then be derived by recalling that

$$x_e(t) = D(\mathcal{G})^T x(t) = \begin{bmatrix} D(\mathcal{G}_\tau)^T \\ T^T D(\mathcal{G}_\tau)^T \end{bmatrix} x(t) = R^T D(\mathcal{G}_\tau)^T x(t).$$

Left-multiplication of (4.26) by $R^T D(\mathcal{G}_\tau)^T$ leads to

$$R^T \dot{x}_\tau(t) = -R^T \left(D(\mathcal{G}_\tau)^T D(\mathcal{G}_\tau) + D(\mathcal{G}_\tau)^T D(\mathcal{G}_\tau) T T^T\right) x_\tau(t),$$

which is the desired result (4.23). The reduced order representation follows directly from the structure of the matrix R.   ■

Theorem 4.4 can be used, in conjunction with (4.15) - (4.16), to show that the cycle states serve as an internal feedback on the dynamics of edges of the spanning tree subgraph of $\mathcal{G}$; this is depicted in Figure 4.2.

4.4 BEYOND LINEARITY

As we have seen so far, the agreement protocol introduced in (3.2) can be extended and examined in several directions, including imposing an orientation on the underlying interaction rule, or allowing the underlying network to switch during the protocol's evolution. However, the overarching assumption in our analysis up to now has been the linearity of the interaction rule. In this section, we explore another venue–closely related to

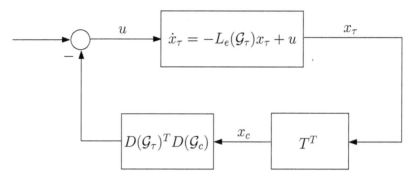

Figure 4.2: Edge dynamics as a feedback structure between spanning tree edges and cycle edges

Lyapunov theory–for analyzing situations when nonlinear elements are included in the general setup of the protocol, defined over connected graphs. This machinery is based on the passivity approach to nonlinear system analysis and design in combination with the edge Laplacian formalism of the previous section. We refer the reader to Appendix A.3 for more on passivity theory; however, we here review the basic setup.

Consider the nonlinear system

$$\dot{z}(t) = f(z(t), u(t)), \quad y(t) = z(t), \tag{4.27}$$

where f is locally Lipschitz (see Appendix A.1) and $f(0,0) = 0$. Then (4.27) is called *passive* if there exists a continuously differentiable, positive semidefinite function V, referred to as the *storage function*, such that

$$u(t)^T y(t) \geq \dot{V}(t) \tag{4.28}$$

for all t; if $\dot{V}$ in (4.28) can be replaced by $\dot{V} + \psi(z)$ for some positive definite function ψ, then we call the system *strictly passive*; in our case, since the output of the system is its state, (4.27) could also be referred to as *output* strictly passive. A storage function is called *radially unbounded* if $V(x) \to \infty$ whenever $\|x\| \to \infty$.

The following theorem is one of the key results in passivity theory.

Theorem 4.5. *Suppose that (4.27) is output strictly passive with a radially unbounded storage function. Then the origin is globally asymptotically stable.*

To demonstrate the utility of this "passivity theorem" in the context of the agreement protocol, consider the interconnection of Figure 4.3, with an integrator in the forward path and the edge Laplacian of a spanning tree, in the feedback path, encoding the edge agreement (4.10). Note that z in this case denotes the vector of edge states x_e (4.9). Then, with respect to the quadratic storage function $V(z(t)) = (1/2)z(t)^T z(t)$, and in reference to Figure 4.3, one has

$$
\begin{aligned}
u(t)^T y(t) &= u(t)^T z(t) \\
&= -z(t)^T L_e(\mathcal{G}_\tau) z(t) + u(t)^T z(t) + z(t)^T L_e(\mathcal{G}_\tau) z(t) \\
&= \dot{V}(z(t)) + z(t)^T L_e(\mathcal{G}_\tau) z(t),
\end{aligned}
$$

implying that the system is strictly (output) passive with a storage function that is radially unbounded. This observation, in turn, makes the convergence analysis for the edge agreement over a spanning tree fall under the domain of Theorem 4.5. Hence, $z(t) \to 0$ as $t \to \infty$, and convergence to the agreement subspace of the "node" states follows.

The connection between the agreement protocol and Theorem 4.5 can be used to extend the basic setup of the agreement protocol in various directions, one of which is the following.

Corollary 4.6. *Suppose that for a connected network of interconnected agents, the edge states evolve according to $\dot{x}_e = -f(\mathcal{G}, x_e)$, where $f : \mathbf{G}_n \times \mathbf{R}^m \to \mathbf{R}^m$ (with $\mathbf{G}_n$ being the set of all graphs on n nodes) satisfies*

$$
x_e^T f(\mathcal{G}, x_e) > 0
$$

for all $x_e \neq 0$, when $\mathcal{G}$ is connected. Then the corresponding node states converge to the agreement subspace.

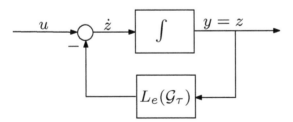

Figure 4.3: Edge agreement over a spanning tree as an output strictly passive system

It is now tempting to extend the agreement protocol to fit one of the many passivity-type results in nonlinear systems theory. One path in this direction would be to base the analysis and design on the following result.

Theorem 4.7. *Consider the feedback connection shown in Figure 4.4, where the time-invariant passive system $G_1 : \dot{z}(t) = f(z(t), u_1(t)), y_1(t) = z(t)$*

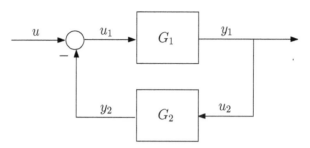

Figure 4.4: Feedback configuration for Theorem 4.7

has a storage function V and the time-invariant memoryless function G_2 is such that $u_2^T y_2 \geq u_2^T \phi(u_2)$ for some function ϕ. Then the origin of the closed loop system (with $u = 0$) is asymptotically stable if $v^T \phi(v) > 0$ for all $v \neq 0$.

To illustrate the ramification of Theorem 4.7, suppose that following the integrator block in Figure 4.3, there exists a nonlinear operator ψ such that for some positive definite functional $V(z)$, one has $\psi(z) = \nabla V(z)$. Then

$$\dot{V}(t) = \nabla V^T \dot{z}(t) = \psi(z)^T \dot{z}(t), \qquad (4.29)$$

implying that the forward path of the feedback configuration shown in Figure 4.5(a) is passive with a storage function V and the function $\phi(v)$ in Theorem 4.7 can be chosen as $\lambda_2(\mathcal{G})v$ for a connected graph. Hence, the asymptotic stability of origin with respect to the edge states $x_e(t)$ can be implied by invoking Theorem 4.7. The more general case of this result for a connected network is also immediate using (4.18) stating that

$$L_e(\mathcal{G}) = R^T L_e(\mathcal{G}_\tau)R,$$

where $\mathcal{G}$ is an arbitrary connected graph. This relationship suggests the loop transformation depicted in Figure 4.5, keeping in mind that passivity of the forward path does not change under post- and premultiplication by matrices R^T and R; Theorem 4.7 can now be invoked under this more general setting.

An example that demonstrates the utility of the above observations for multiagent systems pertains to the Kuramoto model of n coupled oscillators interacting over the network $\mathcal{G}$ as

$$\dot{\theta}_i(t) = k \sum_{j \in N(i)} \sin(\theta_j(t) - \theta_i(t)), \quad i = 1, 2, \ldots, n. \qquad (4.30)$$

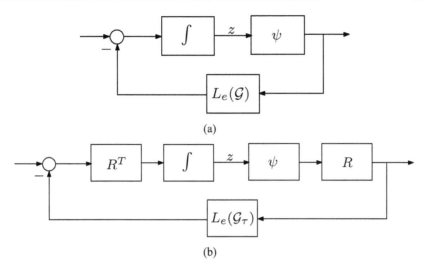

(a)

(b)

Figure 4.5: Loop transformation between feedback connection with edge Laplacian over arbitrary connected graphs (a) to one over spanning trees (b)

In (4.30), the constant k denotes the coupling strength between the oscillators, which for the purpose of this section is assumed to be positive. The nonlinear interaction rule (4.30) can compactly be represented as

$$\dot{\theta}(t) = -kD(\mathcal{G})\sin(D(\mathcal{G})^T\theta(t)), \tag{4.31}$$

where $\theta(t) = [\theta_1(t), \theta_2(t), \ldots, \theta_n(t)]^T$, and we have adopted the convention that when $w = [w_1, \ldots, w_n]^T \in \mathbf{R}^n$, then

$$\sin(w) = [\sin(w_1), \ldots, \sin(w_n)]^T.$$

The edge perspective of §4.3 now leads to

$$D(\mathcal{G})^T\dot{\theta}(t) = -kL_e(\mathcal{G})\sin(D(\mathcal{G})^T\theta(t)); \tag{4.32}$$

hence the Kuramoto model (4.30) can be represented as

$$\dot{z}(t) = -kL_e(\mathcal{G})\sin(z(t)), \tag{4.33}$$

where $z(t) = D(\mathcal{G})^T\theta(t)$. In order to mold the stability analysis of the Kuramoto model (4.30) in the context of passivity theory, we write

$$\dot{z}(t) = -kR^T L_e(\mathcal{G}_\tau)R\sin(z(t)), \tag{4.34}$$

where $L_e(\mathcal{G}_\tau)$ is the edge Laplacian of a spanning tree of $\mathcal{G}$, and hence a positive definite matrix, or when viewed as a dynamic system, a strictly passive element. Now let $V(z(t)) = \mathbf{1}^T(\mathbf{1} - \cos(z(t)))$ be a candidate storage function for the Kuramoto model (4.34). In this case, $V(z) > 0$ for all nonzero $z \pmod{2\pi}$, $V(0) = 0 \pmod{2\pi}$, and $\psi(z) = \sin(z)$, in reference to the identity (4.29) and Fig. 4.5(b). Using the passivity machinery, combined with the edge Laplacian formalism, we thus conclude that for the Kuramoto model over a connected graph, the synchronization state is asymptotically stable.[5]

SUMMARY

In this chapter, we extended the agreement protocol to the case when the network topology undergoes structural changes. What this means is that edges may appear and disappear over time. The key observation is the use of LaSalle's invariance principle in conjunction with the notion of "union" of graphs, or digraphs, over a finite time interval that need to be connected (for graphs) or have a rooted out-branching (for digraphs) to ensure the convergence of the protocol. We then explored the edge Laplacian, which provides the means for viewing the agreement subspace as the origin of an alternate coordinate system. The edge Laplacian also facilitates the extension of the linear agreement protocol to the interconnection of nonlinear elements where certain passivity properties are ensured in the system by the presence of a connected network.

NOTES AND REFERENCES

The use of common Lyapunov functions for studying the agreement protocol over switching networks appeared in the works of Jadbabaie, Lin, and Morse [124] and Olfati-Saber and Murray [182]. The general case of convergence, in the discrete-time case, for the iteration $x(k+1) = W(k)x(k)$ when $W(k)$ is stochastic matrix at each time index k, has long been examined in theory of Markov chains under the heading of inhomogeneous products of non-negative matrices [213],[249]. Convergence analysis of such iterations when they are induced by an underlying switching network was studied by Bertsekas and Tsitsiklis in the context of distributed computation [22] and by Jadbabaie, Lin, and Morse [124], who were motivated by Viscek's model of collective motion of self-driven particles [238], as well as by Tanner, Jadbabaie, and Pappas [228] and Lin, Broucke, and Francis [147]. Extension to the nonlinear setting was pioneered by Moreau [162]; see also Slotine and Wang [219]. The study of

---

[5]By synchronization we refer to the case when $\theta_1 = \theta_2 = \cdots = \theta_n \bmod 2\pi$.

multiagent systems using passivity theory has been pursued by Arcak [10]. Theorems 4.5 and 4.7 are found in Chapter 6 of Khalil [131]. A nice extension of the agreement protocol examined under quantization is found in [43].

SUGGESTED READING

We suggest the work of Moreau [162] and Cortés [53] for examining the agreement protocol under a more general setup over undirected and directed time-varying interconnection topologies. For passivity theory we recommend the books by Khalil [131] and Brogliato, Lozano, Maschke, and Egeland [35].

EXERCISES

Exercise 4.1. What is the relation between the eigenvalues of A and the eigenvalues of its powers? Conclude that if $\lim_{k \to \infty} A^k \neq \infty$, then all eigenvalues of A belong to the unit circle in the complex plane. What can you say about the eigenvalues of the matrix $e^{-L(\mathcal{D})}$ when $\mathcal{D}$ contains a rooted out-branching?

Exercise 4.2. Examine the argument following Corollary 4.2. Provide an analysis for why the value of the Lyapunov function $V(x) = \max_i x_i - \min_i x_i$ has to decrease at every iteration if x does not belong to the agreement subspace, that is, when $V(x) > 0$. Plot $V(x)$ for a representative digraph on five nodes containing a rooted out-branching running the agreement protocol.

Exercise 4.3. Consider the system

$$\dot{\theta}_i(t) = \omega_i + \sum_{j \in N(i)} \sin(\theta_j(t) - \theta_i(t)), \quad \text{for } i = 1, 2, \ldots, n, \quad (4.35)$$

which resembles the agreement protocol with the linear term $x_j - x_i$ replaced by the nonlinear term $\sin(x_j - x_i)$. For $\omega_i = 0$, simulate (4.35) for $n = 5$ and various connected graphs on five nodes. Do the trajectories of (4.35) always converge for any initialization? How about for $\omega_i \neq 0$? (This is a "simulation-inspired question" so it is okay to conjecture!)

Exercise 4.4. Show that the set $[-\pi/2, \pi/2]$ is positively invariant for the edge representation of the Kuramoto model (4.33) when $k \geq 0$, that is, when $z(0) \in [-\pi/2, \pi/2]$, $z(t) \in [-\pi/2, \pi/2]$ for all $t \geq 0$.

Exercise 4.5. Provide an example for an agreement protocol on a digraph that always converges to the agreement subspace (from arbitrary initialization), yet does not admit a quadratic Lyapunov function of the form $\frac{1}{2}x^T x$, that testifies to its asymptotic stability with respect to the agreement subspace.

Exercise 4.6. Consider a matrix $A = [a_{ij}] \in \mathbf{R}^{n \times n}$ with entries that are non-negative. Associate with the matrix A, the digraph $\mathcal{D}$ as follows. Let $V(\mathcal{D})$ be $\{1, 2, \ldots, n\}$ and $ji \in E(\mathcal{D})$ if and only if $a_{ij} > 0$. Show that if the digraph associated with the matrix A is strongly connected, then for some positive integer m, $\sum_m A^m$ has all positive entries.

Exercise 4.7. Let $\lambda_1, \ldots, \lambda_n$ be the ordered eigenvalues of the graph Laplacian associated with an undirected graph. We have seen that the second eigenvalue λ_2 is important both as a measure of the robustness in the graph, and as a measure of how fast the protocol converges. Given that our job is to build up a communication network by incrementally adding new edges (communication channels) between nodes, it makes sense to try and make λ_2 as large as possible.

Write a program that iteratively adds edges to a graph (starting with a connected graph) in such a way that at each step, the edge (not necessarily unique) is added that maximizes λ_2 of the graph Laplacian associated with the new graph. In other words, implement the following algorithm:

Step 0: Given G0 a spanning tree on n nodes. Set k=0
Step 1: Add a single edge to produce Gnew from Gk such that
           lambda2(Gnew) is maximized. Set k=k+1, Gk=Gnew
Repeat Step 1 until Gk=Kn

for $n = 10, 20$, and 50. Did anything surprising happen?

Exercise 4.8. Consider n agents placed on the line at time t, with agent 1 at position 2Δ, agent 2 at position Δ, and the remaining agents at the origin. An edge between agents exists if and only if $|x_i - x_j| \leq \Delta$. Compute where the agents will be at time $t + \delta t$, for some small δt, under the agreement protocol. In particular, for what values of n is the graph connected at time $t + \delta t$?

Exercise 4.9. Does the convergence of the edge states to the origin in an arbitrary graph imply that the node states converge to the agreement set?

Exercise 4.10. Consider the scenario where the relative states in the agreement protocol over a tree are corrupted by a zero mean Gaussian noise with identity covariance. How can (4.24) be modified to reflect this? The $\mathcal{H}_2$ norm of the system $\dot{x}(t) = Ax(t) + Bw(t)$ can be calculated by finding $(\mathbf{trace}\ X)^{1/2}$ where $AX + XA^T + BB^T = 0$; this norm measures how a Gaussian noise w is amplified in the system shown in the figure below. Show that the $\mathcal{H}_2$ norm of the noisy agreement problem over a tree is proportional to the number of edges in the graph. How about the case when the underlying graph is a cycle?

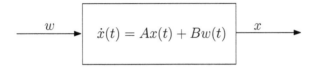

Chapter Five

Probabilistic Analysis of Networks and Protocols

"Always be a little improbable."
— Oscar Wilde

In this chapter we show the extent by which the previously described tools and techniques for reaching agreement on graphs can be adapted to situations where the network contains stochastic elements. In particular, we will discuss the case of random graphs, in which the existence of an edge between a pair of vertices at any given time is driven by a Bernoulli random process. Such models are useful, for instance, for modeling intermittently faulty communication channels. A related topic, also included in this chapter, is how to reach agreement in the presence of noise. We conclude the chapter with a brief overview of other probabilistic models of networks.

Allowing the underlying network in the agreement protocol to switch among a finite number of topologies can be by design or necessity. In the latter case, the analysis that allows us to ascertain that the protocol retains its convergence properties can be categorized as *robustness analysis*. In this chapter, we first consider yet another facet associated with the robustness of the agreement protocol, this time, by allowing random failures in the edges of the network.

5.1 RANDOM GRAPHS

In the Erdős-Rényi model of random graphs on n vertices, the existence of an edge between a pair of vertices in the set $V = \{1, \ldots, n\}$, is determined randomly, independent of other edges, with probability $p \in (0, 1]$. (The case $p = 0$ corresponds to the static, empty graph.) The sample space of all such random graphs will be denoted by $\mathbf{G}(n, p)$. Note that the value of edge probability p is assumed to be same for all potential edges of $\mathcal{G} \in \mathbf{G}(n, p)$; relaxing this assumption however is possible. This edge probability can be fixed, or in more interesting scenarios, a function of the order of the graph; hence the notation $p(n)$ is often used to specify the edge probability.

In the random graph model, all graph parameters are interpreted prob-abilistically. Thus one considers the "expected" number of edges in the random graph $\mathcal{G} \in \mathbf{G}(n, p)$, as the number of edges of $\mathcal{G} \in \mathbf{G}(n, p)$ is a random variable. To further clarify this point, let us consider the probability that $\mathcal{G} \in \mathbf{G}(n, p)$ has exactly m edges. Since there are $\binom{n}{2}$ potential edges among n vertices, the probability that we have any of the m potential edges is

$$p^m (1 - p)^{\binom{n}{2} - m};$$

however there are $\binom{n}{m}$ possible choices for those m edges and hence, the probability of having exactly m edges in $\mathcal{G} \in \mathbf{G}(n, p)$ is

$$\binom{n}{m} p^m (1 - p)^{\binom{n}{2} - m}.$$

Moreover, the expected number of edges in $\mathcal{G} \in \mathbf{G}(n, p)$ is $p\binom{n}{2}$: this fol-lows from the fact that if we let $X_{ij} = 1$ when there is an edge between vertices i and j and $X_{ij} = 0$ if otherwise, then the expected value for the number of edges in the random graph is

$$\mathbf{E} \left\{ \sum_{ij \in E(\mathcal{G})} X_{ij} \right\} = \sum_{ij} \mathbf{E} \left\{ X_{ij} \right\} = \sum_{ij} p = p \binom{n}{2}.$$

The above observation implies that the size of a random graph in $\mathbf{G}(n, p)$ has a binomial distribution: it is the sum of Bernoulli random variables, each taking on values 0 and 1 with probabilities $q = 1 - p$ and p, respectively. This simple observation proves to be useful in showing other results for a number of graph parameters, as the next proposition states.

Proposition 5.1. *The expected number of vertices of degree k for $\mathcal{G} \in$* $\mathbf{G}(n, p)$ *is*

$$n \binom{n - 1}{k} p^k q^{n - 1 - k}.$$

Proof. The probability that a vertex in $\mathcal{G} \in \mathbf{G}(n, p)$ has degree k is $p^k q^{n-1-k}$. There are $\binom{n-1}{k}$ choices for edges to be incident on this vertex; the statement of the proposition now follows by linearity of the expectation operator.

∎

Another probabilistic notion often employed in the theory of random graphs is that of *almost all graphs*. In this venue, one considers a particular graph property $\mathcal{P}$ (for example, connectedness) and examines the asymptotic behavior of

$$\mathbf{Pr}\{\mathcal{G} \in \mathbf{G}(n,p) \text{ has the property } \mathcal{P}\}$$

as $n \to \infty$. If this probability tends to 1, then we say that *almost all graphs in* $\mathbf{G}(n,p)$ *have property* $\mathcal{P}$.

One powerful way to show that a property holds for almost all graphs is to find the expected value or the variance of a random variable associated with $\mathbf{G}(n,p)$ and then invoke one of the *concentration inequalities* in probability. We demonstrate this via an example. First recall that the Markov inequality states that if X is a random variable that only takes non-negative values, then

$$\mathbf{Pr}\left\{ X \geq t \right\} \leq \frac{\mathbf{E}\{X\}}{t}.$$

In particular, if X is integer-valued, then $\mathbf{E}\{X\} \to 0$ as $n \to \infty$ implies that

$$\mathbf{Pr}\{X = 0\} \to 1 \quad \text{as } n \to \infty.$$

Using the Markov inequality, we can state the following observation.

Theorem 5.2. *If* p *is constant, then almost all graphs in* $\mathbf{G}(n,p)$ *are connected and have diameter 2.*

Proof. Let X_{ij} be a random variable on $\mathbf{G}(n,p)$ such that

$$X_{ij} = \begin{cases} 1 & \text{if } v_i \text{ and } v_j \text{ have no common neighbors,} \\ 0 & \text{otherwise.} \end{cases} \tag{5.1}$$

Then $\mathbf{Pr}\{X_{ij} = 1\} = (1-p^2)^{n-2}$ since if two vertices do share a neighbor, none of the other $n-2$ vertices do. Letting $X = \sum_{ij} X_{ij}$, where indices i and j run over all distinct pairs of vertices, we obtain

$$\mathbf{E}\{X\} = \binom{n}{2}(1 - p^2)^{n-2},$$

which, when p is fixed, approaches zero as $n \to \infty$. By the Markov inequality, it now follows that for a fixed p, the probability that a pair of vertices do not have a common neighbor approaches zero as $n \to \infty$. Hence, the probability that the graph has diameter 2, and hence is connected, goes to 1 as $n \to \infty$. ∎

Results such as Theorem 5.2 are abundant in the theory of random graphs–often quite elegant and surprising. The crucial step in the proof of many such results is the "right" choice of the random variable and the application of the "right" concentration inequality (such as the Markov inequality). For example, the expected value of the vertex degree in a random graph is $p(n-1)$; using the Chernoff bound (see Appendix A.4) one can show that

$$|d(v_i) - pn| \le 2(\sqrt{pn \log n} + \log n),$$

which in our asymptotic notation can be written as

$$|d(v_i) - \mathbf{E}\{d(v)\}| = \mathcal{O}(\sqrt{n \log n})$$

for almost all graphs. We will use this asymptotic bound in §5.2.2.

5.2 AGREEMENT OVER RANDOM NETWORKS

We now assume that the underlying network in the agreement protocol (3.2) is in fact a random graph, when sampled at particular time intervals. Specifically, we assume that during a given time interval, there is a probability $q \in [0,1)$ of loosing the information-exchange link between a pair of agents, that is, an edge between a pair of vertices exists with probability $p \in (0,1]$, where $p = 1 - q$. Moreover, as the random model requires, we consider the situation where the failure probability on one edge is independent of that on others.

Having embedded a random network in the dynamic system (3.2), analogous to the switching networks of §4.1.2, we proceed to consider its evolution in a sampled-data setting. Thus, we consider an arbitrary sampling of the time axis at intervals $\delta > 0$, and monitor the trajectory of $z(k) = x(k\delta)$ expressed by

$$z(k+1) = e^{-\delta L_k} z(k), \quad k = 0, 1, \ldots, \tag{5.2}$$

where L_k is the Laplacian matrix of the random graph as realized at time $k\delta$. We denote by $\mathbf{L}(n,p)$ as the set of Laplacian matrices associated with random graphs on n nodes with probability of link failures $1 - p$. Thereby, in (5.2), for each k, one has $L_k \in \mathbf{L}(n,p)$. We will assume that during the time interval $[k\delta, (k+1)\delta)$, the dynamics of the system is governed by L_k, that is, in between the sample times, the graph does not change.

Let us start gathering the ingredients that we will need for applying the *stochastic version of LaSalle's invariance principle* to the agreement protocol over random graphs.

5.2.1 Agreement via the Stochastic Version of LaSalle's Invariance Principle

We start this section with an observation.

Corollary 5.3. *Suppose that for every realization $L \in \mathbf{L}(n,p)$ one has*

$$z^T \left(I - e^{-\delta L} \right) z = 0; \tag{5.3}$$

then $z \in \mathcal{A}$, where $\mathcal{A}$ is defined in (3.5).

Proof. Note that $I - e^{-\delta L}$ is positive semidefinite; hence (5.3) is equivalent to $(I - e^{-\delta L})z = 0$. The proof of the corollary is now a direct consequence of the Perron-Frobenius theorem (see Appendix), as one can choose a connected graph with the Laplacian $L \in \mathbf{L}(n,p)$ such that the corresponding $e^{-\delta L}$ is positive. In this case, the vector z satisfying (5.3) has to be in the span of the eigenvector of $e^{-\delta L}$ corresponding to the unit eigenvalue and, thus in $\mathcal{A}$. ∎

For the proof of convergence to the agreement set $\mathcal{A}$ for the trajectories governed by (5.2), one can invoke a variety of techniques ranging from the ergodic behavior of positive matrices (or non-negative matrices in the case of digraphs) to those based on Lyapunov techniques. Our exposition, in line with our aspiration for consistency and reinforcing our bulldozer analogy at the beginning of Chapter 4, will be based on the *stochastic version of the Lyapunov technique.*

First, we need to recall some definitions for convergence in a probabilistic setting which will be used shortly; for the distinction between various forms of convergence in the probabilistic setting, we refer the reader to the Appendix; see also Notes and References to this chapter. The following mode of convergence is often considered the strongest type of convergence for a random sequence.

Definition 5.4. *A random sequence $\{z(k)\}$ in $\mathbf{R}^n$ converges to z^* with probability 1 (w.p.1) if, for every $\epsilon > 0$,*

$$\mathbf{Pr} \left\{ \sup_{k \geq N} \| z(k) - z^* \| \geq \epsilon \right\} \to 0 \quad as \quad N \to \infty.$$

Similarly, for $\mathcal{A} \subseteq \mathbf{R}^n$, we write

$$\{z(k)\} \to \mathcal{A} \quad w.p.1$$

if, for every $\epsilon > 0$,

$$\mathbf{Pr}\left\{\sup_{k \geq N} \mathbf{dist}\,(z(k), \mathcal{A}) \geq \epsilon\right\} \to 0 \quad \text{as} \quad N \to \infty,$$

where $\mathbf{dist}\,(z, \mathcal{A}) = \inf_{y \in \mathcal{A}} \|y - z\|.$

In view of our observation in (3.9),

$$\mathbf{dist}\,(z(k), \mathcal{A}) = \left\| z(k) - \frac{1^T z(0)}{1^T 1} 1 \right\|$$

$$= \left(z(k)^T z(k) - \frac{(1^T z_0)^2}{1^T 1} \right)^{1/2}$$

$$= \left(\frac{1}{n} z(k)^T \widehat{L} z(k) \right)^{1/2},$$

where $\widehat{L} = L(K_n)$ is the graph Laplacian for the complete graph K_n. Note that

$$\widehat{L} = nI - 11^T.$$

The main result of this section is as follows.

Proposition 5.5. *The trajectory of the random dynamical system (5.2) with matrix* $L \in \mathbf{L}(n, p)$ *converges to the agreement set* $\mathcal{A}$ *(3.5) w.p.1.*

Proof. Consider the function

$$V(z(k)) = \frac{1}{n} z(k)^T \widehat{L} z(k) = \frac{1}{n} \sum_{i \neq j} \|z_i(k) - z_j(k)\|^2;$$

then one has

$$V(z(k)) = z(k)^T z(k) - \frac{c^2}{n} \quad \text{when} \quad 1^T z(k) = c.$$

Consider the evolution of the quantity $\mathbf{E}\,[V(z(k+1)) - V(z(k)) \,|\, z(k)]$, which, when restricted to the trajectories of the random dynamical system (5.2), assumes the form

$$\mathbf{E}\{ z(k)^T (e^{-2\delta L} - I) z(k) \,|\, z(k)\}.$$

Define $\mathbf{E}\,\{ e^{-2\delta L} - I \} = -C$. We know that for all $L \in \mathbf{L}(n, p)$, the spectrum of $e^{-\delta L}$ is as

$$e^{-\lambda_n(\mathcal{G})}, e^{-\lambda_{n-1}(\mathcal{G})}, \ldots, e^{-\lambda_2(\mathcal{G})}, 1,$$

and the matrix C is positive semidefinite. Hence $V(z(k))$ is a supermartingale (see Appendix A.4). Invoking the stochastic version of LaSalle's invariance principle, we conclude that

$$\mathbf{E}\{z(k)^T(e^{-2\delta L} - I)z(k)\} \to 0 \quad \text{as} \quad k \to \infty,$$

and $z(k)^T\left(\mathbf{E}\{e^{-2\delta L} - I\}\right)z(k) \to 0$ w.p.1. The set

$$\mathcal{I} = \{z \mid z^T(\mathbf{E}\{I - e^{-2\delta L}\})z = 0\} \tag{5.4}$$

is an invariant set for the dynamical system (5.2). Moreover, we note that

$$\mathbf{E}\{I - e^{-2\delta L}\} = \sum_{i=1}^{g(n)} p_i(I - e^{-2\delta L_i}),$$

where $g(n) = 2^{\binom{n}{2}}$ is the cardinality of the set of graphs on n vertices and L_i is the Laplacian matrix associated with the ith graph in this set. Since for all $i = 1, \ldots, g(n)$, the matrix $I - e^{-2\delta L_i}$ is positive semidefinite and $p_i > 0$, one must have that for all $z \in \mathcal{I}$ and realizations $L \in \mathbf{L}(n, p)$,

$$z^T\left(I - e^{-2\delta L}\right)z = 0.$$

Consequently, in view of Corollary 5.3, the largest invariant set contained in $\mathcal{I}$ (5.4) is nothing but the agreement subspace. ∎

5.2.2 Rate of Convergence

Although determining the rate of convergence of the agreement protocol over a switching network is rather involved, the convergence analysis of the protocol over random networks benefits from a combination of stochastic Lyapunov theory and random matrix theory. In this direction, let $\{\widehat{z}(k)\}_{k \geq 1}$ be the projection of $\{z(k)\}_{k \geq 1}$ onto the subspace orthogonal to the agreement subspace $\mathcal{A}$. Thus, for all $z(k) \in \mathbf{R}^n$, one has $\widehat{z}(k)^T \mathbf{1} = 0$. Now we monitor the behavior of the Lyapunov function in Proposition 5.5 along the projected trajectories onto this subspace, $\mathcal{A}^\perp$. Of course, as before,

$$\mathbf{E}\{V(\widehat{z}(k+1)) - V(\widehat{z}(k)) \mid \widehat{z}(k)\} = \widehat{z}(k)^T \mathbf{E}\{e^{-2\delta L} - I\}\widehat{z}(k).$$

As the vector $\mathbf{1}$ is the eigenvector corresponding to the largest eigenvalue of all matrix exponentials $e^{-2\delta L}$ with $L \in \mathbf{L}(n, p)$, one has

$$\widehat{z}(k)^T \mathbf{E}\{e^{-2\delta L} - I\}\widehat{z}(k) \leq (\lambda_{n-1}(\mathbf{E}\{e^{-2\delta L}\}) - 1)\,\widehat{z}(k)^T\widehat{z}(k), \tag{5.5}$$

where $\lambda_{n-1}(\mathbf{E}\{e^{-2\delta L}\})$ denotes the *second largest* eigenvalue of the matrix $\mathbf{E}\{e^{-2\delta L}\}$. In particular, we note that the inequality

$$\lambda_{n-1}(\mathbf{E}\{e^{-2\delta L}\}) < 1$$

guarantees that the Lyapunov function V is a supermartingale with a bounded decrease in its value at each time step. In order to gain some insight into the behavior of the quantity

$$\lambda_{n-1}(\mathbf{E}\{e^{-2\delta L}\})$$

for random graphs, we note that

$$\lambda_{n-1}(\mathbf{E}\{e^{-2\delta L}\}) = \max_{\|z\|=1, z \perp \mathbf{1}} \sum_{i=1}^{g(n)} p_i\, z^T e^{-2\delta L_i} z$$

$$\leq \mathbf{E}\{e^{-2\delta \lambda_2(L)}\}.$$

Let us define the quantity $\alpha(n, p, \delta) = \mathbf{E}\{e^{-2\delta \lambda_2(L)}\}$. Since for a subset of indices i,

$$0 < e^{-2\delta \lambda_2(L_i)} < 1 \quad \text{(when } \mathcal{G} \text{ is connected)},$$

and for the complement subset

$$e^{-2\delta \lambda_2(L_i)} = 1 \quad \text{(when } \mathcal{G} \text{ is disconnected)},$$

one has

$$0 < \alpha(n, p, \delta) < 1.$$

Thus

$$\mathbf{E}\{V(\hat{z}(k+1)) - V(\hat{z}(k)) \mid \hat{z}(k)\} \leq (\alpha(n, p, \delta) - 1)\, \|\hat{z}(k)\|^2,$$

and for all $\gamma > 0$,

$$\mathbf{Pr}\left\{\sup_{k \geq N} \hat{z}(k)^T \hat{z}(k) \geq \gamma \right\} \leq \frac{\alpha(n, p, \delta)^N}{\gamma}\, \hat{z}(0)^T \hat{z}(0).$$

The rate of convergence of the agreement protocol on a random network is hence dictated by the quantity $\alpha(n, p, \delta)$ as well as by $\lambda_{n-1}(\mathbf{E}\{e^{-2\delta L}\})$.

5.2.3 Convergence of Random Agreement over Large-scale Networks

Let us make a few remarks on the rate of convergence of the agreement protocol over "large" random networks. As evident from the previous section, in particular the inequality (5.5), the convergence of this protocol is dictated by[1]

$$\mathbf{E}\{e^{-2\delta \lambda_2(L)}\} \quad \text{and} \quad \lambda_{n-1}(\mathbf{E}\{e^{-2\delta L}\})$$

---

[1]The quantity $\lambda_{n-1}(\mathbf{E}\{e^{-2\delta L}\})$ points out the importance of the spectral properties of an "average" graph for the convergence of the protocol on a random network.

as $n \to \infty$. In the rest of this section, we provide some insight into the behavior of $\lambda_2(L)$ as a function of n and p; we denote this eigenvalue by $\lambda_2(n, p)$.

As $L(\mathcal{G})$ for $\mathcal{G} \in \mathbf{G}(n, p)$ is a *random matrix*, it is natural to explore the relevance of the spectral theory of random matrices in the context of random graphs. The celebrated Wigner's semicircle law is a central result in this direction. In the following, we will be referring to $[A]_{kj}$ as a_{ij}.

Theorem 5.6. *Let A be a symmetric $n \times n$ matrix where the entries a_{ij} for $i \geq j$ are independent real-valued random variables with finite moments. The entries a_{ij} with $i > j$ are required to have an identical distribution function; moreover, the entries a_{ii} possess the same distribution. Consider the quantity*

$$W_n(x) = \frac{\text{number of eigenvalues} \leq x}{n}.$$

Assuming that the variance of a_{ij} is σ^2, one has

$$\lim_{n \to \infty} W_n(2\sigma\sqrt{n}\,x) = W(x) \quad \text{in probability},$$

where W is an absolutely continuous distribution function with density

$$W(x) = \begin{cases} \frac{2}{\pi}\sqrt{1 - x^2} & \text{for} \quad |x| \leq 1, \\ 0 & \text{for} \quad |x| > 1, \end{cases}$$

namely, a semicircular distribution.

Wigner's semicircle law provide a limiting distribution for the location of the eigenvalues of a random symmetric matrix. However, for our purpose, the following ramification of the semicircle law proves to be particularly helpful; see Notes and References for pointers to the proof of the semicircle law.

Corollary 5.7. *Let A be a symmetric $n \times n$ matrix where the entries a_{ij} for $i \geq j$ are independent real-valued random variables with finite moments. The entries a_{ij} with $i > j$ are required to have identical distribution functions; moreover the entries a_{ii} possess the same distribution. Let*

$$\lambda_n \geq \lambda_{n-1} \geq \cdots \geq \lambda_1$$

be the ordered eigenvalues of A. Then, for any $\epsilon > 0$, one has

$$\lambda_{n-1}(A) = \mathcal{O}(n^{\frac{1}{2}+\epsilon}) \quad \text{in probability}.$$

Using Corollary 5.7, a bound on the second smallest eigenvalue of the graph Laplacian can be derived as follows. Recall that $L(\mathcal{G}) = \Delta(\mathcal{G}) - A(\mathcal{G})$, where $\Delta(\mathcal{G})$ is the degree matrix for the graph and $A(\mathcal{G})$ is its adjacency matrix. Thus

$$L(\mathcal{G}) = (-A(\mathcal{G}) + p(n-1)I) + (\Delta(\mathcal{G}) - p(n-1)I),$$

and therefore for any $\epsilon > 0$, one has[2]

$$\lambda_2(n,p) = pn + \mathcal{O}(n^{\frac{1}{2}+\epsilon}) \quad \text{in probability.} \tag{5.6}$$

The bound (5.6) can in fact be sharpened as follows: let $p \in (0,1]$ and for any $\epsilon \in (0,2)$, one has

$$\lim_{n \to \infty} \mathbf{Pr}\left\{pn - f_\epsilon^+(n) < \lambda_2(n,p) < pn - f_\epsilon^-(n)\right\} = 1, \tag{5.7}$$

where

$$f_\epsilon^+(n) = \sqrt{(2+\epsilon)\,p\,(1-p)\,n \log n}$$

and

$$f_\epsilon^-(n) = \sqrt{(2-\epsilon)\,p\,(1-p)\,n \log n}.$$

We note that the inequalities in (5.7) indicate that for fixed p and large values of n, $\lambda_2(n,p)$ is an increasing function of n. Moreover, one has

$$\mathbf{Pr}\left\{e^{-\lambda_2(L)} > e^{-pn + f_\epsilon^-(n)}\right\} \to 1 \quad \text{as} \quad n \to \infty.$$

As a direct consequence of this observation, one can state that for the agreement protocol (5.2), *the rate of convergence is improved, at least linearly, for random networks of larger order when link probabilities are fixed.* This observation can also be interpreted in terms of the improved probabilistic *robustness* properties of large random networks as they pertain to the agreement protocol; see Figure 5.1. In fact, when the probability of edge failures in the random graph is fixed, larger networks exhibit better convergence rates for the agreement protocol.

---

[2]In fact, it can be shown that $\lim_{n \to \infty}(\lambda_n(A)/n) = p$.

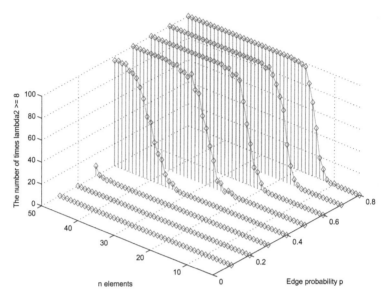

Figure 5.1: The number of times, out of 100 trials, that the inequality $\lambda_2(n, p) \geq 8$ holds for each values of n and p; $2 \leq n \leq 50$, $0.1 \leq p \leq 0.8$

5.3 AGREEMENT IN THE PRESENCE OF NOISE

In this section, we consider a slightly modified discrete version of the agreement protocol where the interagent, relative information exchange is corrupted by noise. Although this is topically distinct from the random graph discussion of the previous section, it requires similar stochastic analysis tools, justifying its inclusion in this chapter.

In particular, assuming that $z_i(k)$ denotes the state of node i at time step k, we consider the evolution of the discrete time system

$$z_i(k+1) = z_i(k) - \gamma(k)\, u_i(k), \quad k = 1, 2, \ldots, \tag{5.8}$$

where $\gamma(k) > 0$ is a time-varying step size and $u_i(k)$ is the control input or update direction at time k for vertex i. The update direction for i is assumed to depend on a noisy measurement of its relative state with respect to its neighbors as

$$u_i(k) = \sum_{j \in N(i)} (z_i(k) - z_j(k) + \eta_{ji}(k)), \qquad (5.9)$$

where $N(i)$ denotes the set of neighbors of node i, and $\eta_{ji}(k)$ is the noise on edge ij at time k. We assume that all η_{kj} are independent, uncorrelated, and Gaussian distributed with zero mean and variance σ^2. Note that in order to guarantee convergence of the protocol, it is imperative to include a time-dependent multiplicative term such as $\gamma(k)$ in (5.8), as otherwise there would not be an asymptotic noise rejection mechanism.

The control input (5.9) can be conveniently written in terms of the Laplacian and adjacency matrices of the network graph as

$$u_i(k) = \sum_{j=1}^n [L(\mathcal{G})]_{ij}\, x_j(k) + \sum_{j=1}^n [A(\mathcal{G})]_{ji}\, \eta_{ji}(k). \qquad (5.10)$$

Note that the second term on the right-hand side of (5.10) is the total noise input for vertex i at time k; it will be denoted by $w_i(k)$. Observe that for all i and k, the $w_i(k)$ are zero mean, independent, uncorrelated Gaussian distributed random variables. The variance of $w_i(k)$ can be computed using the following fact.

Lemma 5.8. *Let M be a positive integer-valued random variable. For independent identically distributed random variables $Y_1, Y_2, \ldots,$ with mean μ and variance σ^2, the mean and variance of the sum $Y = Y_1 + Y_2 + \cdots + Y_M$, are $\mathbf{E}\{Y\} = \mu\, \mathbf{E}\{M\}$ and*

$$\mathbf{var}\{Y\} = \mu^2\, \mathbf{var}\{M\} + \sigma^2\, \mathbf{E}\{M\}. \qquad (5.11)$$

Since all η_{ji} are Gaussian distributed with zero mean and variance σ^2, it follows from (5.11) that for all k,

$$\mathbf{var}\{w_i(k)\} = \mathbf{E}\{|w_i(k)|^2\} = \sigma^2\, \mathbf{E}\{d(v)\}.$$

We note that for random graphs for example, $\mathbf{E}\{d(v)\} = p(n-1)$. The discrete time protocol in (5.8) can now be expressed as

$$\begin{aligned} z(k+1) &= z(k) - \gamma(k)(L(\mathcal{G})z(k) + w(k)) \\ &= (I - \gamma(k)L(\mathcal{G}))z(k) - \gamma(k)w(k), \end{aligned} \qquad (5.12)$$

where $w(k) = [w_1(k), \ldots, w_n(k)]^T$. It is apparent from (5.12) that convergence of the agreement protocol is dependent on the choice of the step size $\gamma(k)$. For a fixed γ and a connected graph, however, the variance of vertex i's state at any time cannot be less than $\gamma \sigma^2 \mathbf{E}\left[d(v)\right]$ (this follows from the fact that $\mathbf{var}\left[w_i(k)\right] = \sigma^2 \, \mathbf{E}\left[d(v)\right]$) and the agreement protocol fails to converge w.p.1. It is interesting to note that for a given $\gamma(k)$, a higher average node degree implies higher variance in $u_i(k)$ and possibly slower convergence of the protocol. This is in contrast to the noise-free case, where a higher node degree often implies greater information sharing and faster convergence.

One approach by which the variance of the state vector can be driven to zero is to adopt a time-varying step size in the agreement protocol (5.12) that satisfies the following conditions:

$$\lim_{k \to \infty} \gamma(k) = 0, \quad \sum_{k=1}^{\infty} \gamma(k) = \infty, \quad \text{and} \quad \sum_{k=1}^{\infty} \gamma^2(k) < \infty. \quad (5.13)$$

Note that the second condition is necessary to allow for a sufficient number of updates for the protocol to converge. As we will show shortly, an additional condition on $\gamma(k)$–related to the underlying graph Laplacian–is required to ensure convergence of the agreement protocol over noisy networks.

5.3.1 Convergence Analysis

In this section we consider the probabilistic convergence of the agreement protocol (5.12) in the presence of noise. One of the crucial constructs in this venue is the notion of the *pseudogradient*. The pseudogradient inequality is defined implicitly in Proposition 5.9 below (via the inequality (5.16)). First let us consider the quadratic function

$$V(z(k)) = \frac{1}{2} \, z(k)^T \, L(\mathcal{G}) \, z(k), \quad (5.14)$$

admitting the gradient $\nabla V(z(k)) = L(\mathcal{G})z(k)$. Note that $\nabla V(z(k))$ is Lipschitz continuous with a constant $\lambda_n(\mathcal{G}) > 0$ for nonempty graphs; that is,

$$\| L(\mathcal{G}) \, z(k_1) - L(\mathcal{G}) \, z(k_2) \|_2 \le \lambda_n(\mathcal{G}) \, \| z(k_1) - z(k_2) \|_2 \quad (5.15)$$

since the spectral norm of $L(\mathcal{G})$ is equal to $\lambda_n(\mathcal{G})$. We now show that the process (5.12) is a strong pseudogradient with respect to $V(z(k))$ defined above.

Proposition 5.9. *For a connected graph,* $u(k) = L(\mathcal{G}) \, z(k) + w(k)$ *is a strong pseudogradient of* $V(z(k))$ *(5.14), that is,*

$$\nabla V(z(k))^T \, \mathbf{E}\{u(k) \,|\, z(k)\} \ge \beta \, V(z(k)), \quad (5.16)$$

if $0 < \beta \leq 2\lambda_2(\mathcal{G})$, *where* $\lambda_2(\mathcal{G})$ *is the second smallest eigenvalue of the graph Laplacian* $L(\mathcal{G})$.

Proof. Observe that

$$\mathbf{E}\{u(k) \,|\, z(k)\} = \mathbf{E}\{L(\mathcal{G})z(k) + w(k) \,|\, z(k)\} = L(\mathcal{G})\,z(k),$$

since $\mathbf{E}\{w(k) \,|\, z(k)\} = \mathbf{E}\{w(k)\} = 0$. For the inequality (5.16) to hold, it suffices to ensure that for all $z(k)$,

$$z(k)^T \left[L(\mathcal{G})^2 - \frac{\beta}{2}\, L(\mathcal{G}) \right] z(k) \geq 0.$$

The last inequality certainly holds if $L(\mathcal{G})^2 - (\beta/2)\, L(\mathcal{G})$ is positive semidefinite for some $\beta > 0$. Since the spectrum of $L(\mathcal{G})^2$ is $\{\lambda_i^2(\mathcal{G}) : 1 \leq i \leq n\}$, with $\lambda_i(\mathcal{G})$ being the ith smallest eigenvalue of $L(\mathcal{G})$, it follows that $L(\mathcal{G})^2 - (\beta/2)\, L(\mathcal{G})$ is positive semidefinite for $\beta \leq 2\lambda_2(\mathcal{G})$. Since $\lambda_2(\mathcal{G}) > 0$ for a connected graph, the strong pseudogradient inequality (5.16) is satisfied for any $\beta \in (0, 2\lambda_2(\mathcal{G})]$. ∎

We now state an observation, followed by a lemma that is essential to the subsequent convergence analysis.

Proposition 5.10. *Let* $\mathcal{G}$ *be a connected graph. Suppose that the trajectory of (5.12) is such that, for the quadratic function (5.14),* $V(z(k)) \to 0$ *w.p.1. Then* **dist** $(z(k), \mathcal{A}) \to 0$ *w.p.1.*

Lemma 5.11. *Consider the sequence of non-negative random variables*

$$\{V(k)\}_{k \geq 0} \quad \textit{with} \quad \mathbf{E}\{V(0)\} < \infty.$$

Let

$$\mathbf{E}\{V(k+1) \,|\, V(0), \ldots, V(k)\} \leq (1 - c_1(k))\, V(k) + c_2(k), \quad (5.17)$$

with $c_1(k)$ *and* $c_2(k)$ *satisfying*

$$0 \leq c_1(k) \leq 1, \ c_2(k) \geq 0, \ \sum_{k=0}^{\infty} c_2(k) < \infty, \ \sum_{k=0}^{\infty} c_1(k) = \infty$$

and

$$\lim_{k \to \infty} \frac{c_2(k)}{c_1(k)} = 0. \qquad (5.18)$$

Then $V(k) \to 0$ *w.p.1.*

Proof. We only provide a sketch of the proof; see notes and references for pointers to the full proof. First, one constructs the auxiliary non-negative sequence

$$U(k) = V(k) + \sum_{j=k}^{\infty} c_2(j).$$

Next, using the first two conditions on $c_1(k)$ and $c_2(k)$, it can be shown that $\mathbf{E}\{U(k+1)\} \le U(k)$. Therefore, the sequence $\{U(k)\}_{k \ge 0}$ is a non-negative supermartingale and converges to some non-negative random variable U^* w.p.1. On the other hand, since $\{c_2(k)\}_{k \ge 0}$ is summable, it follows that the sequence $\{V(k)\}_{k \ge 0}$ converges to some non-negative random variable V^* w.p.1. Using the conditions on $c_1(k)$ and $c_2(k)$, it then follows that

$$\mathbf{E}\{V(k)\}_{k \ge 0} \to 0 \quad \text{w.p.1.}$$

The convergence statements

$$V(k) \to V^* \quad \text{and} \quad \mathbf{E}\{V(k)\} \to 0,$$

both w.p.1, can now be used to prove that $V^* = 0$.   ∎

We now arrive at the main result of this section.

Proposition 5.12. *For a connected graph, the trajectory of the system (5.12) converges to the agreement set $\mathcal{A}$ w.p.1 if the conditions in (5.13) hold and for all $k \ge 1$, $\gamma(k) \le 2/\lambda_n(\mathcal{G})$.*

Proof. Using the quadratic function (5.14), one has

$$
\begin{aligned}
V(z(k+1)) &= \frac{1}{2} z(k+1)^T L(\mathcal{G}) z(k+1) \\
&= \frac{1}{2} (z(k) - \gamma(k) u(k))^T L(\mathcal{G}) (z(k) - \gamma(k) u(k)) \\
&= \frac{1}{2} z(k)^T L(\mathcal{G}) z(k) - \gamma(k) (z(k)^T L(\mathcal{G})) u(k) \\
&\quad + \frac{\gamma(k)^2}{2} u^T(k) L(\mathcal{G}) u(k) \\
&\le V(z(k)) - \gamma(k) \nabla V(z(k))^T u(k) \\
&\quad + \frac{\gamma(k)^2}{2} \lambda_n(\mathcal{G}) \|u(k)\|^2,
\end{aligned}
\tag{5.19}
$$

since

$$u(k)^T L(\mathcal{G})u(k) = \|u(k)^T L(\mathcal{G})u(k)\|$$
$$\leq \|L(\mathcal{G})\| \, \|u(k)\|^2$$
$$= \lambda_n(L(\mathcal{G})) \, \|u(k)\|^2.$$

Concurrently,

$$\mathbf{E}\{u(k)\|^2 \mid z(k)\} = \mathbf{E}\{(L(\mathcal{G})z(k) + w(k))^T \, (L(\mathcal{G})z(k) + w(k)) \mid z(k)\}$$
$$= z(k)^T L(\mathcal{G})^2 \, z(k) + \mathbf{E}\{w^T(k)w(k)\}$$
$$= (z(k)^T L(\mathcal{G}))(L(\mathcal{G})z(k)) + \sum_{i=1}^{n} \mathbf{var}\{w_i(k)\}$$
$$= \nabla V(z(k))^T \, \mathbf{E}\{u(k) \mid z(k)\} + \sigma^2 \sum_{i=1}^{n} d(v_i)$$
$$= \nabla V(z(k))^T \, \mathbf{E}\{u(k) \mid z(k)\} + \left(n\sigma^2\right) \bar{d}(\mathcal{G}), \qquad (5.20)$$

where $d(v_i)$ denotes the degree of node i in the graph and

$$\bar{d}(\mathcal{G}) = \frac{1}{n} \sum_{i=1}^{n} d(v_i)$$

is the average vertex degree in $\mathcal{G}$. Since the sequence $\{z(k)\}$ is a Markov process,[3] we can take the conditional expectation of (5.19) with respect to $z(k)$ and use (5.20) to find an upper bound for

$$\mathbf{E}\{V(z(k+1)) \mid z(k)\}$$

as

$$\mathbf{E}\{V(z(k+1)) \mid z(k)\} \leq V(z(k)) - \gamma(k) \, \nabla V(z(k))^T \, \mathbf{E}\{u(k) \mid z(k)\}$$
$$+ \frac{\lambda_n(\mathcal{G})\gamma(k)^2}{2} \, \mathbf{E}\{\|u(k)\|^2 \mid z(k)\}$$
$$\leq V(z(k)) + c_2(k)$$
$$- \left(\gamma(k) - \frac{\lambda_n(\mathcal{G})\gamma(k)^2}{2}\right) \nabla V(z(k))^T \, \mathbf{E}\{u(k) \mid z(k)\},$$

where $c_2(k)$ is defined by

$$c_2(k) = \frac{n\sigma^2 \, \gamma(k)^2 \, \lambda_n(\mathcal{G}) \, \bar{d}(\mathcal{G})}{2}. \qquad (5.21)$$

---

[3] A random process where the distribution of the process at a given time conditioned on its previous time history is identical to its distribution conditioned only on its immediately preceding time step.

We next invoke the strong pseudogradient property (5.16) and observe that

$$\mathbf{E}\left\{V(z(k+1))\,|\,z(k)\right\} \leq (1 - \beta\gamma(k) + \frac{\beta\gamma(k)^2\,\lambda_n(\mathcal{G})}{2})V(z(k)) + c_2(k)$$
$$= (1 - c_1(k))V(z(k)) + c_2(k), \qquad (5.22)$$

where

$$c_1(k) = \beta\,\gamma(k)\left(1 - \frac{\gamma(k)\,\lambda_n(\mathcal{G})}{2}\right). \qquad (5.23)$$

It is now straightforward to show that when $\gamma(k)$ satisfies conditions (5.13) and

$$\gamma(k) \leq 2/\lambda_n(\mathcal{G})$$

for all k, $c_1(k)$ and $c_2(k)$ satisfy

$$0 \leq c_1(k) \leq 1,\ c_2 \geq 0,\ \sum_{k=1}^{\infty} c_1(k) = \infty,\ \sum_{k=1}^{\infty} c_2(k) < \infty$$

and

$$\lim_{k\to\infty} \frac{c_2(k)}{c_1(k)} = 0.$$

These conditions in turn, allow us to invoke Lemma 5.11 and conclude that $V(z(k)) \to 0$ w.p.1. By Proposition 5.10, the statement of the proposition now follows.   ∎

We remark that since $\mathbf{E}\{w(k)\} = 0$ the expected value of the limiting state is also the mean of the initial states of the vertices in the network. In particular, for all $k \geq 0$, one has

$$\mathbf{1}^T z(k+1) = \mathbf{1}^T z(k) - \gamma(k)\,\mathbf{1}^T(Lz(k)+w(k)) = \mathbf{1}^T z(k) - \gamma(k)\,\mathbf{1}^T w(k);$$

hence, for all $k \geq 0$,

$$\mathbf{E}\{\mathbf{1}^T z(k+1)\} = \mathbf{E}\{\mathbf{1}^T z(k)\}.$$

Let us emphasize that in order to guarantee the convergence of the protocol, the step size $\gamma(k)$ needs to satisfy $\gamma(k) \leq 2/\lambda_n(\mathcal{G})$ (Proposition 5.12) in addition to conditions (5.13). A function which satisfies these criteria is

$$\gamma(k) = \frac{2}{\lambda_n(\mathcal{G})\,k}, \qquad (5.24)$$

since $\sum_{k=1}^{\infty}(1/k) = \infty$, and $\sum_{k=1}^{\infty}(1/k^2) = \pi^2/6$. Alternately, one can use a switching step size of the form

$$\gamma_{\text{sw}}(k) = \begin{cases} 2/\lambda_n(\mathcal{G}) & \text{if } k \leq \lceil \lambda_n(\mathcal{G})/2 \rceil, \\ 1/k & \text{otherwise,} \end{cases} \qquad (5.25)$$

where $\lceil x \rceil$ denotes the least integer upper bound for $x \in \mathbf{R}$.

Example 5.13. *To demonstrate the importance of the graph parameters on the convergence of the agreement protocol, we consider the protocol on a graph of one hundred nodes that has been realized as follows. We assign each vertex a coordinate in the unit square that has been generated by a uniform distribution and let two vertices be neighbors of each other if their distance is less than or equal to 0.35 (this corresponds to a "random geometric graph"–more on this model in the next section). We let the noise in the corresponding agreement protocol be a zero mean Gaussian noise with $\sigma^2 = 0.1$. The initial states for these simulations have been uniformly randomly distributed between 180 and 220 degrees. The right panel of Figure 5.2 demonstrates that violating $\gamma(k) \leq 2/\lambda_n(\mathcal{G})$ for all $k \geq 1$ compromises the convergence properties of the protocol.*

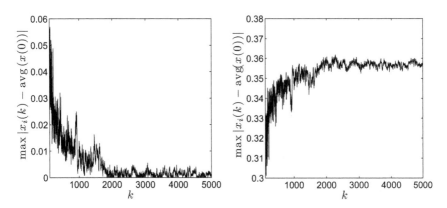

Figure 5.2: Performance of the agreement protocol, measured in terms of the worst case deviation from the steady state over all nodes, on a realization of a random geometric graph on 100 vertices; left panel: $\gamma(k) = \gamma_{\text{sw}}(k)$ (5.25), right panel: $\gamma(k) = 1/k$.

5.4 OTHER PROBABILISTIC MODELS OF NETWORKS

In this section, we provide a brief introduction to other probabilistic models of networks besides the Erdős-Rényi random graph model studied earlier in this chapter.

5.4.1 Random Geometric Graphs

Another probabilistic graph structure arises when n independent and identically distributed points (that is, vertices) on the unit square $[0, 1]^2$ are considered. In such a setting, let r be a positive real number–often referred to as the range or radius of a vertex–and let z_i denote the position of vertex i in the plane.

A random geometric graph on $[0, 1]^2$, denoted by $\mathcal{G}(n, r)$, is defined on the vertex set $V(\mathcal{G}) = [n]$ and the edge set $E(\mathcal{G}) = \{ij : i, j \in [n], i \neq j\}$, such that $ij \in E(\mathcal{G})$ if $\| z_i - z_j \| \leq r$. Due to this "distance-dependent edge existence property," random geometric graphs have been used for modeling several real-world networks, among them wireless ad hoc and sensor networks, biological networks, and social networks.

A random geometric graph on a Poisson point process is known as a Poisson geometric graph. By definition, the following properties hold for a Poisson geometric graph: (1) for every region S in the unit square $[0, 1]^2$, the number of nodes in S, $N(S)$, is Poisson distributed with parameter $n|S|$, where n is the total number of nodes in $[0, 1]^2$ and $|S|$ is the area of the region S, that is,

$$\mathbf{Pr}\left\{ N(S) = k \right\} = \left((n|S|)^k / k! \right) e^{-n|S|};$$

and (2) for every finite collection of disjoint regions in the unit square $[0, 1]^2$, $\{S_1, S_2, \ldots, S_m\}$, the random variables $N(S_1), N(S_2), \ldots, N(S_m)$ are independent. An example of a Poisson random geometric graph on fifteen vertices for three values of the range parameter is shown in Figure 5.3. The above properties can be employed to derive the distribution of vertex degrees in a Poisson geometric graph, as expressed by the following lemma.

Lemma 5.14. *Assume that $n \gg 1$ and $\pi r^2 \ll 1$ for a Poisson random geometric graph $\mathcal{G}(n, r)$ on $[0, 1]^2$. Let $d(v_i)$ denote the degree of vertex i, located in subregion S of area πr^2. Then $d(v_i)$ is Poisson distributed with parameter $n\pi r^2$. Therefore, for all i, $\mathbf{E}\{d(v_i)\} = n\pi r^2$.*

Our next observation pertains to probabilistic connectivity of Poisson geometric graphs as a function of the vertex range r.

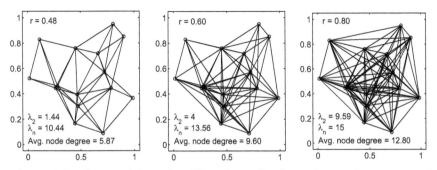

Figure 5.3: Connectivity of a 15-node realization of a random geometric graph $\mathcal{G}(15, r)$ for $r = 0.48$, $r = 0.60$, and $r = 0.80$

Lemma 5.15. *Consider a Poisson random geometric graph $G(n, r)$ in the plane. For any real number α, $\mathbf{Pr}\{\, \mathcal{G}(n, r) \text{ is connected}\,\} \geq e^{-e^{-\alpha}}$, when $n\pi r^2 > \ln(n) + \alpha$.*

Lemma 5.15 provides guidelines for choosing the "range" of vertices in a random geometric graph to ensure a certain probability of connectivity for the resulting network. For example, if we would like the probability of network connectivity to be at least 0.99, then one can set $\alpha = 4.7$. For a 10-node connected network, $r > 0.473$ would suffice; for 20- and 100-node networks, one must have $r > 0.35$ and $r > 0.173$, respectively.

5.4.2 Small-world and Scale-free Networks

Although in this chapter we offered a glimpse into certain probabilistic models of networks, for example, random and random geometric graphs, the area of network modeling, particularly for social and sensor networks, is a rich area of research. For example, the *small-world network* offers a model that exhibits two features that are often observed in social networks: (1) most vertices have low pairwise distance, resembling an average short distance between a pair of nodes in a random graph; and (2) there is a higher probability that two vertices will be connected directly to one another if they have another neighboring vertex in common. The second feature is referred to as the clustering effect; the *clustering coefficient of the network*, often normalized to be in the unit interval, captures, for example, how likely it is for two individuals who have a common friend in a social network to be friends themselves. A particular example of the small-world network was offered by Watts and Strogatz, where a few edges in a k-regular lattice are chosen and randomly "rewired," in the sense that one of their end points is moved to a new random vertex (however, the resulting graph should remain a simple graph even after this random rewiring). Another model of interest in many applications is the scale-free network, where certain nodes in the network are highly connected whereas most nodes have low degrees. In particular, the underlying degree distribution is assumed to follow a power law relation

$$p(k) = \mathbf{Pr} \{ \text{a random node in the network has } k \text{ neighbors} \} \approx k^{-\gamma},$$

where γ is typically between 2 and 3. This network model has been used to mimic a number of physical, social, and biological networks. A particular mechanism for generating scale-free networks is the *preferential attachment* as proposed by Barabási and Réka [16]. In this proposed procedure for generating scale-free networks, the probability that a node attains a new neighbor is an increasing function of its degree in the network.

SUMMARY

In this chapter, we provided an introduction to the probabilistic aspects of network protocols. Our emphasis has been on two extensions of the basic Lyapunov approach for the analysis of the agreement protocol–one for the case when the underlying network is random, and the other for when the protocol is corrupted by noise.

We concluded the chapter by an overview of other probabilistic network models, including random geometric graphs, and small-world and scale-free networks.

NOTES AND REFERENCES

Random graphs constitute an active area of research at the intersection of combinatorics, graph theory, and probability. They are often on the list of networks that are collectively referred to as *complex networks* in physics and engineering literature.

The model that we have used in this chapter for "random agreement" is often called the Erdős-Rényi model of random networks [76],[99], due to the pioneering work of Erdős and Rényi [76]. The use of notions from stochastic stability, namely, supermartingales, for analysis of agreement over random graphs was introduced by Hatano and Mesbahi [113]. Extensions to random digraphs and using the theory of stochastic matrices can be found in the works of Porfiri and Stilwell [194], Wu [250], and Bruneau, Joye, and Merkli [39]. Cogburn [51] and Tahbaz-Salehi and Jadbabaie [227] present necessary and sufficient conditions for almost sure convergence for the product of random stochastic matrices, which in turn has a direct interpretation in terms of agreement over random networks.

Our discussion of the behavior of $\lambda_2(\mathcal{G})$ when $\mathcal{G}$ is a random graph is based on the works of Juhász [128] and Juvan and Mohar [129], which are inspired by the semicircle law for symmetric matrices; see Wigner [245], Arnold [12], and particularly Füredi and Komlós [94].

Section 5.3 is based on the work of Hatano, Das, and Mesbahi [114], which closely parallels Chapter 2 of the book by Polyak [193] as well as §7.8 of the book by Bertsekas and Tsitsiklis [22], extended to the case when part of the gradient is governed by the graph Laplacian. Lemma 5.11 is from Chapter 2 of [193]. The randomized gossip algorithm of Exercise 5.10 is discussed in the paper by Boyd, Ghosth, Prabhakar, and Shah [34].

SUGGESTED READING

We suggest the books of Janson, Luczak, and Rucinski [125] and Bollobás [27] for much more on random graphs. There are a number of excellent books on probability theory; we recommend the one by Shiryaev [216]. For an overview of various probabilistic models of networks we recommend Newman, Barabási, and Watts [175]. This reference not only collects some of the influential papers in the general area of networks, but also provides concise tutorials for each network model.

EXERCISES

Exercise 5.1. Consider the backward product of independent identically distributed stochastic matrices as

$$H_k = P_k P_{k-1} \cdots P_1.$$

Derive conditions for convergence

$$\lim_{k \to \infty} H(k) \to \mathbf{1}v^T \quad \text{in the mean,}$$

for some vector v.

Exercise 5.2 Regenerate, even approximately, Figure 5.1.

Exercise 5.3. Prove Proposition 5.10.

Exercise 5.4. What is the expected number of edges in $\mathcal{G} \in \mathbf{G}(n, p)$?

Exercise 5.5. Propose a random model for balanced directed graphs and extend the analysis of §5.2 to this proposed model.

Exercise 5.6. A set of vertices in the graph is called an independent set if none of the vertices are neighbors of each other. Find a bound on the probability that $G(n, p)$ has k independent vertices, where $2 \leq k \leq n$.

Exercise 5.7. Prove Lemma 5.14.

Exercise 5.8. Comment on the conditions listed in Equation 5.13 and why they might be required for the convergence of the noisy agreement protocol.

Exercise 5.9. Examine the behavior of the noisy agreement protocol on a graph with n vertices, with the fixed step size $\gamma(k) = 1/n$. Do the agents' states converge–in some probabilistic sense–to the agreement subspace? Examine the covariance of the limiting state and whether it is dependent on the structure of the underlying graph.

Exercise 5.10. Consider a weighted digraph on n nodes, where the weight p_{ij} on edge (i, j) represents the probability that vertex i communicates with vertex j. Let P be the corresponding stochastic matrix and assume that P is such that it has a unique eigenvalue with unit magnitude. Consider next the *gossip algorithm*, where at time index k, a node in the digraph awakes with probability $1/n$, and averages its value at time index k with only one of its neighbors–with a probability that is dictated by the weight on the corresponding edge.

(a) Construct the matrix W such that the update rule for the above gossip algorithm can be expressed as

$$x(k + 1) = W x(k), \quad k = 0, 1, 2, \ldots,$$

where $x(k) = [x_1(k), x_2(k), \ldots, x_n(k)]$.

(b) Show that for this gossip algorithm, when

$$k \geq 3 \log(1/\epsilon) / \log \lambda_2(W)^{-1},$$

one has

$$\mathbf{Pr}\{\frac{\|x(k) - (1/n)x(0)^T \mathbf{11}\|}{\|x(0)\|} \geq \epsilon\} \leq \epsilon,$$

where $\lambda_2(W)$ is the second largest (in magnitude) eigenvalue of W.

Exercise 5.11. Consider the noisy agreement protocol (5.12) over a random graph $\mathcal{G} \in \mathbf{G}(n, p)$. Simulate this protocol for various values of n, p, and σ, with the step size specified as (5.24). What observations can you make on the relationship between the convergence properties of the protocol on one hand, and the range of values for these three parameters, on the other?

PART 2
MULTIAGENT NETWORKS

Chapter Six

Formation Control

> "Philosophy is like trying to open a safe with a combination lock:
> each little adjustment of the dials seems to achieve nothing,
> only when everything is in place does the door open."
> — Ludwig Wittgenstein

Formations can be loosely characterized as geometrical patterns to be realized by a multiagent team. Formations appear in a number of biological systems, such as the well-known V-shape, employed by geese and other large migratory birds that is thought to reduce the drag force on individual birds while ensuring sufficient interagent visibility. In this chapter, we discuss various issues related to the specification and execution of formations. We also show how the agreement protocol and its extensions can be used to obtain relative state-based coordination strategies, as well as a nonlinear, distance-based strategy for formation control.

Formation control is one of the first problems one typically addresses when controlling multiple mobile agents. In this chapter, we present this topic by first discussing how formations can be specified, and then proceed by presenting a suite of graph-based formation control strategies.

Regardless of the particulars of a given target formation problem, these problems all share the general property of involving moving the agents in such a way that they satisfy a particular *shape* or *relative state* and a certain aspect of assigning roles (targets in the shape or the relative state) to individual agents. In fact, formation control problems can be defined by a shape or a relative state, as well as an *assignment* component. One can think of the first component as dictating what the formation should "look" like, and the second component as codifying which agent should take on what role in the formation.

6.1 FORMATION SPECIFICATION: SHAPES

We start with the specification of formation shapes and the notion of graph rigidity.

6.1.1 Shapes

Let D be a set of relative, desired interagent distances, that is,

$$D = \{d_{ij} \in \mathbf{R} \mid d_{ij} > 0, \ i, j = 1, \ldots, n, \ i \neq j\},$$

with $d_{ij} = d_{ji}$, and where we assume that D is a *feasible formation*. Feasibility means that the formation can in fact be realized, that is, that there are points $\xi_1, \ldots, \xi_n \in \mathbf{R}^p$ ($p = 2$ corresponds to the planar case, while $p = 3$ encodes 3D formations) such that

$$\|\xi_i - \xi_j\| = d_{ij} \quad \text{for all } i, j = 1, \ldots, n, \ i \neq j.$$

An example of a feasible and an infeasible formation shape specification over three agents is shown in Figure 6.1.

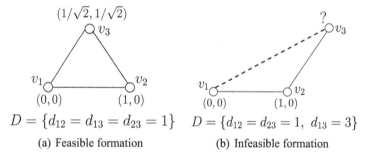

(a) Feasible formation   (b) Infeasible formation

Figure 6.1: Two formations specified through desired interagent distances. The left figure corresponds to a case where all three distances are equal to one, resulting in an equilateral triangle. The right figure shows a case where it is impossible to place three agents in the plane–or in any other Euclidean space for that matter–such that the desired interagent distances are realized.

By a *scale invariant* formation D we understand any set of distances D' such that

$$D' = \alpha D$$

for any $\alpha \in \mathbf{R}_+$. This type of formation specification makes sense in applications where the environment is moderately cluttered and a scaled contraction or expansion of the formation may be needed to negotiate the environment. An example of this is shown in Figure 6.2, where three agents are to traverse a narrow passage while maintaining a desired triangular shape. The way this is achieved is by scaling the formation throughout the maneuver so that a tighter triangular formation is employed while going through the narrow passage. It should be noted that an alternative approach to this is to switch among a collection of formations, which will be the topic of §6.5.

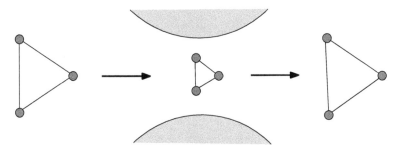

Figure 6.2: Three agents are traversing a narrow passage by contracting the formation to ensure that the team clears the obstacles while maintaining the desired triangular shape.

The most common type of formation specification, D, is one in which scaling is not allowed, that is, the formation specification is exactly given in terms of interagent distances. One way in which such a formation specification can be encoded is through a weighted graph $\mathcal{G}_f = (V, E_f, w)$, where the subscript f denotes "formation." As before, V is the set of vertices in the formation, E_f is the set of edges that specify what interagent distances are defined by the formation, and the edge weight w encodes the distance specifications in the sense that $w(v_i v_j) = d_{ij}$, where $w : E_f \to \mathbf{R}_+$.

Note that it is tempting to let $\mathcal{G}_f$ be complete, that is, to specify an edge between every pair of vertices. However, it is often desirable to produce more parsimonious graphs and only keep the edges needed to ensure that the proper shape is maintained. This is the case when each edge corresponds to an explicit task, that is, the establishment and maintenance of a particular interagent separation. Hence, in order to avoid overloading the agents with such tasks, edges that are not strictly necessary should be avoided. Similarly, if some edges are significantly longer than others, they require longer range sensing or higher power communication capabilities. Therefore, it would be cost-effective to remove those edges, if this is at all possible.

It is sometimes the case that the formation is only allowed to be translationally invariant, that is, the rotation is no longer free. In this case, we need to incorporate this in the specification as well. In other words, rather than using D as the formation specification, one could use

$$\Xi = \{\xi_1, \dots, \xi_n\}, \ \xi_i \in \mathbf{R}^p, \ i = 1, \dots, n,$$

where $\|\xi_i - \xi_j\| = d_{ij}, \ i, j = 1, \dots, n, \ i \neq j$. Any collection of points $x_1, \dots, x_n$ in $\mathbf{R}^p$ is thus said to satisfy the formation specification if

$$x_i = \xi_i + \tau \ \text{ for all } i = 1, \dots, n$$

for some arbitrary translation $\tau \in \mathbf{R}^p$. We summarize these different formation specifications in the following table.

Interagent distances $D = \{d_{ij} = d_{ji} \geq 0, \ i, j = 1, \dots, n, \ i \neq j\}$		
formation	specification	interpretation
scale invariant	D	$\|x_i - x_j\| = \alpha d_{ij}$ for some $\alpha > 0$
rigid	D	$\|x_i - x_j\| = d_{ij}$
translational invariant	Ξ	$x_i = \xi_i + \tau$ for some $\tau \in \mathbf{R}^p$

6.1.2 Rigidity

Although we have already implicitly hinted at the concept of graph *rigidity*, this concept has a clear and concise graph theoretic interpretation. Graph rigidity is the study of formation graphs for which the only permissible motions, while maintaining proper edge distances, are rigid motions. Graphs that satisfy these properties are promising candidates for specifying formations in that they describe the formation shape using only interagent distance specifications.

Assume that a formation has been specified by the formation graph $\mathcal{G}_f = (V, E_f, w)$, where $w : E_f \to \mathbf{R}_+$ associates a feasible, desired interagent distance to each agent pair in the formation graph. Now, given a set of feasible points Ξ, we define a *framework* as $\mathcal{G}(\Xi) = (\Xi, \mathcal{G}_f)$, where Ξ and $\mathcal{G}_f$ are as previously defined. We can also define a *trajectory* of a framework $\mathcal{G}(\Xi)$ as the set of continuous states, $x_1(t), \dots, x_n(t)$, with initial conditions $x_i(0) = \xi_i(0), \ i = 1, \dots, n$, and $t \geq 0$. A trajectory represents the *motion* of a multiagent network that is initially in the desired target formation; we say that such a motion is *edge-consistent* if $\|x_i(t) - x_j(t)\|$ is constant for all $\{v_i, v_j\} \in E_f$.

Using this terminology, we call the trajectory *rigid* if the distances between every pair of states $x_i(t)$ and $x_j(t)$, not just between states corresponding to nodes that are adjacent in the formation graph, remain constant. Thus, a rigid trajectory represents a *rigid motion* of the network, starting from the target formation, during which all interagent distances are maintained.

Definition 6.1 (Rigid and Flexible Frameworks). *A framework is* rigid *if and only if all edge-consistent trajectories of the framework are rigid trajectories. If a framework is not rigid, we refer to it as* flexible.

It is clear that rigid frameworks represent rigid formations, that is, formations whose shape can be maintained rigidly while only maintaining the desired interagent distances. Figure 6.3 provides examples of rigid and flexible frameworks.

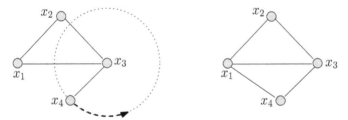

Figure 6.3: The left figure corresponds to a flexible framework in which agent 4 can move anywhere on the dotted circle (thus changing its distances to agents 1 and 2) while still satisfying the distance constraint with agent 3. The right figure corresponds to a rigid framework in the planar case.

Note that rigidity is defined for frameworks, that is, graphs together with configurations, rather than graphs alone. In fact, in order to properly define graph rigidity, we need the additional notion of *infinitesimal rigidity*. Assume that the trajectories $x_i(t)$ associated with each vertex in the framework are differentiable functions. Since we have defined an edge-consistent trajectory of $\mathcal{G}(\Xi)$ to be one such that the distance between points $x_i(t)$ and $x_j(t)$ remains constant along the trajectory, it follows that

$$(\dot{x}_i(t) - \dot{x}_j(t))^T (x_i(t) - x_j(t)) = 0, \quad \text{for all } \{v_i, v_j\} \in E_f. \quad (6.1)$$

The assignment of constant instantaneous velocities $\dot{x}_i = u_i$ that satisfy (6.1) at $t = 0$ is described as an *infinitesimal motion* of the framework. Let $u = [u_1^T, u_2^T, \ldots, u_n^T]^T$ be such an infinitesimal motion, applied at the configuration points $\xi_1, \ldots, \xi_n$. The relation (6.1) can then be represented in matrix form as

$$R(\mathcal{G}(\Xi))u = 0, \tag{6.2}$$

where $R(\mathcal{G}(\Xi))$ is known as the *rigidity matrix*. We note that the rigidity matrix has card(E_f) rows and pn columns, where p is the dimension of the state of the agents, that is, $\xi_i \in \mathbf{R}^p$, $i = 1, \ldots, n$.

Definition 6.2. *A framework $\mathcal{G}(\Xi)$ is infinitesimally rigid if $R(\mathcal{G}(\Xi))u = 0$ for all infinitesimal motions u.*

The special case of planar agents, that is, where $p = 2$, is particularly well understood.

Theorem 6.3. *A framework with $n \geq 2$ points in $\mathbf{R}^2$ is infinitesimally rigid if and only if* **rank** $R(\mathcal{G}(\Xi)) = 2n - 3$.

The following theorem establishes the relationship between infinitesimal rigidity and rigidity in any dimension.

Theorem 6.4. *Infinitesimal rigidity implies rigidity.*

We note, however, that rigidity does not imply infinitesimal rigidity, which is a fact that will be pursued in Exercise 6.2.

It is clear that the rigidity of a framework $\mathcal{G}(\Xi)$ depends on both the topology (the graph) and the configuration. For a given formation graph $\mathcal{G}_f$, we can think of a framework $\mathcal{G}(\Xi) = (\Xi, \mathcal{G}_f)$ as a particular *realization* of $\mathcal{G}_f$, and we define a *generically rigid graph* as follows.

Definition 6.5. *A graph is* generically rigid *if it has an infinitesimally rigid realization.*

Note that generic rigidity is a property of a graph, not of a framework. Therefore, we refer to generically rigid graphs as *rigid graphs*. If $\mathcal{G}_f$ is a rigid graph, and $\mathcal{G}(\Xi) = (\Xi, \mathcal{G}_f)$ is infinitesimally rigid, we say that Ξ is a *generic configuration* for $\mathcal{G}_f$, and that $\mathcal{G}(\Xi)$ is a *generic realization*.

The configuration $\Xi = \{\xi_1, \ldots, \xi_n\}$ defines a point in $\mathbf{R}^{pn}$. There are, however, a number of such points that are generic realizations of a particular formation graph. The following lemma describes the set of generic configurations for a generically rigid graph.

Lemma 6.6. *If $\mathcal{G}_f$ is a generically rigid graph, then the set of all generic configurations for $\mathcal{G}_f$ is a dense, open subset of $\mathbf{R}^{pn}$.*

This implies that, for a generically rigid graph $\mathcal{G}_f$, any configuration Ξ' that is a realization of $\mathcal{G}_f$ can be well approximated by a generic configuration Ξ such that $\mathcal{G}(\Xi) = (\Xi, \mathcal{G}_f)$ is infinitesimally rigid and, therefore, rigid by Theorem 6.4.

It is clear that adding edges to a rigid graph cannot affect its rigidity, which seems to imply that there is a minimum number of edges needed to produce a rigid graph. We define a *minimally rigid graph* as follows.

Definition 6.7. *A graph is* minimally rigid *if it is rigid but does not remain rigid after the removal of a single edge.*

The following theorem provides necessary and sufficient conditions for a graph to be minimally rigid in the planar case.

Theorem 6.8. *A graph with $n \geq 2$ vertices in $\mathbf{R}^2$ is minimally rigid if and only if*

    (1) it has $2n - 3$ edges, and
    (2) each induced subgraph of $n' \leq n$ vertices has no more than $2n' - 3$ edges.

As a consequence, through rigidity, we have a handle on favorable shape specifications for formation control applications if the formations are allowed to be both translationally and rotationally invariant. Moreover, we note that all that is needed in order to enforce a rigid formation is the ability for the individual agents to measure and maintain interagent distances.

6.2 FORMATION SPECIFICATION: RELATIVE STATES

Translationally invariant formations can be directly specified by a set of desired relative states in the formation configuration space, as opposed to a set of relative distances among the agents, as examined in §6.1. For example, suppose we want to specify that a group of three point masses in $\mathbf{R}^3$ keep a particular relative position in space. Denoting the position of point mass i as $x_i \in \mathbf{R}^3$, this specification can be accomplished by defining the vector

$$z(t) = [\,(x_1(t) - x_2(t))^T, (x_2(t) - x_3(t))^T\,]^T \in \mathbf{R}^6$$

and then specifying a reference relative state vector z_{ref} for the desired formation. In this case, since we are using vector specification as opposed to a shape specification, the formation configuration is completely specified. For example, in the above scenario, the vector z_{ref} implicitly specifies *all* interagent relative states as, for example,

$$x_1(t) - x_3(t) = (x_1(t) - x_2(t)) + (x_2(t) - x_3(t)).$$

The configuration specification of formations provides the flexibility of completely specifying them either relatively or inertially. By an inertial frame, we refer to a coordinate axis that is assumed to be nonaccelerating and nonrotating.[1] For example, specifying the formation inertially can be accomplished by letting x_o be the coordinates of a fictitious point mass with respect to the inertial frame, and then specifying the formation of the three point masses by defining

$$z(t) = [\,(x_o(t) - x_1(t))^T, (x_1(t) - x_2(t))^T, (x_2(t) - x_3(t))^T\,]^T \in \mathbf{R}^9$$

and specifying the corresponding z_{ref}. Note that x_o can be set as the origin of the inertial frame as well. In the latter case, this specification is equivalent to specifying the inertial vector $[\,x_1^T\ x_2^T\ x_3^T\,]^T \in \mathbf{R}^9$.

Desired formation configurations can conveniently be encoded using the incidence matrix of the graph. Thus for the example above one can define the vector z as

$$z(t) = D(\mathcal{D})^T x(t),$$

where

$$x(t) = [\,x_1(t)^T, x_2(t)^T, x_3(t)^T\,]^T$$

and $\mathcal{D}$ is a directed path graph on three nodes with the incidence matrix

$$D(\mathcal{G}) = \begin{bmatrix} 1 & 0 \\ -1 & 1 \\ 0 & -1 \end{bmatrix} \otimes I,$$

with I as the 3×3 identity matrix and $\otimes$ denoting the Kronecker product. Subsequently, the formation can be specified in its configuration space by specifying the vector z_{ref}. More generally, a formation on n agents, each of whose state evolves in $\mathbf{R}^p$, can be specified by choosing a spanning subgraph of the directed complete graph,[2] denoted by $\mathcal{D}$, defining $z(t) = (D(\mathcal{D})^T \otimes I)x(t)$, with I being the $p \times p$ identity matrix, and then setting z_{ref} as the desired formation configuration. We refer to the formation

---

[1]With respect to another inertial frame, say, one that is attached to a distant star!

[2]The directed version of the complete graph is a digraph where every edge is replaced with two directed edges, each with different end points.

specification via weakly connected digraphs on n vertices as a *relative state specification* (RSS).

Analogous to shapes, relative formation configurations can be specified in different, yet equivalent, ways. As an example, suppose that the formation has been specified via two distinct spanning digraphs, $\mathcal{D}_j$ and $\mathcal{D}_d$, on n vertices. In order to show the equivalence between these two RSS, we seek a linear transformation T_{dj} such that

$$T_{dj}\, D(\mathcal{D}_j)^T = D(\mathcal{D}_d)^T. \tag{6.3}$$

We refer to the transformations between distinct pairs of RSS as T transformations. We now proceed to gain a better insight into the form of various T transformations among distinct classes of (weakly connected) digraphs on the same vertex set. Our discussion revolves around three canonical cases. These cases include transformations (a) from an RSS to one of its subgraphs, (b) from a spanning directed tree RSS to any other RSS, and (c) between two arbitrary weakly connected RSS.

(a) Transformation from an arbitrary RSS to one of its subgraphs. First consider the scenario where the desired RSS, $\mathcal{D}_d$, is a subgraph of another RSS, $\mathcal{D}_j$. Given that $\mathcal{D}_d$ and $\mathcal{D}_j$ have m_d and m_j edges, respectively, the transformation T_{dj}, satisfying $D(\mathcal{D}_d)^T = T_{dj} D(\mathcal{D}_j)^T$, is an $m_d \times m_j$ matrix. Consider next the decomposition

$$T_{dj} = [\widehat{T}_{dj}\ \ \widetilde{T}_{dj}\,] \in \mathbf{R}^{m_d \times m_j}, \tag{6.4}$$

where $\widehat{T}_{dj} \in \mathbf{R}^{m_d \times m_d}$ and $\widetilde{T}_{dj} \in \mathbf{R}^{m_d \times m_j - m_d}$, and the corresponding rearranging of the incidence matrix $D(\mathcal{D}_j)$ is

$$D(\mathcal{D}_j) = [\, D(\mathcal{D}_d)\ \ D(\mathcal{D}_{j/d})\,];$$

hence

$$\widehat{T}_{dj} = I \quad \text{and} \quad \widetilde{T}_{dj}\, D(\mathcal{D}_{j/d})^T = 0.$$

The trivial solution for $\widetilde{T}_{dj}$ is the zero matrix while the general matrix solution consists of rows that belong to the null space of $D(\mathcal{D}_{j/d})$.

(b) Transformation from a directed spanning tree to an arbitrary RSS. The second canonical case corresponds to the transformation T_{dj} where $\mathcal{D}_j$ is a spanning tree and the target digraph $\mathcal{D}_d$ is an arbitrary RSS. In this case, one can make the following observation.

Proposition 6.9. *Let $\mathcal{D}_j$ be a directed spanning tree RSS and $\mathcal{D}_d$ be an arbitrary RSS. Then*

$$T_{dj} = D(\mathcal{D}_d)^T D(\mathcal{D}_j)\, [\, D(\mathcal{D}_j)^T\, D(\mathcal{D}_j)\,]^{-1}, \tag{6.5}$$

where $D(\mathcal{D}_j)$ and $D(\mathcal{D}_d)$, represent, respectively, the incidence matrices associated with RSS $\mathcal{D}_j$ and $\mathcal{D}_d$.[3]

Proof. From the matrix equation (6.3) it follows that $D(\mathcal{D}_j) T_{dj}^T = D(\mathcal{D}_d)$, and hence

$$\{D(\mathcal{D}_j)^T D(\mathcal{D}_j)\} T_{dj}^T = D(\mathcal{D}_j)^T D(\mathcal{D}_d).$$

Since **rank** $D(\mathcal{D}_j) = n - 1$, the matrix product $D(\mathcal{D}_j)^T D(\mathcal{D}_j)$ is invertible, and (6.5) follows.   ∎

(c) Transformation between two arbitrary weakly connected RSS. There are at least two approaches to the characterization of a T transformation between two arbitrary weakly connected RSS, $\mathcal{D}_d$ and $\mathcal{D}_j$. (1) Transform the given digraph $\mathcal{D}_j$ to a spanning tree subgraph $\mathcal{D}_k$ (which corresponds to a Case 1 scenario above), followed by the transformation from $\mathcal{D}_k$ to the RSS $\mathcal{D}_d$ given in Proposition 6.9, and letting $T_{dj} = T_{dk} T_{kj}$. (2) Complete the cycles of $\mathcal{D}_j$ to obtain a complete digraph $\mathcal{D}_c$, and choose an appropriate subgraph of $\mathcal{D}_c$ (re-orient edges if necessary) that corresponds to $\mathcal{D}_d$; then let $T_{dj} = T_{dc} T_{cj}$. An explicit formula for the T transformation T_{dj} between two arbitrary weakly connected RSS is facilitated by the following lemma.

Lemma 6.10. *The collection of $n - 1$ rows of the incidence matrix for a weakly connected digraph, corresponding to its $n - 1$ signed characteristic vectors of single vertex cuts, are linearly independent.*

As a direct consequence of Proposition 6.9, we arrive at the following corollary.

Corollary 6.11. *Let $\mathcal{D}_j$ and $\mathcal{D}_d$ represent two arbitrary weakly connected RSS. Then*

$$T_{dj} = \widehat{D}(\mathcal{D}_d)^T \{\widehat{D}(\mathcal{D}_j)\widehat{D}(\mathcal{D}_j)^T\}^{-1}\widehat{D}(\mathcal{D}_j), \qquad (6.6)$$

where the rows of $\widehat{D}(\mathcal{D}_j)$ and $\widehat{D}(\mathcal{D}_d)$ correspond, respectively, to the $n - 1$ signed characteristic vectors of single vertex cuts of RSS $\mathcal{D}_j$ and $\mathcal{D}_d$.

Proof. The proof follows from the observation that $T_{dj} D(\mathcal{D}_j)^T = D(\mathcal{D}_j)^T$ if and only if $T_{dj} \widehat{D}(\mathcal{D}_j)^T = \widehat{D}(\mathcal{D}_j)^T$; moreover, the matrix $\widehat{D}(\mathcal{D}_j)$ is full row-rank.   ∎

---

[3]The identity (6.5) hints at the fact that the transformation T_{jd} can be viewed as the projection of the network $\mathcal{D}_d$ along the edge space of the RSS $\mathcal{D}_j$ [71].

6.3 SHAPE-BASED CONTROL

In this section, we move from specifying formations to actually achieving them. In particular, we focus our attention on the problem of driving a collection of mobile agents to a rotationally invariant formation, encoded through the formation graph $\mathcal{G}_f = (V, E_f)$, together with an associated target location set Ξ, as in §6.1. The reason for starting with this formulation is that it supports a solution in terms of linear formation control algorithms. In fact, for such a setup, it will turn out that we can directly use the agreement protocol, as discussed in Chapter 3.

As before, let $x_i \in \mathbf{R}^p$ denote the position of agent i. What should be achieved by the formation control protocol is that for some $\tau \in \mathbf{R}^p$, $x_i = \xi_i + \tau$, for all $i = 1, \ldots, n$. For this, we encode the actual agent network through the graph $\mathcal{G} = (V, E)$, which we will refer to as the interaction graph. This graph may be dynamic, that is, have the edge set E change over time as agents change their adjacency relation as they move around in the environment, or static.

Regardless of whether the graph is static or dynamic, what we want the formation control to achieve is to drive the agents in such a way that:

(R1) $\|x_i(t) - x_j(t)\|$ converges asymptotically to d_{ij} for all i, j such that $\{v_i, v_j\} \in E_f$.

(R2) If the interaction graph $\mathcal{G}(t)$ is dynamic, it should converge to a static graph that is a supgraph of the desired graph $\mathcal{G}_f$ (without weights) in finite time. In other words, what we want is that $E_f \subseteq E(t)$ for all $t \geq T$, for some finite $T \geq 0$.

6.3.1 The Static Case

We first consider the situation in which the interaction graph is static. For the formation control problem to be solvable in the sense of (R1) above, we must assume that (R2) is trivially satisfied, that is, $E_f \subseteq E$. If that is indeed the case, we can define τ_i as the displacement of x_i from the target location $\xi_i \in \Xi$. In other words, we let

$$\tau_i(t) = x_i(t) - \xi_i, \quad i = 1, \ldots, n.$$

Now, by reaching agreement over the τ_i, we would have that $x_i - \xi_i = \tau$ for a constant displacement vector τ, which would mean that the translationally invariant formation control problem has been solved. In this direction, we simply let

$$\dot{\tau}_i(t) = - \sum_{j \in N_f(i)} (\tau_i(t) - \tau_j(t)).$$

Here $N_f(i)$ is the set of nodes adjacent to v_i in the formation graph $\mathcal{G}_f$, that is,

$$N_f(i) = \{j \in \{1, \ldots, n\} \mid \{v_i, v_j\} \in E_f\}.$$

But, noting that for all t, $\dot{\tau}_i(t) = \dot{x}_i(t)$ as well as $\tau_i(t) - \tau_j(t) = x_i(t) - x_j(t) - (\xi_i - \xi_j)$, leads us to the distributed linear formation control strategy,

$$\dot{x}_i(t) = - \sum_{j \in N_f(i)} (x_i(t) - x_j(t)) - (\xi_i - \xi_j). \qquad (6.7)$$

By virtue of the convergence of the agreement protocol (as long as the formation graph is connected), we have in fact solved the formation control problem. An example of this is seen in Figure 6.4, in which ten agents are driven to an equidistant circular formation.

Theorem 6.12. *Consider the connected target formation graph $\mathcal{G}_f$ given by (V, E_f) and a set of target locations Ξ. If the static interaction graph $\mathcal{G} = (V, E)$ satisfies $E_f \subseteq E$, then the protocol (6.7) will asymptotically drive all agents to a constant displacement of the target positions, that is, for all i,*

$$x_i(t) - \xi_i \to \tau$$

as $t \to \infty$.

6.3.2 The Dynamic Case

If the interaction graph is dynamic, then as a direct consequence of Theorem 6.12, as long as for all $t \geq 0$, $E_f \subseteq E(t)$ (with $E(t)$ being the dynamic edge set associated with the interaction graph), the protocol (6.7) is still just dealing with the "static" formation graph. We state this observation as a corollary.

Corollary 6.13. *Given a connected, target formation graph $\mathcal{G}_f = (V, E_f)$, the protocol (6.7) will asymptotically drive all agents to a constant displacement of the target positions if for all $t \geq 0$, $E_f \subseteq E(t)$.*

It may not be the case that $E_f \subseteq E(t)$ for all $t \geq 0$. For example, if the interaction graph is a Δ-disk proximity graph, that is, one in which $\{v_i, v_j\} \in E(t) \Leftrightarrow \|x_i(t) - x_j(t)\| \leq \Delta$, we might have to be slightly

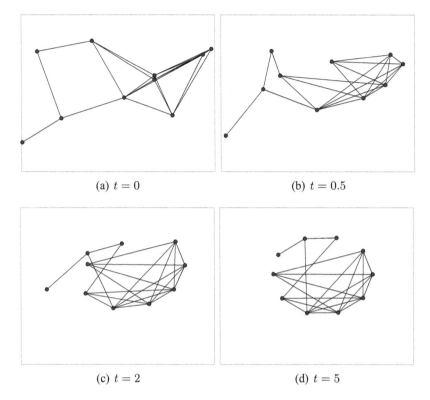

(a) $t = 0$ (b) $t = 0.5$

(c) $t = 2$ (d) $t = 5$

Figure 6.4: A collection of 10 agents that execute the formation control strategy (6.7) in order to reach an equidistant circular formation

more creative. For instance, one could first use the agreement protocol discussed in Chapter 3 to solve the rendezvous problem in order to achieve a complete graph. Once that graph is achieved (provided that no desired edge distances are greater than Δ), one can switch to the protocol (6.7). A hybrid control strategy that implements this is shown in Figure 6.5, with a particular example in Figure 6.6.

When using this strategy, the transition from rendezvous to formation control happens when a complete graph is achieved, which can be accomplished based on a simple single-hop communication rule. First, each agent i that can verify that $\text{card}(N_i) = n - 1$, that is, when it is interacting with all other agents in the formation, sends out a signal denoting completeness. The agents will know that the graph is complete when they receive a similar signal back from all other agents in the formation.

It should be noted that this strategy only works if the assumptions in Corollary 6.13 hold, that is, if $E_f \subseteq E(t)$ for all times after the transition from "Rendezvous" to "Formation Control." To enforce this in all situations

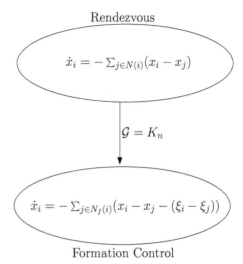

Rendezvous

$$\dot{x}_i = -\sum_{j \in N(i)}(x_i - x_j)$$

$$\mathcal{G} = K_n$$

$$\dot{x}_i = -\sum_{j \in N_f(i)}(x_i - x_j - (\xi_i - \xi_j))$$

Formation Control

Figure 6.5: The agents start out executing the standard agreement protocol. Once a complete graph has been obtained ($\mathcal{G} = K_n$), they switch to a linear formation control algorithm that only takes into account the adjacency relation, as specified by E_f.

is, unfortunately, something that cannot be achieved by linear means alone. Instead, nonlinear control strategies must be employed, a topic which will be discussed in Chapter 7, where the problem of controlling a collection of mobile robots, with limited sensing and communications capabilities, is investigated.

6.4 RELATIVE STATE-BASED CONTROL

In this section, we consider control of formations specified via relative states among their agents. This will be first pursued for linear formation control for agents with single, double, and linear time-invariant dynamics. We then proceed to explore how Laplacian-based potential or navigation functions can be employed to synthesize various formation control algorithms for a group of unicycles.

6.4.1 Linear Formation Control

We start our discussion by examining linear formation control laws for one-dimensional single and double integrator agent models, as well as those specified by a linear time-invariant model. In this direction we assume,

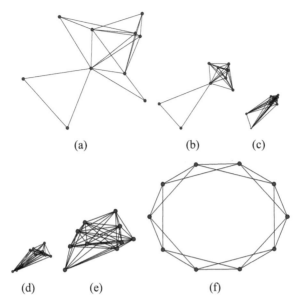

(a)   (b)   (c)

(d)   (e)   (f)

Figure 6.6: A progression is shown where the first three figures (a), (b), and (c) correspond to the execution of the rendezvous control strategy until the dynamic interaction graph is a complete graph. The following three figures (d), (e), and (f) show how the complete graph is changed to the desired formation using only local information.

without loss of generality, that the desired formation has been specified in the relative configuration space via a spanning tree digraph $\mathcal{D}$ (see §6.2). Moreover, we assume that the relative state measurement corresponding to a directed edge in $\mathcal{D}$ is available to both of its end vertices.

The model for single integrator agents is specified as

$$\dot{x}_i(t) = u_i(t), \quad i = 1, 2, \ldots, n, \tag{6.8}$$

where u_i denotes the admissible control input for agent i.

Given the spanning tree digraph $\mathcal{D}$ employed for the formation specification, let

$$z(t) = D(\mathcal{D})^T x(t),$$

and let z_{ref} be the constant reference relative position for the n integrators consistent with $\mathcal{D}$. The formation error at time t is therefore

$$e(t) = z_{\text{ref}} - z(t),$$

and hence

$$\dot{e}(t) = -D(\mathcal{D})^T u(t).$$

Next, consider the state feedback controller of the form

$$u(t) = kD(\mathcal{D})e(t) \tag{6.9}$$

for $k > 0$, which is essentially a *proportional control* with gain $D(\mathcal{D})$. The resulting closed loop system is now given by

$$\dot{e}(t) = -kL_e(\mathcal{D})\,e(t),$$

where $L_e(\mathcal{D}) = D(\mathcal{D})^T D(\mathcal{D})$ is the edge version of the graph Laplacian for the digraph $\mathcal{D}$ as discussed in Chapter 2, §2.3.4. However, since the edge Laplacian is positive definite for a spanning tree, it follows that

$$\lim_{t \to \infty} e(t) = 0,$$

and the n single integrators (6.8), adopting the control law (6.9), asymptotically achieve the desired relative position specified by z_{ref}. In fact, applying the controller (6.9) for the formation control of n single-integrators results in the closed loop system

$$\dot{x}(t) = -kL(\widetilde{\mathcal{D}})x(t) + kD(\mathcal{D})z_{ref}, \tag{6.10}$$

where $k > 0$, and $\widetilde{\mathcal{D}}$ is the disoriented digraph of $\mathcal{D}$. Note that for $z_{ref} = 0$, (6.10) reduces to the (scaled) agreement protocol (3.2).

Next, consider the relative state dynamics for a network of double integrators,

$$\ddot{x}_i(t) = u_i(t), \quad i = 1, 2, \ldots, n, \tag{6.11}$$

where u_i denotes the admissible control input for agent i. It is assumed that the desired relative positions and velocities for these double integrators have been specified via a spanning tree digraph $\mathcal{D}$ as

$$[\, z_{ref}(t)^T, \dot{z}_{ref}(t)^T \,]^T.$$

By setting $e(t) = z_{ref}(t) - D(\mathcal{D})^T x(t)$, and assuming that $\ddot{z}_{ref}(t) = 0$, it follows that

$$\ddot{e}(t) = -D(\mathcal{D})^T \ddot{x}(t) = -D(\mathcal{D})^T u(t).$$

Now, let the state feedback controller be of the form

$$u(t) = k[D(\mathcal{D})\ D(\mathcal{D})]\begin{bmatrix} e(t) \\ \dot{e}(t) \end{bmatrix},\qquad\qquad(6.12)$$

for some $k > 0$, which is essentially a proportional-derivative (PD) control law with position and velocity error gains specified by $D(\mathcal{D})$. The resulting closed loop system is then of the form

$$\begin{bmatrix} \dot{e}(t) \\ \ddot{e}(t) \end{bmatrix} = \begin{bmatrix} 0 & I \\ -kL_e(\mathcal{D}) & -kL_e(\mathcal{D}) \end{bmatrix}\begin{bmatrix} e(t) \\ \dot{e}(t) \end{bmatrix},$$

where, once again, $L_e(\mathcal{D})$ is the edge Laplacian of the digraph $\mathcal{D}$. The characteristic equation for the matrix governing the error dynamics, that is,

$$A_{cl} = \begin{bmatrix} 0 & I \\ -kL_e(\mathcal{D}) & -kL_e(\mathcal{D}) \end{bmatrix},$$

is

$$\det(\lambda I - A_{cl}) = \det(\lambda^2 I + (\lambda + 1)kL_e(\mathcal{D})) = 0.$$

Since $\lambda = -1$ does not satisfy this equation, it is not an eigenvalue of A_{cl}. The eigenvalues of A_{cl} thus satisfy

$$\det(\lambda^2/(\lambda + 1)\,I + kL_e(\mathcal{G})) = 0.$$

Denoting the eigenvalues of $-kL_e(\mathcal{D})$ by μ, one has that, for each i,

$$\mu_i = \lambda_i^2/(\lambda_i + 1),$$

and hence

$$\lambda_i = \frac{1}{2}\left(\mu_i \pm \sqrt{\mu_i^2 + 4\mu_i}\right).$$

However, since for $k > 0$, $-kL_e(\mathcal{D})$ is negative definite when $\mathcal{G}$ is a spanning tree, $\mu_i < 0$ for all i and consequently, the matrix A_{cl} is Hurwitz. This, on the other hand, guarantees that

$$\lim_{t\to\infty}\begin{bmatrix} e(t) \\ \dot{e}(t) \end{bmatrix} = 0,$$

ensuring that the formation of n double integrators (6.11) under the formation control protocol (6.12) achieves the desired relative position and velocity specified by z_{ref} and $\dot{z}_{\text{ref}}$.

The closed loop system with the above control law now assumes the form

$$\frac{d}{dt}\begin{bmatrix} z(t) \\ \dot{z}(t) \end{bmatrix} = \mathcal{E}(\mathcal{D})\begin{bmatrix} z(t) \\ \dot{z}(t) \end{bmatrix} + \mathcal{F}(\mathcal{D})\begin{bmatrix} z_{\text{ref}}(t) \\ \dot{z}_{\text{ref}}(t) \end{bmatrix}, \qquad (6.13)$$

where

$$\mathcal{E}(\mathcal{D}) = \begin{bmatrix} 0 & I \\ -kL_e(\mathcal{D}) & -kL_e(\mathcal{D}) \end{bmatrix}, \quad \mathcal{F}(\mathcal{D}) = \begin{bmatrix} 0 & 0 \\ kL_e(\mathcal{D}) & kL_e(\mathcal{D}) \end{bmatrix}.$$

If we change our point of view from the closed loop dynamics on the edges to the corresponding dynamics on the vertices, we obtain

$$\frac{d}{dt}\begin{bmatrix} x(t) \\ \dot{x}(t) \end{bmatrix} = \mathcal{L}(\widetilde{\mathcal{D}})\begin{bmatrix} x(t) \\ \dot{x}(t) \end{bmatrix} + \mathcal{D}(\mathcal{D})\begin{bmatrix} z_{\text{ref}}(t) \\ \dot{z}_{\text{ref}}(t) \end{bmatrix}, \qquad (6.14)$$

where

$$\mathcal{L}(\widetilde{\mathcal{D}}) = \begin{bmatrix} 0 & I \\ -kL(\widetilde{\mathcal{D}}) & -kL(\widetilde{\mathcal{D}}) \end{bmatrix}, \quad \mathcal{D}(\mathcal{D}) = \begin{bmatrix} 0 & 0 \\ kD(\mathcal{D}) & kD(\mathcal{D}) \end{bmatrix},$$

and $L(\widetilde{\mathcal{D}})$ is, once again, the graph Laplacian of the disoriented $\mathcal{D}$ and $k > 0$. This can be further unwrapped as

$$\ddot{x}(t) = -kL(\widetilde{\mathcal{D}})x(t) - kL(\widetilde{\mathcal{D}})\dot{x}(t) + kD(\mathcal{D})z_{\text{ref}}(t) + kD(\mathcal{D})\dot{z}_{\text{ref}}(t).$$

We now consider yet another class of formations, namely those whose agents have an internal dynamics described by

$$\dot{x}_i(t) = ax_i(t) + bu_i(t), \quad i = 1, 2, \ldots, n,$$

where $a, b \in \mathbf{R}$. The main point of our discussion below is that when these agents interact over a network to achieve a formation specified by the desired relative states, the stability of the formation's "relative dynamics" is not only a function of the dynamics of each agent, in this case, parameterized by scalars a and b, but also the structure of the underlying interaction graph. In this direction, let $z(t) = D(\mathcal{D})^T x(t)$ for a spanning digraph $\mathcal{D}$, which is assumed to be consistent with the underlying formation relative sensing geometry, possibly after the application of an appropriate T transformation discussed in §6.2. Hence

$$\dot{z}(t) = az(t) + bD(\mathcal{D})^T u(t).$$

Now by letting

$$u(t) = kD(\mathcal{D})(z_{\text{ref}} - z(t)) \quad \text{for some } k \in \mathbf{R},$$

one obtains the closed loop system for the formation relative dynamics as

$$\dot{z}(t) = (aI - kbL_e(\mathcal{D}))z + kbL_e(\mathcal{D})z_{\text{ref}}.$$

To assess the stability of relative dynamics of these agents, it suffices to consider the eigenvalues of the matrix

$$aI - kbL_e(\mathcal{D})$$

which assume the form

$$a - \lambda_i(\mathcal{G})kb, \quad i = 2, \dots, n,$$

where $\lambda_i(\mathcal{G})$'s are the eigenvalues of $L_e(\mathcal{D})$, as well as being the nonzero eigenvalues of $L(\mathcal{G})$. Hence the relative dynamics of the formation for the proposed formation control is stable if and only if for all nonzero eigenvalues of the graph Laplacian one has[4]

$$a - \lambda_i(\mathcal{G})kb < 0 \quad \text{for all} \quad i > 2.$$

6.4.2 Control of Unicycles

The ideas from the previous sections can be extended for coordination and synchronization of multiple identical planar unicycles interacting over an information-exchange network. Unicycles are convenient models in a wide range of applications, including those found in aerospace (unmanned aerial vehicles) and biology (fish locomotion). Viewing the position of the unicycle i in $\mathbf{R}^2$, with coordinates $[x_i, y_i]^T$, it becomes convenient to view its coordinates in $\mathbf{C}$, represented by the complex number

$$r_i(t) = x_i(t) + j\, y_i(t) \quad \text{for } t \geq 0,$$

where $j = \sqrt{-1}$. Now, since

$$\dot{x}_i(t) = v_i \cos \theta_i(t), \; \dot{y}_i(t) = v_i \sin \theta_i(t), \text{ and } \dot{\theta}_i(t) = \omega_i(t),$$

with v_i denoting the speed of the unicycle, we can conveniently represent the kinematics of n unicycles in $\mathbf{C}$, assuming that $\omega_i(t) = u_i(t)$, via

---

[4]As before, assuming a spanning tree digraph for the relative sensing geometry whose measurements are accessible to both end vertices of the corresponding directed edge.

$$\dot{r}_i(t) = v_i e^{j\theta_i(t)}, \quad \dot{\theta}_i(t) = u_i(t), \quad i = 1, 2, \ldots n. \qquad (6.15)$$

By normalizing the speed of the unicycle as $v_i(t) = 1$, it becomes evident that its dynamics can be studied on the unit disk in the complex plane; see Figure 6.7. Let us now consider a group of identical unicycles, whose

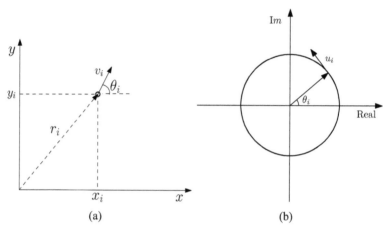

Figure 6.7: Planar unicycle coordinates: (a) in Cartesian coordinates, (b) in complex plane.

angular state and control input at time t can be represented as vectors

$$\theta(t) = [\,\theta_1(t), \theta_2(t), \ldots, \theta_n(t)\,]^T \quad \text{and} \quad u(t) = [\,u_1(t), u_2(t), \ldots, u_n(t)\,]^T.$$

We also adopt the notation

$$e^{j\theta(t)} = [\,e^{j\theta_1(t)}, e^{j\theta_2(t)}, \ldots, e^{j\theta_n(t)}\,]^T. \qquad (6.16)$$

Our goal is to explore (undirected) local interaction rules among the multiple unicycles that lead to coordinated behavior among them. The unicycle dynamics, in the meantime, offer a set of coordinated behaviors that are desirable in applications, yet are unique with respect to their linear counterparts. These behaviors include (1) synchronization, where the heading angles for the unicycles assume a common value, (2) balanced behavior, where the center of mass of the evolution of the unicycles remain constant, (3) spacing, where the unicycles rotate around a prespecified center(s), and (4) symmetrical phase patterns, where the unicycles rotate about a given center with a certain regularity in their phase differences. Let us proceed to use the representation of the unicycle kinematics on the unit disk in the complex plane (6.15) in conjunction with Laplacian-based potentials, to address how the above four types of behavior can emerge from local interaction rules implemented over a graph.

In what follows, it becomes convenient to think of $\dot{r}_i(t) = e^{j\theta_i}$ as the "state" of agent i. First, we synthesize a *navigation function* for unicycle coordination along the following constructs. For positive integer m, define the mth order average state,

$$p_m(\theta) = \frac{1}{nm}\mathbf{1}^T e^{jm\theta},$$

and the mth order potential $U_m : [0, 2\pi]^n \to \mathbf{R}_+$,

$$U_m(\theta) = \frac{n}{2}|p_m(\theta)|^2 = \frac{1}{2nm^2}(e^{jm\theta})^* \mathbf{11}^T e^{jm\theta}, \qquad (6.17)$$

where $(e^{jm\theta})^*$ is the complex conjugate of $e^{jm\theta}$. Thus, for example, $p_1(\theta)$ is the velocity of the center of mass of the n unicycles as

$$p_1(\theta) = \frac{1}{n}\sum_i \dot{r}_i(t) = \frac{d}{dt}\left(\frac{1}{n}\sum_i r_i(t)\right),$$

and $U_1(\theta)$ can be considered as the kinetic energy of this center of mass. It is also convenient to define, for each positive integer m, the "average" angle ψ_m such that

$$p_m(\theta) = |p_m(\theta)|e^{j\psi_m}.$$

The utility of the potential function U_m (6.17) is now made explicit through the following observation.

Proposition 6.14. *For each positive integer m, the unique minimum of the potential U_m (6.17) corresponds to the case when*

$$p_m(\theta) = 0. \qquad (6.18)$$

The unique maximum of $U_m(\theta)$ (6.17), on the other hand, corresponds to the case when for distinct pairs of unicycles i and j, one has

$$\theta_i = \theta_j \qquad \mathrm{mod}\ (2\pi/m).$$

Proof. Since $U_m(\theta) \geq 0$ for all θ, it follows that it achieves its global minimum value of zero when $p_m(\theta) = 0$. On the other hand, the maximum is achieved when all phases are aligned, in the sense that

$$m\theta_i = \psi_m \quad \mathrm{mod}\ 2\pi.$$

It remains to show that the other critical points of $U_m(\theta)$, that is, those that make $\partial U_m/\partial \theta_i = 0$ for $i = 1, \ldots, n$, *do not* correspond to either a minimum or a maximum of $U_m(\theta)$. First, we note that all critical points of $U_m(\theta)$ satisfy

$$\frac{\partial U_m(\theta)}{\partial \theta_i} = \left\langle p_m(\theta), je^{jm\theta_i} \right\rangle$$

$$= |p_m(\theta)| \left\langle e^{j\psi_m}, je^{jm\theta_i} \right\rangle = 0, \quad i = 1, 2, \ldots, n. \quad (6.19)$$

If $p_m(\theta) \neq 0$, condition (6.19) expresses that at the critical configuration, some of the unicycles, say $n - r$ of them, are aligned with ψ_m, and r of them are 180 degrees apart from ψ_m, where without loss of generality we have assumed that $0 \leq r < n/2$.[5] In the meantime, since at these critical points

$$m|p_m(\theta)| = 1 - (2r/n),$$

it follows that, when all unicycles are aligned with ψ_m, the potential $U_m(\theta)$ reaches its maximum. For the critical points of $U_m(\theta)$ one has

$$\frac{\partial^2 U_m(\theta)}{\partial \theta_i^2} = \frac{1}{n} - m \left\langle p_m(\theta), e^{jm\theta_i} \right\rangle$$

$$= \frac{1}{n} - m \cos(\psi_m - m\theta_i)|p_m(\theta)|. \quad (6.20)$$

Since when $r > 0$, one has $m|p_m(\theta)| = 1 - (2r/n) > 1/n$, the expression (6.20) takes negative values when $m\theta_i = \psi_m$ and positive values when $m\theta_i = \psi_m + \pi$, indicating that the critical points, other than the minimum and maximum states identified in the statement of the proposition, are in fact saddle points.   ∎

The following definitions once again reinforce which coordinated behaviors are of particular interest in our discussion.

Definition 6.15. *The phase vector θ is called a balanced configuration of order m when*

$$p_m(\theta) = 0$$

and a synchronized configuration of order m when for all distinct pairs of unicycles i and j,

$$\theta_i = \theta_j \qquad \mod (2\pi/m).$$

---

[5]Note that multiplication of a complex number by j corresponds to rotating it in the complex plane by 90 degrees.

When $m = 1$ the balanced and synchronized configurations of order m are referred to as balanced and synchronized configurations, respectively.

As the maximum and minimum of $U_1(\theta)$ correspond to the synchronized and balanced configurations, respectively, it is natural to propose the gradient control law

$$u_i(t) = -k\nabla_i U_1(\theta) = -k\left\langle p_1(\theta(t)), je^{j\theta_i(t)} \right\rangle$$

$$= -\frac{k}{n}\sum_{j=1}^{n}\sin(\theta_j(t) - \theta_i(t)), \quad i = 1, 2, \ldots, n,$$

which steers the unicycle group, when $k > 0$, toward the minimum of $U_1(\theta)$, or the balanced configuration, and, when $k < 0$, toward the maximum of $U_1(\theta)$ or synchronization; see Figure 6.8. In both cases, the critical points that do not correspond to the minimum or maximum of $U_1(\theta)$ are unstable.

Since the control law above requires information exchange among all unicycles, we would like to consider to what extent this control law can be adapted to the case when the interunicycle information exchange is dictated by an underlying–not necessary complete–undirected network $\mathcal{G}$. In this venue, since the Laplacian over the complete graph is

$$L(K_n) = nI - \mathbf{1}\mathbf{1}^T, \tag{6.21}$$

it follows that the potential $U_m(\theta)$ (6.17) is in fact

$$U_m(\theta) = \frac{n}{2m^2} - \frac{1}{2nm^2}(e^{jm\theta})^* L(K_n) e^{jm\theta},$$

suggesting that we should consider the critical points of the Laplacian-based potential

$$W_m(\theta) = \frac{1}{2}(e^{jm\theta})^* L(\mathcal{G}) e^{jm\theta} = \frac{1}{2}\left\langle e^{jm\theta}, L(\mathcal{G})\, e^{jm\theta} \right\rangle, \tag{6.22}$$

for synthesizing distributed control laws that operate over arbitrary connected graphs. Recall that for a connected network $\mathcal{G}$, the null space of $L(\mathcal{G})$ is characterized by the agreement subspace. Hence, the minimum of potential W_m (6.22) corresponds to the case when

$$e^{jm\theta(t)} = e^{j\theta_o}\mathbf{1},$$

for some $\theta_o \in [0, 2\pi)$, resulting in the potential value of zero. Thereby, following the gradient control law

$$u_i(t) = -k\frac{\partial W_m(\theta)}{\partial \theta_i} = mk\sum_{j\in N(i)}\sin m(\theta_j(t) - \theta_i(t)), \tag{6.23}$$

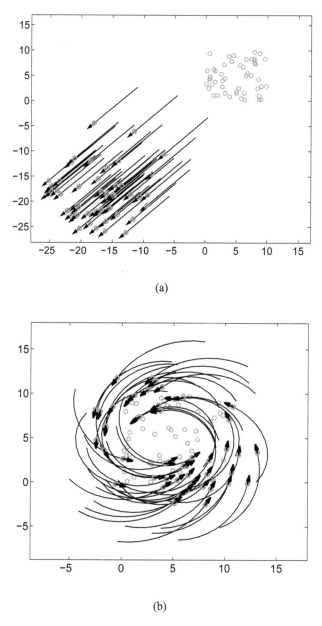

(a)

(b)

Figure 6.8: The gradient control law, steering the unicycle group toward (a) synchronization, where $\theta = \theta_o \mathbf{1}$, and (b) a balanced configuration, where $p_1(\theta) = 0$

when $k > 0$ and invoking LaSalle's invariance principle, ensures that the unicycles are steered toward the synchronized configuration of order m, which is the minimum of the Laplacian-based quadratic potential $W_m(\theta)$ (6.22).

It is tempting to conjecture that the global maximum of the potential $W_m(\theta)$ (6.22) corresponds to the balanced configuration of order m for the group of unicycles interacting over an arbitrary connected network. However, this potential can have multiple local maxima, and in fact, nonunique global maxima, some of which, do not correspond to the balanced configuration of order m for the unicycles. In the meantime, when the network is assumed to be a circulant graph $\mathcal{G}$, the function $W_m(\theta)$ (6.22) is in fact globally maximized when $e^{jm\theta(t)}$ is the eigenvector corresponding to the maximum eigenvalue of $L(\mathcal{G})$. When this occurs, we have $\mathbf{1}^T e^{jm\theta(t)} = 0$ and hence the unicycle group, while following the gradient control (6.23) with $k < 0$, steers the unicycle group to a balanced configuration of order m. We state the above observations as a theorem; see notes and references.

Theorem 6.16. *Let $\mathcal{G}$ be a connected graph. Then the global minimum of $W_m(\theta)$ (6.22) is the synchronized configuration of order m. Moreover, if the interaction network is a circulant graph, then the global maximum of $W_m(\theta)$ (6.22) is a balanced configuration of order m. In either case, a gradient law of the form (6.23) provides a distributed control strategy to attain these configurations with $k > 0$ for reaching synchronization, and $k < 0$ for reaching a balanced configuration.*

In addition to synchronization and balanced configurations that deal with phases of the unicylces, another facet of their coordination involves their spacing. In order to gain an insight into controlling this aspect of their motion, consider the case when $u_i(t) = \omega_o$, for all i in (6.15), where ω_o is a nonzero constant. In this scenario, the unicycles will traverse circles centered at

$$c_i(t) = r_i(t) + \frac{j}{\omega_o} e^{j\theta_i(t)}, \quad i = 1, 2, \ldots, n,$$

each with radius

$$\rho_o = \frac{1}{|\omega_o|}.$$

The reader is invited to verify that the direction of the rotation of the unicycles depends on the sign of ω_o. In order to consider the *agreement on the center of rotation* for all unicycles, it is convenient to introduce a new variable

$$q_i(t) = -j\omega_o c_i(t) = e^{j\theta_i(t)} - j\omega_o r_i(t), \quad i = 1, 2, \ldots, n, \quad (6.24)$$

which simultaneously encapsulates information on the heading and the center of rotation for the ith unicycle.

Theorem 6.17. *Consider the unicycle group interacting over a connected graph $\mathcal{G}$, and construct the potential*

$$S(q(t)) = \frac{1}{2}\langle q(t), L(\mathcal{G})q(t)\rangle,$$

where $q(t) = [q_1(t), \ldots, q_n(t)]^T$ and each $q_i(t)$ is defined as in (6.24). Then $S(q)$ reaches its global minimum when $q(t) = q_o\mathbf{1}$ for some $q_o \in \mathbf{C}$. Moreover, the gradient flow

$$u_i(t) = \omega_o + k\left\langle [L(\mathcal{G})]_{i.}q(t), je^{j\theta_i(t)}\right\rangle \tag{6.25}$$

for $k < 0$ and $\omega_o \neq 0$ steers the group of unicycles toward agreement on their centers of rotation as well as on their respective phases (mod 2π).[6]

The proof of this theorem, which not surprisingly involves LaSalle's invariance principle, is left as an exercise.

The Laplacian-based potential (6.22) can also be utilized to exercise more control on the phase patterns that the unicycle group exhibits. In this direction, let an (η, n)-pattern for a group of unicycles, with η as a divisor of n, be a symmetric arrangement of n phases in η clusters, uniformly spaced around the unit circle. Hence, the $(1, n)$-pattern is the synchronized configuration while the (n, n)-pattern is the splay state–in this case, each unicycle has a distinct phase, spread evenly on the unit disk.

Lemma 6.18. *Let $L(\mathcal{G})$ be the Laplacian of a circulant graph on n vertices. For any positive integer m, an (η, n)-pattern is a critical point of the potential*

$$W_m(\theta) = \frac{1}{2}\left\langle e^{jm\theta}, L(\mathcal{G})e^{jm\theta}\right\rangle.$$

Proof. Recall from Chapter 2 that the matrix of eigenvectors of $L(\mathcal{G})$ is the Fourier matrix. That is, for the angle $\bar{\theta}$ characterizing the (η, n)-pattern, $e^{jm\bar{\theta}}$ is an eigenvector of $L(\mathcal{G})$. Thus $\bar{\theta}$ is a critical point of $W_m(\theta)$. ∎

We conclude this section with an observation on the means by which a group of unicycles can be driven to symmetric patterns via local interactions over an undirected graph.

---

[6]The notation $[L(\mathcal{G})]_{i.}$ signifies the ith row of the matrix $L(\mathcal{G})$.

Theorem 6.19. *Consider a connected circulant graph $\mathcal{G}$. If $\bar{\theta}$ character-izes an (η, n)-symmetric pattern with η as a divisor of n, then it is a local minimum of the potential*

$$\widehat{W}(\theta) = -\frac{1}{2} \sum_{m=1}^{\eta} k_m \left\langle e^{jm\theta}, L(\mathcal{G})e^{jm\theta} \right\rangle, \qquad (6.26)$$

where $k_m > 0$ for $m = 1, \ldots, \eta - 1$, and

$$k_\eta < -\sum_{m=1}^{\eta-1} k_m.$$

Hence, the potential (6.26) can be used to synthesize a distributed gradient-based control law that steers the unicycles to a formation with an (η, n)-pattern.

Proof. The proof involves the linearization of the gradient of the potential $\widehat{W}(\theta)$ (6.26) at $\bar{\theta}$ which leads to a form built around a matrix that is essentially a weighted Laplacian for the underlying circulant graph. The complete proof is left as an exercise. ∎

We emphasis that the convergence of the gradient algorithm proposed in Theorem 6.19 is only local in nature.

6.5 DYNAMIC FORMATION SELECTION

6.5.1 The Centralized Case

Given that we know how to achieve target formations, the next question is what formations to use in the first place. As already stated, one can easily envision that the agents should be spread out when navigating and explor-ing free space, while a more tight formation is preferred when negotiating cluttered environments, as illustrated in Figure 6.2. What this implies is that it could potentially be beneficial to let the team switch between different formations in reaction to environmental changes.

To this end, we can define a *formation error* with respect to each possible formation under consideration,

$$\mathcal{E}^k : \mathbf{R}^p \times \cdots \times \mathbf{R}^p \to \mathbf{R}_+ \cup \{0\},$$

with smaller values indicating a smaller error, for example,

$$\mathcal{E}^k(x_1, \ldots, x_n) = \sum_{i=1}^{n} \sum_{j=1}^{n} \omega_{ij}^k \left(\|x_i - x_j\|^2 - (d_{ij}^k)^2 \right)^2. \qquad (6.27)$$

Here the superscript k indicates the kth formation as specified through D_k, $d_{ij}^k = d_{ji}^k$ is the desired distance between agents i and j, and $\omega_{ij}^k = \omega_{ji}^k > 0$ is a weight that corresponds to the relative importance of enforcing the correct distance between agents i and j. This construction ensures that $\mathcal{E}^k$ is positive semidefinite as well as $\mathbf{E}^k(x_1, \ldots, x_n) = 0$ only if the desired formation is perfectly achieved, that is, when $\|x_i - x_j\| = d_{ij}^k$, for all $i \neq j$.

Now, given a collection of potentially useful formations, with instantaneous formation errors $\mathcal{E}^1(t), \mathcal{E}^2(t), \ldots, \mathcal{E}^M(t)$, we can let the system execute formation j whenever $\mathcal{E}^j < \mathcal{E}^i$, for all $i \neq j$. In other words, we will always choose the formation with the smallest error.

An example of this strategy, where the agents switch between a line and a triangular formation, is shown in Figure 6.9, together with the corresponding error functions $\mathcal{E}_L$ and $\mathcal{E}_T$; here the subscripts L and T denote line and triangle formations, respectively. In this example, we let the agent travel towards the goal (located at $(9, 9)$). For this, they need to move through a narrow passage defined by the objects located at $(2, 3), (3, 2), (4, 3)$ and $(3, 4)$. A potential-based obstacle avoidance behavior deforms the triangle so that $\mathcal{E}_L < \mathcal{E}_T$ and, accordingly, we make a transition to the line formation. When the agents arrive at the obstacle located at $(5, 5)$ the obstacle avoidance behavior puts the leading agent out of its current heading, resulting in $\mathcal{E}_T < \mathcal{E}_L$, and the agents make a transition back to the triangular formation.

Now, it should be noted that this approach requires that all agents have perfect knowledge of the (relative) positions of all other agents, that is, that the interaction graph is a complete graph. If this is not the case, the total formation errors cannot be directly computed by each agent. To overcome this, we need to develop a decentralized, distributed version of the formation selection algorithm, which is the topic of the next subsection.

6.5.2 The Decentralized Case

If global information is not available to all agents, we can introduce $\mathcal{E}_i^k(t)$ as a measure of the kth formation error, as perceived locally by agent i, for example,

$$\mathcal{E}_i^k(t) = \sum_{j \in N(i)} \omega_{ij}^k \big(\|x_i(t) - x_j(t)\|^2 - (d_{ij}^k)^2\big)^2. \qquad (6.28)$$

It now seems like a reasonable strategy somehow to propagate and compare these local errors throughout the network. To this end, we define $\zeta_i^k(t)$ as an estimate of the kth *global* formation error, as estimated by agent i at time t. (We note here that $\mathcal{E}_i^k(t)$ is a purely instantaneous evaluation of how

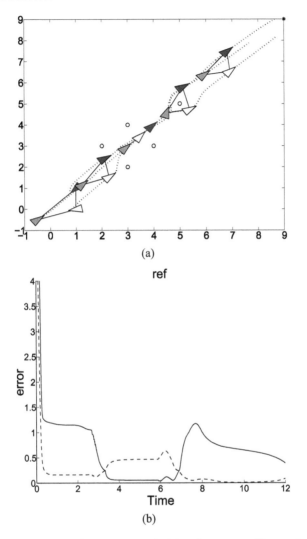

(a)

(b)

Figure 6.9: Three triangular agents switching between a line and a triangular formation in order to avoid the circular objects along the way (top), and their corresponding error functions (bottom), with $\mathcal{E}_L$ (line formation) solid and $\mathcal{E}_T$ (triangular formation) dashed

well formation k is being maintained, as perceived by agent i, while $\zeta_i^k(t)$ is a local, individual estimate of the global performance of the team.)

One way in which the agreement protocol of Chapter 3 can be used now is to let the agents communicate ζ_i^k to their neighbors, and then select the formation with the smallest global formation error. In other words, we could let

$$\begin{cases} \zeta_i^k(0) = \mathcal{E}_i^k(0), \\ \dot{\zeta}_i^k(t) = -\sum_{j \in N_i} (\zeta_i^k(t) - \zeta_j^k(t)), & \text{for } i = 1, \dots, n, \ k = 1, \dots, M. \end{cases}$$

$$(6.29)$$

How the agents should, in fact, act on this global error estimate is not self-evident, and in the following paragraphs we will discuss some possible strategies. These strategies involve selecting the formation with the smallest

1. instantaneous local error;
2. instantaneous global error estimate;
3. asymptotic global error estimate.

It should be noted that there is nothing that fundamentally dictates the "right" approach to this problem. Rather, this choice should be application driven. In essence, any dynamic formation selection mechanism is a hybrid control strategy in that it involves switching discretely between different "continuous" behaviors, that is, formation controllers. These transitions can be defined in different ways depending on how the performance, or formation error, is established for the network.

The most obvious and direct way in which a decentralized formation selection mechanism can be defined is by simply letting each agent select the formation with the smallest instantaneous error. As an example, consider a situation in which we have two possible formations. This formation strategy would thus result in agent i selecting to execute formation 1 (using the dynamics $\dot{x}_i(t) = f_i^1(x(t))$) whenever $\mathcal{E}_i^1 < \mathcal{E}_i^2$, and formation 2 ($\dot{x}_i(t) = f_i^2(x(t))$) if this is not the case, as shown in Figure 6.10.

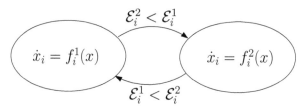

Figure 6.10: A hybrid automaton implementing the strategy in which agent i selects the dynamics $\dot{x}_i(t) = f_i^1(x(t))$ over $\dot{x}_i(t) = f_i^2(x(t))$ if $\mathcal{E}_i^1 < \mathcal{E}_i^2$

This strategy suffers from an obvious flaw in that decisions are made (and then executed) based on local properties, while global performance of the network is not taken into account. An agent switches its selection based on its instantaneous error measurements but does not consider what the other agents are trying to do, with the potential result that multiple agents may be executing completely different formations, which may in turn be nonoptimal from a global perspective yet locally beneficial. To address this issue, it makes sense to take the locally estimated global performance measure $\zeta_i^k(t)$ into account. And, since $\zeta_i^k(t)$ is updated using the agreement protocol, the global error estimate will converge to the average of the initial local errors. Thus, it makes sense to let agent i execute $\dot{x}_i(t) = f_i^k(x(t))$ if $\zeta_i^k(t) < \zeta_i^\ell$, for all $\ell = 1, \ldots, M$, $\ell \neq k$.

But there is a caveat with this strategy in that agents must be able to communicate with each other to transmit their global error estimates. Moreover, a problem that persists is that the agents act instantaneously, with the possible result that they may end up switching at different time instances as well as to different formations. As such, it might be more desirable to develop strategies where the agents spend more time "thinking" than "moving."

By letting the agents wait a certain, prespecified time before transitioning between formations, the (potentially) undesirable, instantaneous nature of the previous strategies is avoided. The main idea is to let the agents "wait" until (6.29) converges before taking action. Since the convergence is asymptotic, this means that, in theory, the agents have to wait an infinite amount of time. However, we can cap this by instead defining a convergence threshold which gives rise to a finite "convergence time" T_{conv}.

Assuming that the network is connected and static, (6.29) can be rewritten in matrix form, which yields $\dot{\xi}^k(t) = -L(\mathcal{G})\xi^k(t)$, where, as before, $L(\mathcal{G})$ is the graph Laplacian of the underlying interaction topology. As shown in Chapter 3, $\xi^k(t)$ asymptotically approaches $\bar{\zeta}^k \mathbf{1}$, where, $\bar{\zeta}^k = \frac{1}{n}\sum_{i=1}^n \zeta_i^k(0)$.

For static, undirected graphs, a direct consequence of Theorem 3.4 in Chapter 3 is that

$$\|\zeta_i^k(t) - \bar{\zeta}^k \mathbf{1}\| \leq \|\zeta_i^k(0) - \bar{\zeta}^k \mathbf{1}\| e^{-\lambda_2(L(\mathcal{G}))t} \quad \text{for all } t \geq 0. \quad (6.30)$$

If $\zeta_i^k(0)$ is bounded through $\|\zeta_i^k(0) - \bar{\zeta}^k \mathbf{1}\| \leq \kappa$ for some known constant $\kappa > 0$, we define a convergence threshold as $\|\zeta_i^k(t) - \bar{\zeta}^k \mathbf{1}\| \leq \varepsilon$ for some $\varepsilon > 0$. It then directly follows that

$$T_{conv} \geq \frac{1}{\lambda_2(L(\mathcal{G})) \ln(\frac{\kappa}{\varepsilon})}. \quad (6.31)$$

If $k = 2$, we can, as before, let $\dot{x}_i(t) = f_i^1(x)$ if $\zeta_i^1(t) < \zeta_i^2(t)$ and $\dot{x}_i(t) = f_i^2(x)$ otherwise, as seen in Figure 6.11. The price one has to pay for achieving simultaneous and delayed (as opposed to instantaneous) transitions is that the agents must somehow keep track of time in that they are required to wait T_{conv} time units and then select a formation.

If ε is small enough, all agents will not only switch at the same time, but will select the same formation. However, as they move around (waiting for T_{conv}), the environment may change. This observation leads to the incorporation of new data into the global performance estimates of the network.

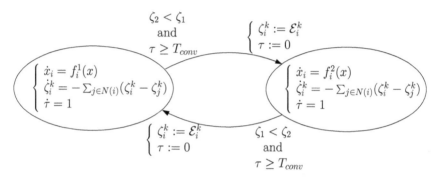

Figure 6.11: A transition between formations occurs when the guard conditions are met ($\zeta_i^k < \zeta_i^\ell$ and $\tau \geq T_{conv}$) at which point the states are reset to $\zeta_i^k = \mathcal{E}_i^k$ (new error estimate) and $\tau = 0$ (waiting time reset).

In the context of formation selection, this new information may simply arrive in the form of an agent's instantaneous local formation error. Thus, we modify the update equation for the local estimate of the global error as,

$$\dot{\zeta}_i^k(t) = - \sum_{j \in N(i)} ((\zeta_i^k(t) - \zeta_j^k(t)) + F(\mathcal{E}_i^k(t), \zeta_i^k(t))),$$

where $F(\mathcal{E}_i^k(t), \zeta_i^k(t))$ represents an insertion of the *current* instantaneous local error, $\mathcal{E}_i^k(t)$, associated with agent i's local perception of how well formation k is being maintained. But, it is not entirely clear how this new, current information should be inserted into the agreement protocol to establish the correct form of the function $F(\mathcal{E}_i^k(t), \zeta_i^k(t))$.

If we let $\mathcal{E}^k(t) = (\mathcal{E}_1^k(t), \dots, \mathcal{E}_n^k(t))^T$, we see that the average error associated with formation k, as perceived by the different agents, is simply $(1/n)\mathbf{1}^T \mathcal{E}^k$. What we would like an estimation algorithm to do is to make the individual beliefs about the global performance of a particular formation approach, this (time-varying) average.

In other words, if $\zeta^k(t) = (\zeta_1^k(t), \ldots, \zeta_n^k)^T$, we would like

$$e_{\zeta^k}(t) = \zeta^k(t) - \frac{1}{n}\mathbf{1}\mathbf{1}^T \mathcal{E}^k(t) \to 0 \quad \text{as} \quad t \to \infty. \qquad (6.32)$$

The first approach to the distributed performance estimation problem is to use a *proportional estimator*, such as

$$\begin{cases} \dot{w}_i^k(t) = -\gamma w_i^k(t) - \displaystyle\sum_{j \in N(i)} (\zeta_i^k(t) - \zeta_j^k(t)), \\ \zeta_i^k(t) = w_i^k(t) + \mathcal{E}_i^k(t), \end{cases} \qquad (6.33)$$

for $i = 1, 2, \ldots, n$, where $w_i^k(t)$ is the so-called *estimator state* and $\gamma > 0$ is the rate at which new information is introduced. Again, $\zeta_i^k(t)$ is agent i's local estimate of the global error associated with formation k, it is initialized as before by letting $\zeta_i^k(0) = \mathcal{E}_i^k(0)$.

Rewriting (6.33) yields

$$\dot{w}^k(t) = -(\gamma I + L(\mathcal{G}_f))w^k(t) + L(\mathcal{G}_f)\mathcal{E}^k(t), \qquad (6.34)$$

where $\mathcal{G}_f$ is the underlying formation graph.

Lemma 6.20. *Let $\mathcal{G}_f$ be a connected graph. If $w^k(t)$ satisfies (6.34) then*

$$\frac{1}{n}\mathbf{1}^T w^k(t) = e^{-\gamma t}\frac{1}{n}\mathbf{1}^T w^k(0).$$

Proof. We have

$$\frac{1}{n}\mathbf{1}^T \dot{w}^k(t) = -\gamma \frac{1}{n}\mathbf{1}^T w^k(t) - \frac{1}{n}\mathbf{1}^T L(\mathcal{G}_f)(w^k(t) - \mathcal{E}^k(t)).$$

Now, since $\mathbf{1}^T L(\mathcal{G}_f) = (L(\mathcal{G}_f)\mathbf{1})^T = 0$, we directly get that

$$\frac{1}{n}\mathbf{1}^T w^k(t) = \frac{1}{n}\mathbf{1}^T e^{-\gamma t} w^k(0),$$

and the statement of the lemma follows. ∎

Now, we observe that we can rewrite $e_{\zeta^k}(t)$ in (6.32) as

$$e_{\zeta^k}(t) = \Pi\zeta^k(t) - \frac{1}{n}\mathbf{1}^T w^k(t) = \Pi\zeta^k(t) - e^{-\gamma t}\frac{1}{n}\mathbf{1}^T w^k(0), \quad (6.35)$$

where $\Pi = I - \frac{1}{n}\mathbf{1}\mathbf{1}^T$ is the *disagreement projection* operator. As a consequence, we have that

$$\|e_{\zeta^k}(t)\| \le \frac{1}{\sqrt{n}}|\mathbf{1}^T w^k(0)|e^{-\gamma t} + \|\Pi\zeta^k(t)\|.$$

Moreover, the disagreement vector $\Pi\zeta^k$ satisfies

$$\Pi\dot{\zeta}^k(t) = -(\gamma I + L(\mathcal{G}_f))\Pi\zeta^k(t) + \Pi(\gamma\mathcal{E}^k(t) + \dot{\mathcal{E}}^k(t)),$$

so if we let $V^k(t) = (1/2)\|\Pi\zeta^k(t)\|^2$, we get

$$
\begin{aligned}
\dot{V}^k(t) &= \zeta^k(t)^T\Pi^2\dot{\zeta}^k(t) \\
&= -\gamma\zeta^k(t)^T\Pi^2\zeta^k(t) - \zeta^k(t)^T L(\mathcal{G}_f)\zeta^k(t) \\
&\quad + \zeta^k(t)^T\Pi(\gamma\mathcal{E}^k(t) + \dot{\mathcal{E}}^k(t)) \\
&\leq -(\gamma + \epsilon)V^k(t) + \|\gamma\mathcal{E}(t)^k + \dot{\mathcal{E}}^k(t)\|\sqrt{V^k(t)}.
\end{aligned}
$$

If the input $\mathcal{E}^k(t)$ is varying quickly, we have little hope of the system converging. So, we assume that the input is slowly varying in the sense that there is a $\mu^k > 0$ such that

$$\|\gamma\mathcal{E}^k(t) + \dot{\mathcal{E}}^k(t)\| \leq \mu^k \quad \text{for all } t \geq 0. \tag{6.36}$$

Under this assumption, we have the inequality

$$\dot{V}^k(t) \leq -(\gamma + \epsilon)V^k(t) + \mu^k\sqrt{V^k(t)}.$$

The right-hand side in the previous equation is negative if

$$V^k(t) \geq \frac{(\mu^k)^2}{(\gamma + \epsilon)^2}$$

and $\epsilon > 0$. As a consequence, we have that

$$\|\Pi\zeta^k(t)\| \leq \frac{\sqrt{2}\mu^k}{\gamma + \epsilon} \quad \text{for all } t \geq 0. \tag{6.37}$$

Theorem 6.21. *Let $\mathcal{G}_f$ be a static, connected, undirected graph. If the input is slowly varying in the sense of (6.36), then the proportional estimator in (6.33) bounds the error as*

$$\text{limsup}_{t\to\infty}\|e_{\zeta^k}(t)\| \leq \frac{\sqrt{2}\mu^k}{\gamma + \epsilon}$$

for some $\epsilon > 0$.

The proportional estimator does a reasonable job in terms of stabilization. However, as could be expected, the error does not necessarily decay to zero. For this, we need to introduce an integral action as well.

In this final strategy, the new information is introduced into the individual global estimates using a *proportional-integral estimator* as follows:

$$\begin{cases} \dot{w}_i^k(t) = - \displaystyle\sum_{j \in N(i)} (\zeta_i^k(t) - \zeta_j^k(t)), \\ \dot{\zeta}_i^k(t) = \gamma(\mathcal{E}_i^k(t) - \zeta_i^k(t)) - \displaystyle\sum_{j \in N(i)} (\zeta_i^k(t) - \zeta_j^k(t) + w_i^k(t) - w_j^k(t)) \end{cases}$$

(6.38)

for $i = 1, 2, \ldots, n$. As before, $w_i^k(t)$ is the estimator state for agent i associated with formation k. The advantage of this performance estimator lies in the fact that the input $\mathcal{E}_i^k(t)$ does not directly affect ζ_i^k, and hence it provides better filtering of noisy inputs through the integrating action.

Using the same technique as in the previous section, one can arrive at the following result.

Theorem 6.22. *Let $\mathcal{G}_f$ be a static, connected, undirected graph. If the input is slowly varying in the sense of (6.36), then the proportional-integral estimator specified by (6.38) makes the asymptotic error arbitrarily small by choosing the information rate $\gamma > 0$ sufficiently small.*

6.6 ASSIGNING ROLES

Our discussion in §6.1.1 dealt with specifying the desired shape or target geometry for the formation. Once a collection of agents are to achieve this shape, one has to make a decision as to whether the identities of the individual agents matter. If they do, the formation specification is said to come with an *assignment* (encoded by the fact that the indices in the specification need to correspond to particular indices associated with the actual agents). If they do not, the formation specification is said to be *assignment free*.

The problem of determining who goes where in a formation is equivalent to finding a suitable permutation over $[n] = \{1, \ldots, n\}$, that is, to find a bijection

$$\pi : [n] \rightarrow [n]$$

that solves the formation problem

$$\|x_{\pi(i)} - x_{\pi(j)}\| = d_{ij} \quad \text{for all } i, j = 1, \ldots, n, \ i \neq j.$$

If one associates a cost with the assignment of agent j to target i as $c(i, j)$, for example, the distance agent j has to travel to establish the desired formation, the problem of finding the best assignment becomes

$$\min_{\pi} \sum_{i=1}^{n} c(\pi(i), i),$$

which is a combinatorial optimization problem–with a potentially prohibitive computational requirement for large formations. However, for such linear assignment problems, the optimal assignment can in fact be efficiently computed (with a computational complexity of $\mathcal{O}(n^3)$) using the so-called Hungarian method.

The Hungarian method is initialized as a weighted bipartite graph with $2n$ nodes (corresponding to agents and targets, respectively), given by $\mathcal{G} = (V, E, w)$, where $V = X \cup Y$ (X = agents, Y = targets), $E = X \times Y$, and w_{ij} is the weight associated with edge $\{v_i, v_j\} \in E$. Now, a *matching* (assignment) $M \subseteq E$ is a collection of edges such that $d(i) \leq 1$ for all $v_i \in V$ under M, that is, each vertex is incident to at most one edge in M. We say that a matching M is *complete* if $\text{card}(M) = n$, that is, when every agent has been assigned to a target; the assignment problem thus involves finding the "best" such complete matching. In fact, the Hungarian method maximizes the weights in the matching, so we can, for example, let the weight $w_{ij} = -c(i, j)$ to transform the original assignment problem to that of finding a complete, maximum-weight matching.

In order to solve this, we need to introduce a vertex labeling $\ell : V \to \mathbf{R}$, and we say that the labeling is *feasible* if

$$\ell(x) + \ell(y) \geq w_{xy} \quad \text{for all } x \in X, y \in Y.$$

Based on a feasible labeling, we can form the *equality graph* $\mathcal{G}_\ell(V, E_\ell)$, with the edge set given by $E_\ell = \{\{x, y\} \mid \ell(x) + \ell(y) = w_{xy}\}$, and we let $N_\ell(x) = \{y \in Y \mid \{x, y\} \in E_\ell\}$.

Theorem 6.23 (Kuhn-Munkres). *If ℓ is feasible and $M \subseteq E_\ell$ is complete, then M is a maximum-weight matching.*

Proof. Let $M' \subseteq E$ be a complete matching. Its total weight is then

$$w(M') = \sum_{\{x,y\}\in M'} w_{xy} \leq \sum_{\{x,y\}\in M'} \ell(x) + \ell(y) = \sum_{\{x,y\}\in M} (\ell(x) + \ell(y))$$
$$= w(M).$$

The complete matching M' satisfies $w(M') \leq w(M)$, and hence M is of maximum-weight. $\blacksquare$

It turns out that this result is all one needs to formulate the algorithm that, roughly speaking, starts out with a feasible labeling ℓ together with some matching $M \subseteq E_\ell$. Then the following steps are repeated until M is complete: (1) if possible, increase the size of $M \subseteq E_\ell$, (2) if not, improve ℓ to ℓ' such that $E_\ell \subseteq E_{\ell'}$. Since at each step either M or E_ℓ is increased, the process must terminate with a complete matching, that is, one with a maximum-weight matching. More precisely, the algorithm is as follows:

Step 0
Let $\ell(y) = 0$, for all $y \in Y$, $\ell(x) = \max_y w(x, y)$ for all $x \in X$. (This step provides an initial feasible labeling.)

Step 1
Find the equality graph E_ℓ and pick a matching $M \subseteq E_\ell$. Pick unmatched $x \in X$ and set $S = \{x\}$, $T = \emptyset$.

Step 2
If $N_\ell(S) \neq T$ increase (if possible) the matching M by picking any $y \in N_\ell(s) \backslash T$.
If y is unmatched, add $\{x, y\}$ (where $\{x, y\} \in E_\ell$) to M (and if needed, update M further to keep it as a proper matching since x may already be matched–this is always possible without reducing M) and go to Step 1 unless M is complete, at which point the algorithm is terminated.
If y is matched, for instance, to x', let $S = S \cup \{x'\}$, $T = T \cup \{y\}$. Go to Step 3.

Step 3
If $N_\ell(S) = T$, update the labeling function ℓ using

$$\delta_\ell = \min_{x \in S, \, y \notin T} \{\ell(x) + \ell(y) - w_{xy}\}$$

as

$$\ell'(v) = \begin{cases} \ell(v) - \delta_\ell & \text{if } v \in S, \\ \ell(v) + \delta_\ell & \text{if } v \in T, \\ \ell(v) & \text{otherwise.} \end{cases}$$

Update E_ℓ and go to Step 2.

This *Hungarian algorithm* provides a rather nice example of how graph-based algorithms can be used to overcome seemingly intractable combinatorial problems, and it constitutes the basis for many other assignment algorithms, centralized or distributed.

SUMMARY

In this chapter, we explored the formation control problem. Among the host of topics considered, we discussed different means of specifying formations in terms of their scale, shape, relative state, and target role assignment. The role of graph rigidity, particularly in formations specified by shapes, was also examined. A number of control laws were then investigated, ranging from linear, Laplacian-based controllers to nonlinear formation controllers and controllers for unicycle agents. The question concerning what formation to select in the first place was also examined; we proposed to use an error measure that dictates how well a particular formation has been kept. This measure can then be estimated in a distributed fashion to obtain decentralized formation selection mechanisms.

NOTES AND REFERENCES

The problem of specifying, achieving, and maintaining formations has a rich history and a number of different control strategies have been proposed to this end. Some of the works that we drew inspiration from in this chapter appeared in the works of Lawton, Beard, and Young [144], Beard, Lawton, and Hadaegh [17], Broucke [36], [37], Burkard [42], Desai, Ostrowski, and Kumar [66], [65], Egerstedt and Hu [74], Muhammad and Egerstedt [165], Eren, Belhumeur, Anderson, Morse [77], Jadbabaie, Lin, and Morse [124], and Ögren, Egerstedt and Hu [180]. The basic premise behind graph-based formation control has been explicitly discussed in Olfati-Saber and Murray [181], Dunbar and Muray [72], and by Broucke in [36],[37]. Examples of various ways by which formations can be specified include deviations from desired positions, as was the case in the work of Ögren, Egerstedt, and Hu [180], deviations from desired interrobot distances (for example as in Jadbabaie, Lin, and Morse [124]), or as dissimilarities between graphs encoding the desired and actual formations, as was discussed in Ji and Egerstedt [126].

Our discussion of formation control for single and double integrator agents parallels the work of Ren and Beard [204] and Sandhu, Mesbahi, and Tsukamaki [212], whereas for the linear time invariant agents, we have presented the simplified version of the work of Fax and Murray [85] (which is concerned with the normalized version of the Laplacian) and more specifically, the paper by Lafferriere, Williams, Caughman, and Veerman [140]. The notion of T transformations for formation control was examined in the paper by Sandhu, Mesbahi, and Tsukamaki, which considered them in the context of reconfigurable formation control laws. The section on formation control of unicycles closely follows the work of Sepulchre, Paley, and Leonard [214].

The networked control community in general, and formation control researchers in particular, have drawn significant inspiration from interaction rules in social animals and insects. Compelling examples include Couzin [57], Gazi and Passino [95], and Grünbaum, Viscido, Parrish [107]. In particular, nearest neighbor-based formation control, for example as discussed in Ferrari-Trecate, Buffa, and Gati [89], Ji and Egerstedt [126], Olfati-Saber [185], Martínez, Cortés, and Bullo [151], and McNew and Klavins [153], has a direct biological counterpart, as shown by Couzin in [56],[57].

Graph rigidity and persistence is discussed by Coxeter and Greitzer [58], Eren, Whiteley, Anderson, Morse, and Belhumeur [78], Hendrickx, Anderson, Delvenne, and Blondel [115], Gluck [100], Roth [208], Tay and Whiteley [234], and Laman [141], from which the definitions and results discussed in this chapter were taken. To generate minimally rigid graphs, one can utilize the *pebble game* algorithm by Jacobs and Hendrickson [123], leading to an $\mathcal{O}(n^2)$ algorithm for constructing minimally rigid graphs.

SUGGESTED READING

See Burkard [42] and Kuhn [137] for further details of the linear assignment problem. For an excellent discussion of formation control in biology, we recommend the intriguing paper by Couzin [56]. A good introduction to rigidity is given by Roth in [208], and the basics behind graph-based formation control are presented in an easily digested manner by Lawton, Beard, and Young [144]. For formation control for unicycles we recommend the expository paper by Paley, Leonard, Sepulchre, Grünbaum, and Parrish [189].

EXERCISES

Exercise 6.1. Recall that a condition for rigidity is that

$$(x_i - x_j)^T (\dot{x}_i - \dot{x}_j) = 0, \quad \text{for all } \{v_i, v_j\} \in E.$$

Show that this relation can be rewritten as

$$R(q)\dot{q} = 0,$$

where

$$q = \begin{bmatrix} x_1 \\ \vdots \\ x_n \end{bmatrix};$$

the matrix R is called the *rigidity matrix*. It is known that when $p = 2$ and $x_i \in \mathbf{R}^p$, $i = 1, \ldots, n$, a necessary and sufficient condition for the system to be (generically) rigid is that **rank** $R(q_0) = 2n - 3$, where q_0 is feasible with respect to the edge-distance constraints. For $p = 2$ which of the following formations are (generically) rigid?

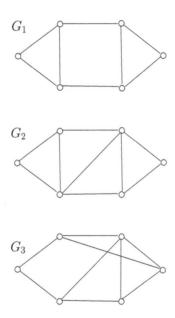

Exercise 6.2. Construct an example of a rigid framework that is not infinitesimally rigid.

Exercise 6.3. Show that if $\mathcal{G}$ contains cycles, the system $\dot{\varepsilon}(t) = -D(\mathcal{G})^T u(t)$ is uncontrollable. However, show that if the initial conditions $\varepsilon(0) = z_0 - z(0)$ satisfy the cycle constraints then the control $u(t) = D(\mathcal{G})\varepsilon(t)$ is stabilizing (with respect to the origin).

Exercise 6.4. Prove Theorem 6.17 using LaSalle's invariance principle.

Exercise 6.5. If a team of robots is to drive in formation while avoiding obstacles as well as progressing toward a goal location, one can, for example, let the individual agent dynamics be given by

$$\dot{x}_i = F_{form} + F_{goal} + F_{obst},$$

where F_{form} is used to maintain formations. However, F_{goal} is used to steer the robot towards a goal and F_{obst} is used to have it avoid obstacles. Find reasonable F_{goal} and F_{obst} and simulate your proposed solution. The final result should look something like the figure below.

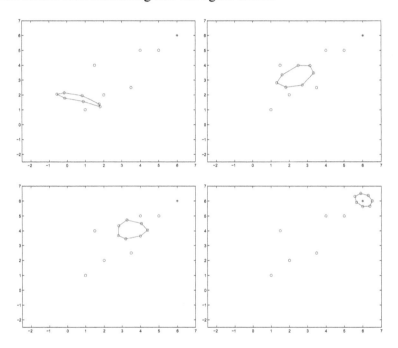

Exercise 6.6. Given a static, undirected, connected graph, let

$$\dot{x}_i = - \sum_{j \in N(i)} (\|x_i - x_j\| - k_{ij})(x_i - x_j),$$

where $k_{ij} = k_{ji}$ is the desired separation between agents i and j. If the desired interagent separations are feasible, show that the dynamics above is locally stable when $\|x_i - x_j\| \approx k_{ij}$.

Exercise 6.7. Using the formation controller in the previous question, explain what happens if the specification is not feasible, that is, no locations exist that satisfy the desired interagent distances.

Exercise 6.8. How does the stability analysis for the formation control in §6.4 of linear time-invariant agents extend to the case when the dimension of the state-space for each agent is larger that one?

Exercise 6.9. Simulate the formation control law in §6.4 for one-dimensional single, double, and linear time invariant agents and provide a simulation example of stable and unstable formations.

Exercise 6.10. Use the algorithm proposed in Theorem 6.19 to simulate convergence to a $(5, 10)$-pattern for a group of 10 unicycles interacting over the cycle graph. Comment on the local nature of the algorithm's convergence by choosing different initial conditions.

Exercise 6.11. Prove Lemma 6.10.

Exercise 6.12. Prove Theorem 6.19.

Exercise 6.13. Find the T transformations between each pair of the digraphs shown below.

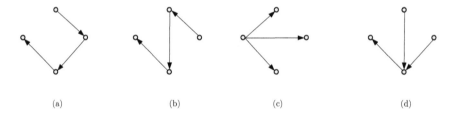

(a)   (b)   (c)   (d)

Chapter Seven

Mobile Robots

This chapter focuses on the question of how to control and coordinate mobile nodes subject to the type of interaction constraints normally associated with mobile robots. In particular, we discuss how the use of Δ-disk proximity graphs as the underlying interaction topologies affects the control design choices. For instance, it is no longer possible to use a linear agreement protocol to solve the rendezvous problem since this protocol may render an initially connected network disconnected. Instead we discuss the use of nonlinear interaction laws for ensuring that the network stays connected. This is discussed for both the rendezvous problem and the formation control problem. Moreover, the issue of graph-based sensor coverage using mobile sensing nodes is discussed in the context of geometrically induced graph-triangulations.

Arguably, a large portion of this book can be thought of as being about teams of networked mobile robots. However, in the previous chapters we have thought of the underlying graph structure as being either static or dynamic without explicit geometric conditions on the existence of edges between vertices. In this chapter, we focus on the situation in which the edges have a direct, geometric interpretation in terms of limiting sensing capabilities, as is the case when the network consists of mobile robots. In particular, we will focus on the case when the graph is a Δ-disk proximity graph, that is, where

$$\{v_i, v_j\} \in E \iff \|x_i - x_j\| \leq \Delta.$$

For example, if the robots are equipped with omnidirectional range sensors, such as sonar rings, they can only detect neighboring robots that are close enough. It should be noted that such graphs are dynamic in nature, as edges may appear or disappear when agents move in or out of the sensing (or communication) range of each other.

7.1 COOPERATIVE ROBOTICS

What makes the multirobot problem challenging is that the agents' movements can no longer be characterized by purely combinatorial interaction conditions. Instead, the coupling between geometry and combinatorics must be taken into account. However, as the *agreement protocol* describes viable means of making networked agents achieve a common value in a decentralized manner, it makes sense to modify this protocol to take the geometric range constraints into account in an explicit manner. We will pursue this endeavor for a suite of problems, including rendezvous and formation control.

To be able to define geometrically constrained agreement protocols, we will need to change the format of the protocol slightly. As the focus of this chapter is on coordination, that is, on interaction models and high-level control strategies rather than on nonlinear vehicle models, we keep the dynamics of each individual agent as a single integrator

$$\dot{x}_i(t) = u_i(t), \ i = 1, \ldots, n, \tag{7.1}$$

where $x_i \in \mathbf{R}^p$ for $i = 1, \ldots, n$. Let us first say a few words about the case when the underlying interaction graph is static. In this case, predefined, fixed links have been established between the agents, and these links are assumed to be available throughout the duration of the movement. We associate a *static interaction graph* (SIG) $\mathcal{G} = (V, E)$ to this network by letting the $V = \{v_1, \ldots, v_n\}$ represent the group, and the static edge set $E \subseteq [V]^2$ is the unordered pairs of agents, with $\{v_i, v_j\} \in E$ if and only if an interaction link exists between agents i and j. Now, what we understand by a limited information, time-invariant, decentralized control law in (7.1) is a control law of the form

$$u_i(t) = \sum_{j \in \mathcal{N}_\sigma(i)} f(x_i(t) - x_j(t)), \tag{7.2}$$

where $\mathcal{N}_\sigma(i)$ is a *subset of* the neighbors of node i in $\mathcal{G}$; the symmetric indicator function $\sigma(i, j) = \sigma(j, i) \in \{0, 1\}$ determines whether or not the information available through edge $\{v_i, v_j\} \in \mathcal{G}$ should be taken into account, with

$$j \in \mathcal{N}_\sigma(i) \Leftrightarrow \{v_i, v_j\} \in E(\mathcal{G}) \text{ and } \sigma(i, j) = 1. \tag{7.3}$$

In other words, just because two nodes are "neighbors" it does not follow that they are "friends." Along the same lines, the decentralized control law

$f(x_i - x_j)$ is assumed to be antisymmetric, that is, for all t,

$$f(x_i(t) - x_j(t)) = -f(x_j(t) - x_i(t)) \quad \text{for all } \{v_i, v_j\} \in E. \quad (7.4)$$

A few remarks about these particular choices of control and indicator functions are in order. First, the reason we only allow the function f in (7.4) to depend on the relative states among interacting agents is that this might be the only type of information available using range-based sensors. In this case, agent i simply measures the position of agent j *relative* to its current state. Second, we insist on having agents be homogeneous in that the same control law should govern the motion of all agents. This restriction is quite natural–and arguably necessary–when considering large-scale networks, where it quickly becomes unmanageable to assign and keep track of individual control laws. Such restrictions have natural consequences. For example, it follows that the centroid of the system (7.1), the average of the agents' states while adopting control laws satisfying (7.2), remains constant during the evolution of the system.

Now, let the p-dimensional position of agent i be given by

$$x_i(t) = [\, x_{i,1}(t), \ldots, x_{i,p}(t) \,]^T, \ i = 1, \ldots, n,$$

and let $x(t) = [\, x_1(t)^T, \ldots, x_n(t)^T \,]^T$. We can then define the component-wise operator as

$$c(x(t), j) = [\, x_{1,j}(t), \ldots, x_{n,j}(t) \,]^T \quad \text{for } j = 1, \ldots, p.$$

Using this notation, the standard agreement protocol from Chapter 3 assumes the form

$$\frac{d}{dt} c(x(t), j) = -L(\mathcal{G}) c(x(t), j), \ j = 1, \ldots, n, \quad (7.5)$$

where $L(\mathcal{G}) = D(\mathcal{G}) D(\mathcal{G})^T$ is the graph Laplacian and $D(\mathcal{G})$ is the incidence matrix of $\mathcal{G}$ associated with one of its orientations. And, as we have seen in Chapter 3, it follows that if $\mathcal{G}$ is connected, then $c(x, j)$ asymptotically approaches **span**$\{1\}$. Moreover, since $c(x, j)^T c(x, j)$ is a Lyapunov function for the system (7.5), for any connected graph $\mathcal{G}$, the control law

$$\frac{d}{dt} c(x(t), j) = -L(\mathcal{G}(t)) c(x(t), j) \quad (7.6)$$

drives the system to **span**$\{1\}$ asymptotically as long as the graph trajectory $\mathcal{G}(t)$ *is connected for all* $t \geq 0$. As such, by applying the control law in (7.5) to a *dynamic interaction graph* (DIG), $\mathcal{G}(t) = (V, E(t))$, where $\{v_i, v_j\} \in E(t)$ if and only if $\|x_i(t) - x_j(t)\| \leq \Delta$, we get a system behavior that seemingly solves the rendezvous problem, that is, the problem of driving all robots to the same location. However, the success of the control law (7.5) hinges on the connectedness of the underlying graph at all times. Unfortunately, this property has to be assumed rather than proved. Figure 7.1 shows an example where connectedness is lost when (7.6) is used to control a system whose network topology is a Δ-disk proximity DIG. In

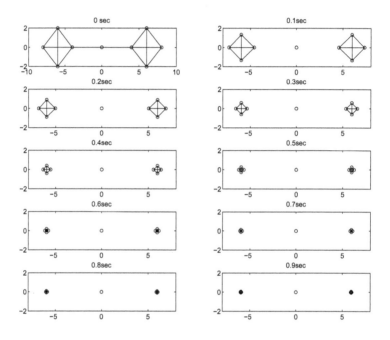

Figure 7.1: A progression where connectedness is lost even though the initial graph is connected ($\Delta = 4$)

the subsequent sections, we will show how the connectedness assumption can be enforced by modifying the control law (7.5) while ensuring that the resulting control laws are still based solely on local information- as is characterized by (7.2).

7.2 WEIGHTED GRAPH-BASED FEEDBACK

In this section, we will restrict the interaction graphs to be static, that is, we will only study the SIG case in which the behavior of the multiagent system is defined through a fixed network topology. In particular, we will show how the introduction of nonlinear edge weights can be used to establish certain invariance properties. To arrive at the desired invariance properties, we will first investigate decentralized control laws of the form

$$\sigma(i, j) = 1,$$
$$f(x_i(t) - x_j(t)) = -w(x_i(t) - x_j(t))(x_i(t) - x_j(t)) \tag{7.7}$$

for all $\{v_i, v_j\} \in E(\mathcal{G})$, where $t \geq 0$ and $w : \mathbf{R}^p \to \mathbf{R}_+$ is a positive, symmetric weighting function that associates a strictly positive and bounded weight to each edge in the SIG. This choice of decentralized control law leads to

$$\dot{x}_i(t) = - \sum_{j \in \mathcal{N}(i)} w(x_i(t) - x_j(t))(x_i(t) - x_j(t)), \tag{7.8}$$

which can be rewritten as

$$\frac{d}{dt} c(x, j) = -D(\mathcal{G}) W(x) D(\mathcal{G})^T c(x, j), \ j = 1, \ldots, p, \tag{7.9}$$

where $W(x) = \mathbf{Diag}([w_1(x), \ldots, w_m(x)]^T) \in \mathbf{R}^{m \times m}$, $m = \text{card}(E)$ is the total number of edges in the graph (the size of the graph), and each edge is identified by one unique index in the set $\{1, \ldots, m\}$.

We can thus define the state-dependent, weighted graph Laplacian as

$$L_w(x) = D(\mathcal{G}) W(x) D(\mathcal{G})^T. \tag{7.10}$$

It is straightforward to establish that as long as the graph is connected, the matrix $L_w(x)$ remains positive semidefinite, with only one zero eigenvalue corresponding to the null space $\mathbf{span}\{\mathbf{1}\}$. Now, given a critical distance δ, we will show that, using appropriate edge weights, the edge lengths never go beyond δ if they start out being less than $\delta - \epsilon$, for some arbitrarily small $\epsilon \in (0, \delta)$. For this, we need to establish some additional notation. In particular, given an edge $\{v_i, v_j\} \in E$, we let $\ell_{ij}(x)$ denote the edge vector between the agents i and j, that is, $\ell_{ij}(x) = x_i - x_j$. Moreover, we define the ϵ-interior of a δ-constrained realization of a SIG $\mathcal{G}$ as

$$\mathcal{D}_{\mathcal{G}, \delta}^{\epsilon} = \{x \in \mathbf{R}^{pn} \mid \|\ell_{ij}\| \leq \delta - \epsilon \text{ for all } \{v_i, v_j\} \in E\}.$$

An edge tension function $\mathcal{V}_{ij}$, can then be defined as

$$\mathcal{V}_{ij}(\delta, x) = \begin{cases} \frac{\|\ell_{ij}(x)\|^2}{\delta - \|\ell_{ij}(x)\|} & \text{if } \{v_i, v_j\} \in E, \\ 0 & \text{otherwise,} \end{cases} \tag{7.11}$$

with

$$\frac{\partial \mathcal{V}_{ij}(\delta, x)}{\partial x_i} = \begin{cases} \frac{2\delta - \|\ell_{ij}(x)\|}{(\delta - \|\ell_{ij}(x)\|)^2}(x_i - x_j) & \text{if } \{v_i, v_j\} \in E, \\ 0 & \text{otherwise.} \end{cases} \tag{7.12}$$

Note that this edge tension function–as well as its derivatives–is infinite when $\|\ell_{ij}(x)\| = \delta$ for some i, j, and as such, it may seem like an odd choice for our formation control law. However, as we will see, we will actually be able to prevent the tension function from reaching infinity; instead, we will examine its behavior on a compact set on which it is continuously differentiable.

Let the total tension energy of $\mathcal{G}$ be defined as

$$V(\delta, x) = \frac{1}{2} \sum_{i=1}^{n} \sum_{j=1}^{n} V_{ij}(\delta, x). \tag{7.13}$$

Lemma 7.1. *Given an initial position $x_0 \in \mathcal{D}_{\mathcal{G}, \delta}^{\epsilon}$, for a given $\epsilon \in (0, \delta)$, if the SIG $\mathcal{G}$ is connected then the set $\Omega(\delta, x_0) = \{x \mid V(\delta, x) \leq V(\delta, x_0)\}$ is an invariant set under the control law*

$$\dot{x}_i(t) = - \sum_{j \in \mathcal{N}(i)} \frac{2\delta - \|\ell_{ij}(x(t))\|}{(\delta - \|\ell_{ij}(x(t))\|)^2} (x_i(t) - x_j(t)). \tag{7.14}$$

Proof. We first note that the control law in (7.14) can be rewritten as

$$\dot{x}_i(t) = - \sum_{j \in \mathcal{N}(i)} \frac{\partial V_{ij}(\delta, x)}{\partial x_i} = -\frac{\partial V(\delta, x)}{\partial x_i} = -\nabla_{x_i} V(\delta, x).$$

This expression may be illdefined since it is conceivable that the edge lengths approach δ; as we will shortly show, this will not happen. In fact, assume that at time τ we have $x(\tau) \in \mathcal{D}_{\mathcal{G}, \delta}^{\epsilon'}$ for some $\epsilon' > 0$. Then the time derivative of $V(\delta, x(\tau))$ is

$$\begin{aligned} \dot{V}(\delta, x(\tau)) &= \nabla_x V(\delta, x(\tau))^T \dot{x}(\tau) \\ &= -\sum_{i=1}^{n} \dot{x}_i(\tau)^T \dot{x}_i(\tau) \\ &= -\sum_{j=1}^{n} c(x(\tau), j)^T L_w(\delta, x(\tau))^2 c(x(\tau), j), \end{aligned} \tag{7.15}$$

where $L_w(\delta, x)$ is given in (7.10), with weight matrix $W(\delta, x)$ (on $\Omega(\delta, x_0)$) as

$$\begin{aligned} W(\delta, x) &= \mathbf{Diag}(w_k(\delta, x)), \quad k = 1, 2, \dots, m, \\ w_k(\delta, x) &= \frac{2\delta - \|\ell_k(x)\|}{(\delta - \|\ell_k(x)\|)^2}, \end{aligned} \tag{7.16}$$

where we have arranged the edges such that subscript k corresponds to edge k. We will use this notation interchangeably with w_{ij} and ℓ_{ij}, whenever it is clear from the context.

Note that for any ϵ' bounded away from 0 from below and from δ from above, and for any $x \in \mathcal{D}^{\epsilon'}_{\mathcal{G},\delta}$, the time derivative of the total tension energy is welldefined. Moreover, for any such x, $\mathcal{V}(\delta, x)$ is non-negative and $\dot{\mathcal{V}}(\delta, x)$ is nonpositive since $L_w(\delta, x)$ is positive semidefinite for all $x \in \Omega(\delta, x_0)$. Hence, in order to establish the invariance of $\Omega(\delta, x_0)$, all that needs to be shown is that, as $\mathcal{V}$ decreases (or at lest does not increase), no edge distances will tend to δ. In fact, since $\mathcal{D}^{\epsilon}_{\mathcal{G},\delta} \subset \mathcal{D}^{\epsilon'}_{\mathcal{G},\delta}$ if $\epsilon > \epsilon'$, we will have established the invariance of $\Omega(\delta, x_0)$ if we can find an $\epsilon' > 0$ such that, whenever the system starts from $x_0 \in \mathcal{D}^{\epsilon}_{\mathcal{G},\delta}$, we can ensure that it never leaves the superset $\mathcal{D}^{\epsilon'}_{\mathcal{G},\delta}$. In this venue, define

$$\hat{\mathcal{V}}_{\epsilon} = \max_{x \in \mathcal{D}^{\epsilon}_{\mathcal{G},\delta}} \mathcal{V}(\delta, x);$$

this maximum always exists and is obtained when all edges are at the maximal allowed distance $\delta - \epsilon$, that is,

$$\hat{\mathcal{V}}_{\epsilon} = \frac{m(\delta - \epsilon)^2}{\epsilon},$$

which is a monotonically decreasing function as ϵ varies in the interval $(0, \delta)$. What we will show next is that we can bound the maximal edge distance that can generate this total tension energy, and the maximal edge length $\hat{\ell}_{\epsilon} \geq \delta - \epsilon$ is one where the entire total energy is contributed from that one single edge. In other words, all other edges have length zero, and the maximal edge length satisfies the identify

$$\hat{\mathcal{V}}_{\epsilon} = \frac{\hat{\ell}^2_{\epsilon}}{\delta - \hat{\ell}_{\epsilon}},$$

that is,

$$\frac{m(\delta - \epsilon)^2}{\epsilon} = \frac{\ell^2_{\epsilon}}{\delta - \ell_{\epsilon}},$$

which implies that

$$\hat{\ell}_{\epsilon} \leq \delta - \frac{\epsilon}{m} < \delta.$$

Hence ℓ_{ϵ} is bounded away from above by δ; moreover it is bounded from above by a strictly decreasing function as ϵ varies in the interval $(0, \delta)$.

Hence, as $\mathcal{V}$ decreases (or at least is nonincreasing), no edge distances will tend to δ, which completes the proof. ∎

The invariance of $\Omega(\delta, x_0)$ shown above now leads to the main SIG theorem.

Theorem 7.2. *Consider a connected SIG, $\mathcal{G}$, with initial condition $x_0 \in \mathcal{D}^\epsilon_{\mathcal{G}, \delta}$ and a given $\epsilon > 0$. Then the multiagent system under the control law (7.14) asymptotically converges to the static centroid $\bar{x}$.*

Proof. The proof of convergence is based on LaSalle's invariance principle (see Appendix A.3). Let $\mathcal{D}^\epsilon_{\mathcal{G}, \delta}$ and $\Omega(\delta, x_0)$ be defined as before. From Lemma 7.1, we know that $\Omega(\delta, x_0)$ is invariant with respect to the dynamics (7.14). We also note that $\mathbf{span}\{1\}$ is $L_w(\delta, x)$-invariant for all $x \in \Omega(\delta, x_0)$. Hence, due to the fact that $\dot{\mathcal{V}}(\delta, x) \leq 0$, with equality only when $c(x(t), j) \in \mathbf{span}\{1\}$, for all $j \in \{1, \dots, p\}$, convergence to $\mathbf{span}\{1\}$ follows.

Next we need to show that the agents converge to the centroid. The centroid at time t is given by

$$\bar{x}(t) = \frac{1}{n} \sum_{i=1}^{n} x_i(t),$$

and the component-wise dynamics of the centroid is

$$\frac{d}{dt}\overline{c(x(t), j)} = \frac{1}{n}\mathbf{1}^T \frac{d}{dt}c(x(t), j) = -\frac{1}{n}\mathbf{1}^T L_w(\delta, x(t))c(x(t), j).$$

Since $\mathbf{1}^T L_w(\delta, x(t)) = (L_w(\delta, x(t))\mathbf{1})^T = 0$ for all t and $x \in \Omega(\delta, x_0)$, we directly have that $\dot{\bar{x}}(t) = 0$, that is, the centroid is static and entirely determined by the initial condition x_0. As such, we can denote the centroid by $\bar{x}_0$. We note that this is a special case of the observation that the centroid is static under any control law of the form (7.2).

Now, let $\bar{\xi} \in \mathbf{R}^n$ be any point on $\mathbf{span}\{1\}$, that is, $\bar{\xi} = (\xi, \dots, \xi)^T$, for some $\xi \in \mathbf{R}$, that is consistent with the static centroid. This implies that

$$\overline{c(x, j)} = \frac{1}{p} \sum_{i=1}^{p} \xi = \xi,$$

and hence ξ has to be equal to the centroid itself. As a consequence, if x_i $(i = 1, \dots, p)$ converged anywhere other than the centroid, we would have a contradiction, and the proof now follows. ∎

We note that the construction we have described above corresponds to adding nonlinear state-dependent weights to the edges in the graph. One could conceivably add weights to the nodes of the graph as well. Unless these weights were all equal, they would violate the general assumption (7.2). For the sake of completeness, however, we briefly discuss this situation in the next few paragraphs.

A node weight can be encoded in the dynamics of the system through the weighting matrix $D(x)$ as

$$\frac{dc(x, j)}{dt} = -D(x)L_w(x)\,c(x, j), \ j = 1, \ldots, p.$$

As long as $D(x)$ is diagonal and positive definite for all x, with the diagonal elements bounded away from zero, one has that, for all $x \in \mathbf{R}^{pn}$,

$$\mathcal{N}(D(x)L_w(x)) = \mathbf{span}\{\mathbf{1}\},$$

and the controller drives the system to the agreement subspace $\mathbf{span}\{\mathbf{1}\}$. However, in this case the positions $x_i \in \mathbf{R}^p$, $i = 1, \ldots, n$, approach the same static point $\bar{x}_D(x_0) \in \mathbf{R}^p$, given by

$$\bar{x}_D(x_0) = \frac{1}{\mathbf{trace}\,(D^{-1}(x_0))} \sum_{i=1}^{n} (d_i^{-1}(x_0))x_{0,i}, \qquad (7.17)$$

where $x_{0,i} \in \mathbf{R}^p$, $i = 1, \ldots, n$, is the initial location of agent i, $d_i(x)$ is the ith diagonal element of $D(x)$. In what follows, we will show that a strategy similar to that discussed in this section can be employed even if the graph is allowed to change over time as the agents move around in their environment.

7.3 DYNAMIC GRAPHS

As already pointed out, during a maneuver, the interaction graph $\mathcal{G}$ may change as the agents move in and out of each others' sensory range. What we focus on in this section is whether stability results, analogous to the static case, can be constructed for the case when $\{v_i, v_j\} \in E$ if and only if $\|x_i - x_j\| \leq \Delta$.

In fact, we intend to reuse the tension energy from the previous section, with the particular choice of $\delta = \Delta$. However, since (7.16) implies that

$$\lim_{\|\ell_k\| \uparrow \Delta} w_k(\Delta, \|\ell_k\|) = \infty,$$

where we use $\uparrow$ to denote the limit as the argument increases, we cannot directly let the interagent tension energy affect the dynamics as soon as two agents form an edge between them, that is, as they move within distance Δ of each other. The reason for this is that we cannot allow infinite tension energies in the definition of the control laws. To overcome this issue, we choose to introduce a certain degree of hysteresis into the system, through the indicator function σ. In particular, we let $\sigma(i, j)$ be given by the "state machine" in Figure 7.2.

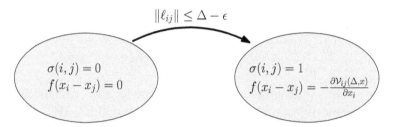

Figure 7.2: The hysteresis protocol for adding interagent tension functions to the total tension function only when agents get within a distance $\Delta - \epsilon$ of each other, rather than when they first encounter each other at a distance Δ

To further elaborate on the state machine in Figure 7.2, we let the total tension energy be affected by an edge $\{v_i, v_j\}$ that has not previously been contributing to the total energy, when $\|\ell_{ij}\| \leq (\Delta - \epsilon)$, where $\epsilon > 0$ is the predefined *switching threshold*. Once the edge is allowed to contribute to the total tension energy, it will continue to do so for all subsequent times.

In other words, for the Δ-disk proximity DIGs we will let

$$
\sigma(i, j)[t^+] = \begin{cases} 0 & \text{if } \sigma(i, j)[t^-] = 0 \text{ and } \|\ell_{ij}\| > \Delta - \epsilon, \\ 1 & \text{otherwise,} \end{cases}
$$

$$
f(x_i - x_j) = \begin{cases} 0 & \text{if } \sigma(i, j) = 0, \\ -\frac{\partial V_{ij}(\Delta, x)}{\partial x_i} & \text{otherwise,} \end{cases} \tag{7.18}
$$

where we have used the notation $\sigma(i, j)[t^+]$ and $\sigma(i, j)[t^-]$ to denote $\sigma(i, j)$ values before and after the state transition in Figure 7.2.

Before we can state the rendezvous theorem for dynamic graphs, we also need to introduce the subgraph $\mathcal{G}_\sigma \subseteq \mathcal{G}$, induced by the indicator function σ,

$$
\mathcal{G}_\sigma = (V(\mathcal{G}), E(\mathcal{G}_\sigma)),
$$

where

$$
E(\mathcal{G}_\sigma) = \{\{v_i, v_j\} \in E(\mathcal{G}) \mid \sigma(i, j) = 1\}.
$$

Theorem 7.3. *Consider an initial position $x_0 \in \mathcal{D}^\epsilon_{\mathcal{G}^0, \Delta}$, where $\epsilon > 0$ is the switching threshold (7.18) and $\mathcal{G}_0$ is the initial Δ-disk DIG. Assume that* *the graph $\mathcal{G}_{0,\sigma}$, induced by the initial indicator function value, is connected. Then, by the control law*

$$u_i(t) = - \sum_{j \in \mathcal{N}_\sigma(i)} \frac{\partial \mathcal{V}_{ij}(\Delta, x)}{\partial x_i}, \qquad (7.19)$$

where $\sigma(i,j)$ is given in (7.18), the group of agents asymptotically converges to **span**$\{1\}$.

Proof. Since from Lemma 7.1 we know that no edges in $\mathcal{G}_\sigma^0$ will be lost, only two possibilities remain, that new edges will or will not be added to the graph during the maneuver. If no edges are added, then we know from Theorem 7.2 that the system will asymptotically converge to **span**$\{1\}$. However, the only graph consistent with $x \in$ **span**$\{1\}$ is $\mathcal{G}_{0,\sigma} = K_n$ (the complete graph on n nodes), and hence no new edges will be added only if the initial graph is complete. If this graph is not complete, at least one new edge will be added. But, since $\mathcal{G}_{0,\sigma}$ is an arbitrary connected graph, and connectivity can never be lost by adding new edges, we conclude that new edges will be added until the indicator induced graph, $\mathcal{G}_\sigma$, is complete, and the system converges asymptotically to **span**$\{1\}$.   ∎

As an example, consider Figure 7.3, showing a collection of agents, influenced by the weighted control law (7.19), with the same initial position as in Figure 7.1. What is different here is–as could be expected–that no links are broken. Figure 7.4 depicts the same situation with the addition of a vertex-weight matrix to the control law, causing the centroid to be no longer static.

7.4 FORMATION CONTROL REVISITED

In the previous sections, we showed a procedure for synthesizing control laws that preserve connectedness while solving the rendezvous problem. In what follows, we will follow the same methodology to solve the distributed formation control problem. By formation control, we understand interagent distance constraints that can be described by a connected edge-labeled graph $\mathcal{G}_d = (V, E_d, d)$, where the subscript d denotes "desired." Here, E_d encodes the desired robot interconnections, that is, whether or not a desired interagent distance is specified between two agents, and the edge labels

$$d : E_d \to \mathbf{R}_+^n$$

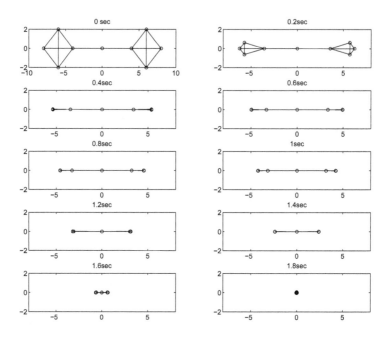

Figure 7.3: A progression is shown where connectedness is maintained during the rendezvous maneuver. Depicted are the positions of the agents and the edges in the DIG as a function of time.

define the desired relative interagent displacements, with $\|d_{ij}\| < \Delta$ for all i, j such that $\{v_i, v_j\} \in E_d$. Given a desired formation, the goal of the distributed formation control is to find a feedback law such that:

F1. The dynamic interaction graph $\mathcal{G}(t)$ converges to a graph that is a supgraph of the desired graph $\mathcal{G}_d$ (without labels) in finite time. In other words, what we want is that $E_d \subseteq E(t)$ for all $t \geq T$, for some finite $T \geq 0$.

F2. The pairwise distances $\|\ell_{ij}(t)\| = \|x_i(t) - x_j(t)\|$ converge asymptotically to $\|d_{ij}\|$ for all i, j such that $\{v_i, v_j\} \in E_d$.

F3. The feedback law utilizes only local information.

Analogous to the treatment of the rendezvous problem, we first present a solution to the formation control problem, and then show that this solution

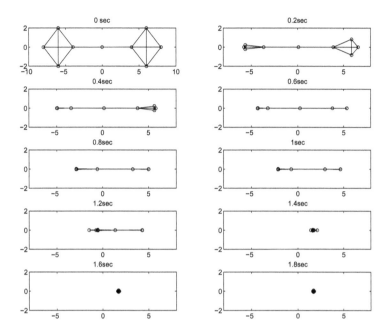

Figure 7.4:   A   progression   where   connectedness   is   maintained during   the   rendezvous   maneuver,   with   vertex   weight   matrix **Diag**$([1\ 1\ 1\ 1\ 1\ 0.5\ 0.5\ 0.5\ 0.5]^T)$

does in fact preserve connectedness as well as guarantee convergence in the sense of (F1) - (F2) above.

The solution will be based on a variation of the previously derived rendezvous controller. In this direction, assume that we have established a set of arbitrary targets $\tau_i \in \mathbf{R}^n$ that are consistent with the desired interagent distances, that is,

$$d_{ij} = \tau_i - \tau_j \quad \text{for all } i, j \text{ such that } \{v_i, v_j\} \in E_d.$$

As in the previous chapter, we can define the displacement from τ_i at time t as

$$y_i(t) = x_i(t) - \tau_i.$$

As before, we let $\ell_{ij}(t) = x_i(t) - x_j(t)$ and $\lambda_{ij}(t) = y_i(t) - y_j(t)$, implying that for all t,

$$\lambda_{ij}(t) = \ell_{ij}(t) - d_{ij}.$$

Now, under the assumption that $\mathcal{G}_d$ is a *spanning graph* of the initial interaction graph $\mathcal{G}$, that is, $V(\mathcal{G}_d) = V(\mathcal{G})$ and $E_d \subseteq E(\mathcal{G})$, we establish the control law

$$\dot{x}_i(t) = - \sum_{j \in N_{\mathcal{G}_d}(i)} \frac{2(\Delta - \|d_{ij}\|) - \|\ell_{ij}(t) - d_{ij}\|}{(\Delta - \|d_{ij}\| - \|\ell_{ij}(t) - d_{ij}\|)^2} (x_i(t) - x_j(t) - d_{ij}).$$

$$(7.20)$$

The reason why this seemingly odd choice for a control law makes sense is that we can, again, use the edge tension function $\mathcal{V}$ to describe the control law. In particular, using the following parameters in the edge tension function

$$\mathcal{V}_{ij}(\delta - \|d_{ij}\|, y) = \begin{cases} \frac{\|\lambda_{ij}\|^2}{\Delta - \|d_{ij}\| - \|\lambda_{ij}\|} & \text{if } \{v_i, v_j\} \in E_d, \\ 0 & \text{otherwise,} \end{cases} \qquad (7.21)$$

we obtain the decentralized control law

$$\sigma(i,j) = 1,$$
$$f(x_i(t) - x_j(t)) = -\frac{\partial \mathcal{V}_{ij}(\Delta - \|d_{ij}\|, y)}{\partial y_i},$$

for all $\{v_i, v_j\} \in E_d$. Along each individual dimension, the dynamics now assumes the form

$$v\frac{dc(x(t), j)}{dt} = \frac{dc(y(t), j)}{dt} = -L_w(\Delta - \|d\|, y(t))c(y(t), j), \ j = 1, \ldots, n,$$

where $L_w(\Delta - \|d\|, y)$ is the graph Laplacian associated with $\mathcal{G}_d$, weighted by the matrix $W(\Delta - \|d\|, y)$, and where we have used the convention that the term $\Delta - \|d\|$ should be interpreted in the following manner:

$$W(\Delta - \|d\|, y) = \mathbf{Diag}(w_k(\Delta - \|d_k\|, y)), \quad k = 1, 2, \ldots, |E_d|,$$
$$w_k(\Delta, -\|d_k\|, y) = \frac{2(\Delta - \|d_k\|) - \|\lambda_k\|)}{(\Delta - \|d_k\| - \|\lambda_k\|)^2}.$$

$$(7.22)$$

Here, again, the index k runs over the edge set E_d. Note that this construction allows us to study the evolution of $y_i(t)$ rather than $x_i(t)$ $(i = 1, \ldots, n)$; we formalize this in the following lemma for static interaction graphs.

Lemma 7.4. *Let the total tension energy function be*

$$\mathcal{V}(\Delta - \|d\|, y) = \frac{1}{2} \sum_{i=1}^{n} \sum_{j=1}^{n} \mathcal{V}_{ij}(\Delta - \|d_{ij}\|, y). \qquad (7.23)$$

If $y_0 \in \mathcal{D}^\epsilon_{\mathcal{G}_d, \Delta - \|d\|}$, with $\mathcal{G}_d$ a (connected) spanning graph, then under the assumption that the interaction graph is static, the set

$$\Omega(\Delta - \|d\|, y_0) = \{y \mid \mathcal{V}(\Delta - \|d\|, y) \leq \mathcal{V}_0\},$$

with V_0 denoting the initial value of the total tension energy function, is an invariant set under the control law in (7.20).

Proof. By the control law (7.20), we have

$$\dot{y}_i = -\sum_{j \in \mathcal{N}_{\mathcal{G}_d}(i)} \frac{\partial \mathcal{V}_{ij}(\Delta - \|d_{ij}\|, y)}{\partial y_i}$$

$$= -\frac{\partial \mathcal{V}(\Delta - \|d\|, y)}{\partial y_i}$$

$$= -\nabla_{y_i} \mathcal{V}(\Delta - \|d\|, y).$$

The nonpositivity of $\dot{\mathcal{V}}$ now follows from the same argument as in (7.15) in the proof of Lemma 7.1. Moreover, for each initial $y_0 \in \mathcal{D}^\epsilon_{\mathcal{G}_d, \Delta - \|d\|}$, the corresponding maximal, total tension energy induces a maximal possible edge length. Following the same line of reasoning as in the proof of Lemma 7.1, the invariance of $\Omega(\Delta - \|d\|, y_0)$ follows. ∎

Note that Lemma 7.4 states that if we can use $\mathcal{G}_d$ as a SIG, $\Omega(\Delta - \|d\|, y_0)$ is an invariant set. In fact, it is straightforward to show that if $\mathcal{G}_d$ is a spanning graph to the initial proximity Δ-disk DIG, then it remains a spanning graph for the graphs $\mathcal{G}(x(t))$ for all $t \geq 0$.

Lemma 7.5. *Given an initial condition x_0 such that $y_0 = (x_0 - \tau_0) \in \mathcal{D}^\epsilon_{\mathcal{G}_d, \Delta - \|d\|}$, with $\mathcal{G}_d$ a connected spanning graph of $\mathcal{G}(x_0)$, the group of autonomous mobile agents adopting the decentralized control law (7.20) are guaranteed to satisfy*

$$\|x_i(t) - x_j(t)\| = \|l_{ij}(t)\| < \Delta \quad \text{for all } t > 0 \text{ and } \{v_i, v_j\} \in E_d.$$

Proof. Consider a pair of agents i and j that are adjacent in $\mathcal{G}_d$, and suppose that $\|\lambda_{ij}\| = \|y_i - y_j\|$ approaches $\Delta - \|d_{ij}\|$. Since $\mathcal{V}_{ij} \geq 0$ for all i, j and $t > 0$, and

$$\lim_{\|\lambda_{ij}\| \uparrow (\Delta - \|d_{ij}\|)} \mathcal{V}_{ij} = \infty,$$

implying that $\mathcal{V} \to \infty$, which contradicts Lemma 7.4. As a consequence, $\|\lambda_{ij}\|$ is bounded away from $\Delta - \|d_{ij}\|$. This means that

$$\|l_{ij}\| = \|\lambda_{ij} + d_{ij}\| \leq \|\lambda_{ij}\| + \|d_{ij}\| < \Delta - \|d_{ij}\| + \|d_{ij}\| = \Delta,$$

and hence edges in E_d are never lost under the control law (7.20). In other words, $\|l_{ij}(t)\| < \Delta$, for all $t \geq 0$, which in turn implies that connectedness is preserved. ∎

We have thereby established that if $\mathcal{G}_d$ is a spanning graph of $\mathcal{G}(x_0)$, then it remains a spanning graph for $\mathcal{G}(x(t))$ for all $t > 0$ (under certain assumptions on x_0), even if $\mathcal{G}(x(t))$ is given by a Δ-disk DIG. In the meantime, since the control law (7.20) only takes pairwise interactions in E_d into account, we can view this dynamic scenario as a static situation, with the SIG given by $\mathcal{G}_d$. What remains to be shown is that the system in fact converges in the sense of the formation control properties (F1) - (F3) as previously defined. That condition F3 (decentralized control) is satisfied follows trivially from the definition of the control law in (7.20). Moreover, we have already established that condition F1 (finite time convergence to the appropriate graph) holds trivially as long as it holds initially, and what remains to be shown is that we can drive the system in finite time to a configuration in which condition F1 holds, after which Lemma 7.5 applies. Moreover, we need to establish that the interagent displacements (defined for edges in E_d) asymptotically converge to the desired relative displacements (F3), which is the topic of the next theorem.

Theorem 7.6. *Under the same assumptions as in Lemma 7.5, for all i, j, the pairwise relative distances $\|\ell_{ij}(t)\| = \|x_i(t) - x_j(t)\|$ asymptotically converge to $\|d_{ij}\|$ for $\{v_i, v_j\} \in E_d$.*

Proof. We first recall that $\mathcal{G}_d$ remains a spanning graph to the DIG. In view of

$$\frac{dc(y, j)}{dt} = -L_w(\Delta - \|d\|, y)c(y, j), \quad j = 1, 2, \ldots, p,$$

Theorem 7.2 ensures that for all $j \in \{1, \ldots, n\}$, $c(y, j)$ will converge to **span**$\{1\}$. What this implies is that all displacements must be the same, that is, that $y_i(t) = \zeta$, for all $i \in \{1, \ldots, n\}$, where $\zeta \in \mathbf{R}^p$. But, this simply means that the system converges asymptotically to a fixed translation away from the target points τ_i, $i = 1, \ldots, n$, that is,

$$\lim_{t \to \infty} y_i(t) = \lim_{t \to \infty} (x_i(t) - \tau_i) = \zeta, \quad \text{for } i = 1, \ldots, n,$$

which in turn implies that

$$\begin{aligned}
\lim_{t \to \infty} \ell_{ij}(t) &= \lim_{t \to \infty} (x_i(t) - x_j(t)) \\
&= \lim_{t \to \infty} (y_i(t) + \tau_i - y_j(t) - \tau_j) \\
&= \zeta + \tau_i - \zeta - \tau_j = d_{ij}
\end{aligned}$$

for all i, j for which $\{v_i, v_j\} \in E_d$, which completes the proof. ∎

7.4.1 Hybrid Rendezvous-to-Formation Control Strategies

The last property that we must establish is whether it is possible to satisfy condition F1, that is, to ensure that the initial Δ-disk proximity DIG converges in finite time to a graph that has $\mathcal{G}_d$ as a spanning tree. If this is achieved, then Theorem 7.6 would be applicable and condition F2 (asymptotic convergence to the correct interagent displacements) would follow. To achieve this, we can use the rendezvous control law developed in the previous section for gathering all agents into a complete graph, of which trivially any desired graph is a subgraph. Moreover, we need to achieve this in such a manner that the assumptions in Theorem 7.6 are satisfied.

Let K_n denote the complete graph on n agents. Moreover, we will use K_n^{Δ} to denote the situation in which the Δ-disk proximity graph is in fact a complete graph, that is, a DIG that is a complete graph in which no pairwise interagent distances is greater than Δ. This notation is potentially confusing as graphs are inherently combinatorial objects while interagent distances are geometric. To be more precise, we will use the notation $\mathcal{G} = K_n^{\Delta}$ to denote the fact that

$$\begin{cases} \mathcal{G} = K_n, \\ \ell_{ij} \leq \Delta \quad \text{for all } i \neq j. \end{cases}$$

The reason for this construction is that, in order for Theorem 7.6 to be applicable, the initial condition has to satisfy

$$y_0 = (x_0 - \tau_0) \in \mathcal{D}_{\mathcal{G}_d, \Delta - \|d\|}^{\epsilon},$$

which is ensured by making ϵ small enough. Moreover, since the rendezvous controller (7.19) asymptotically achieves rendezvous, it will consequently drive the system to K_n^{ϵ} in finite time, for all $0 < \varepsilon < \Delta$. After K_n^{ϵ} is achieved, the controller switches to the controller (7.20), as depicted in Figure 7.5. However, this hybrid control strategy is only viable if the condition that $\mathcal{G} = K_n^{\epsilon}$ is locally verifiable in the sense that the agents can decide for themselves on when the synchronous mode switch should be triggered. In fact, if an agent has $n - 1$ neighbors, all of which are within a distance $\varepsilon/2$, it follows that the maximal separation between two of those neighbors is ε.[1] Hence, when one agent detects this condition, it will trigger a switching signal (involving a one-bit broadcast communication to all its neighbors), and the transition in Figure 7.5 occurs. Note that this might actually occur not at the exact moment when $\mathcal{G}$ becomes K_n^{ϵ}, but rather at a later point. Regardless, we know that this transition will in fact occur in finite time in such a way that the initial condition assumptions of Theorem 7.6 are satisfied.

---

[1] This occurs when the agents are polar opposites on an n-sphere of radius $\varepsilon/2$.

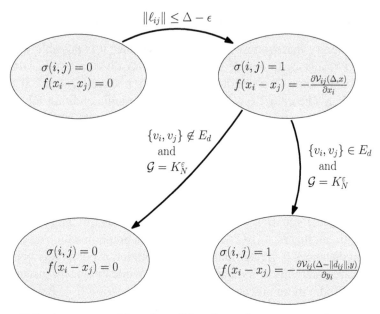

Figure 7.5: A state machine describing how the system undergoes transitions from rendezvous (collection of the agents in a tight, complete graph), to formation control

To illustrate the operation of the formation control strategy specified by (7.20), five agents, starting from a straight line, are to form a pentagonal formation, with $\mathcal{G}_d = C_5$ (the cycle graph with 5 nodes), and the desired interagent distances being $\delta_{ij} = 3.2$ for all $\{v_i, v_j\} \in E_d$. The movement of the group is shown in Figure 7.6.

7.5 THE COVERAGE PROBLEM

In this section, we examine another class of multiagent robotics problem, namely, the *coverage problem*. The coverage problem concerns ensuring that a collection of mobile sensors are placed in such a way that the area under consideration is completely covered by the sensors, that is, such that each location in the area is seen by at least one sensor. An example of this is depicted in Figures 7.7(a) - (b), where (a) corresponds to a situation in which the sensors cover the entire area, under the assumption that the sensors are omnidirectional range sensors. In Figure 7.7(b), coverage is no longer achieved.

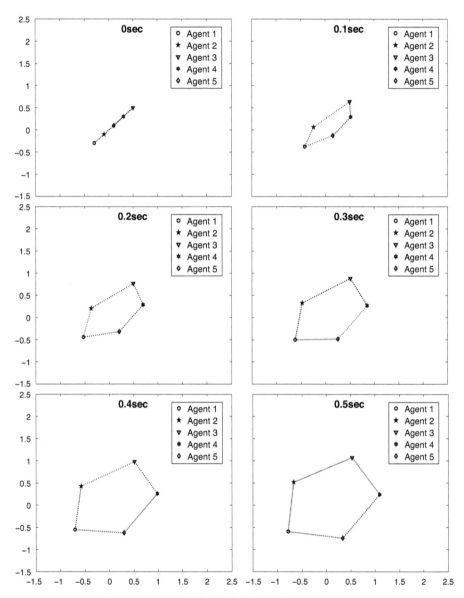

Figure 7.6: Evolution of a formation process

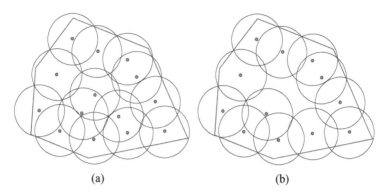

(a)   (b)

Figure 7.7: The left figure shows a sensor coverage scenario in which an enclosed area is completely covered by omnidirectional range sensors. In the right figure, the removal of two sensor nodes results in a hole in the coverage. In the figures, the nodes are shown as dots, while the circle around each node corresponds to the effective sensing area associated with that node.

There is really no clear way in which the coverage problem can be completely decoupled from the geometric properties of the problem. Such geometric inquiries include (i) What is the geometry of the sensing areas? and (ii) What is the geometry of the area in which the sensor nodes are deployed? Despite this seeming obstruction for employing graph-based (and thus combinatorial) methods, it is in fact possible to get close to the geometric formulation of the coverage problem using proximity graphs in conjunction with an interpretation of what "coverage" means in a combinatorial setting.

7.5.1 Triangulations

It is clear that since the coverage problem involves "areas" rather than "points and lines," it may not be enough simply to study vertices and edges of a suitably defined graph in order to get a handle on the coverage problem. Instead we need to get an understanding of how areas are contained by edges through the notion of a *triangulation*.

By a *triangular subgraph*, we mean any subgraph composed of three fully connected vertices. Now, as it is possible to lay out a given graph on the plane in any number of ways, we need to concern ourselves with the problem of selecting a specific such *planar embedding*. In fact, we say that a graph $\mathcal{G}$ is *planar* if there exists an embedding

$$\zeta : V \to \mathbf{R}^2$$

such that when the edges are drawn as straight lines in the plane, no edges intersect except at vertices. An embedded graph $(\mathcal{G}, \zeta)$, that is, a graph together with a corresponding embedding, is a *plane graph* when $\mathcal{G}$ is planar via the realization function ζ. We note that while plane graphs are embedded graphs, not every embedded graph is a plane graph.

The regions of a plane graph bounded by the edges are called *faces* and every finite planar graph has exactly one unbounded face, called the *outer face*.

Definition 7.7. *A plane graph is a* perfect planar triangulation *if the outer face is a cycle and all inner faces are triangles.*

As an example, consider Figure 7.8, where the left figure corresponds to a triangulation while the right figure does not. That this is indeed the case follows from the fact that the graph contains a "hole," that is, one of its faces is not a triangle.

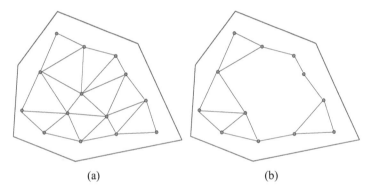

(a)   (b)

Figure 7.8: A perfect planar triangulation (a) and a planar graph that contains a hole (b)

Now, in the context of the coverage problem, the embedding function ζ is simply defining the position $\zeta(v_i)$ of sensor node v_i. Since this is a rather special embedding, we use–as before–$x(v_i)$ to denote the position of node v_i rather than the generic $\zeta(v_i)$. Moreover, the edges in the graph are geometrically induced, that is, they are *proximity graphs*, in that the existence of an edge reflects some underlying fact about the effective sensor ranges. *We will say that a given embedded proximity graph $(\mathcal{G}, x)$ solves the combinatorial coverage problem if it is a perfect, planar triangulation.* It should be noted that this formulation, that is, the formulation of a "combinatorial" coverage problem does not correspond directly to the geometric coverage problem, but it is certainly a close relative. Moreover, it is a formulation that allows us to tackle it using graph theoretic methods.

As we will see in the following sections, simply to use Δ-disk proximity graphs becomes difficult since the combinatorial coverage property is inherently a global problem, defined in terms of global properties of the graph.

As such, the local interactions in a Δ-disk graph must be complemented by more long-distance interactions, which we will pursue through the so-called Gabriel and Voronoi graphs.

7.5.2 Near-Coverage Through Gabriel Graphs

The notion of a Gabriel graph was originally developed in the context of geographic variation analysis. The basic idea behind the Gabriel graph is to generate a proximity graph that is, to a large degree, a nearest neighbor interaction graph. In the meantime, more long-distance interactions may also be included in the structure. This mix of mainly local and a few global interactions provides an effective way of addressing the combinatorial coverage problem via Gabriel graphs.

Consider a graph whose vertices $v_1, \dots, v_n$ correspond to physical, planar agents located at $x_1, \dots, x_n \in \mathbf{R}^2$. The Gabriel graph associated with these agents is given by $\mathcal{G} = (V, E)$, where $\{v_i, v_j\} \in E$ if and only if the interior angle $\angle(x_i, x_k, x_j)$ is acute for all other points x_k, as shown in Figure 7.9. An equivalent way of defining the edge set is to say that $\{v_i, v_j\} \in E$ if and only if the circle of diameter $\|x_i - x_j\|$, containing both points x_i and x_j, does not contain any vertex in its interior. We now

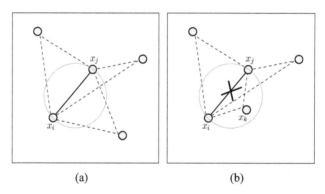

(a)   (b)

Figure 7.9: The acute angle test for establishing edges in the Gabriel graph. In (a), $\{v_i, v_j\} \in E$ since $\angle(x_i, x_k, x_j)$ is acute for all other nodes $v_k \in V$. This is not the case in (b), where, as can be seen, x_k is inside the circle defined by the two points x_i and x_j.

establish some basic results about Gabriel graphs.

Lemma 7.8. *The nearest neighbor edge is always present in the Gabriel graph, that is, if $\|x_i - x_j\| < \|x_i - x_k\|$ for all $k \neq i, j$, then $\{v_i, v_j\} \in E$.*

Proof. Suppose $n \geq 2$ and let $v_i \in V$. Then there exists $v_j \in V$ such that $\|x_i - x_j\| \leq \|x_i - x_k\|$, for all $k \neq i, j$. Now, pick an arbitrary $v_k \in V$.

Since $\|x_i - x_j\| \leq \|x_i - x_k\|$, we have $\angle(x_i, x_k, x_j) \leq \angle(x_i, x_j, x_k)$. Thus $\angle(x_i, x_k, x_j)$ must be acute since we also have that

$$\angle(x_i, x_k, x_j) + \angle(x_i, x_j, x_k) < \pi,$$

and the lemma follows. ∎

Lemma 7.9. *If $\{v_i, v_j\} \notin E$ then there exists v_k such that x_k is closer to both x_i and x_j than x_i and x_j are to each other, that is, $\|x_i - x_k\| < \|x_i - x_j\|$ and $\|x_j - x_k\| < \|x_i - x_j\|$.*

Proof. By definition, if $\{v_i, v_j\} \notin E$ then there exists $v_k \in V$ such that $\angle(x_i, x_k, x_j) \geq \pi/2$. From this, it directly follows that $\|x_i - x_k\| < \|x_i - x_j\|$ and $\|x_j - x_k\| < \|x_i - x_j\|$. ∎

These two lemmas basically state that Gabriel graphs do have a certain nearest neighbor "flair" to them. They are not (in general) disk graphs; they do, however, have other geometric properties. For example, Gabriel graphs are planar; when drawn in the plane, they do not have edges that cross (except of course at the vertices). This is important since as we have previously seen, a perfect planar triangulation certainly is a planar graph as well.

Theorem 7.10. *Any Gabriel graph is planar.*

Proof. The proof is done by contradiction. Suppose $v_i, v_j \in V$, $\{v_k, v_\ell\} \in E$ for some distinct vertices such that the edges $\{v_i, v_j\}$ and $\{v_k, v_\ell\}$ cross each other. Now, consider the quadrilateral defined by the vertices v_i, v_j, v_k, v_ℓ. Certainly at least one of the angles in this quadrilateral must be at least $\pi/2$; without loss of generality, let $\angle(x_\ell, x_i, x_k)$ be such an angle. Then, by definition $\{v_\ell, v_k\} \notin E$, which is a contradiction. ∎

The last missing piece needed to start building up combinatorial coverage structures from Gabriel graphs is to ensure that these graphs are connected.

Theorem 7.11. *Any Gabriel graph is connected.*

Proof. Let V_1 and V_2 be an arbitrary partition of the vertex set V, such that $V_1 \cap V_2 = \emptyset$. To show connectedness, it is enough to show that there is at least one edge between these two vertex sets. Thus, let $v_1 \in V_1$ and $v_2 \in V_2$ be such that $\|x_1 - x_2\| \leq \|x_1' - x_2'\|$ for any other vertices $v_1' \in V_1$ and $v_2' \in V_2$. In other words, v_1 and v_2 are the two closest vertices on the two vertex sets, respectively.

Now, assume that $\{v_1, v_2\} \notin E$. Then, by Lemma 7.9, there is a vertex v_i such that $\|x_1 - x_i\| < \|x_1 - x_2\|$ and $\|x_2 - x_i\| < \|x_1 - x_2\|$. If $v_i \in V_1$, we thus have that v_i is a point in V_1 closer to v_2, which contradicts our selection of v_1 and v_2. Similarly, a contradiction is obtained if $v_i \in V_2$. Thus, $\{v_1, v_2\} \in E$ and the proof follows.   ∎

As a consequence, we have obtained a combinatorial structure with *almost* the correct topology for achieving combinatorial coverage, that is, perfect, planar triangulations. Some examples of Gabriel graphs are given in Figure 7.10. However, it should be noted that even though these structures are quite natural in terms of combining local and global properties, we do not have guarantees that the resulting structures are perfect triangulations. In order to make the Gabriel graphs in Figure 7.10 more appropriate for addressing the coverage problem, we need to move the vertices around, which is the topic of the next few paragraphs.

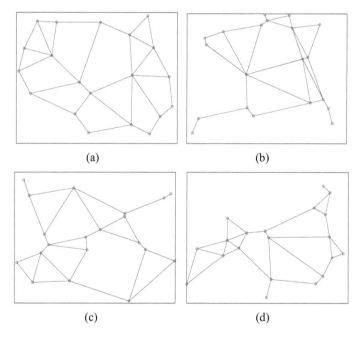

(a)   (b)

(c)   (d)

Figure 7.10: The Gabriel graphs associated with 20 randomly placed nodes. Note that none of these graphs are perfect planar triangulations even though they seem like a good starting point for solving the coverage problem.

Using the notion of a Gabriel graph to determine potentially useful interactions between agents, we can of course move the nodes around in such a way that the desired structure emerges more clearly. What we would like is to ensure a regular structure which can be achieved through an edge potential guaranteeing that desired interagent distances are achieved. In this direction, define the edge potential as

$$U_{ij} = \frac{1}{2}(\|x_i - x_j\| - \Delta)^2 \quad \text{for all } \{v_i, v_j\} \in E, \qquad (7.24)$$

where Δ is the desired interagent distance.

Following the discussion in Chapter 6 on formation control, the corresponding control law becomes

$$\dot{x}_i(t) = - \sum_{j \in N(i)} \nabla_{x_i} U_{ij} = - \sum_{j \in N(i)} \frac{(\|x_i(t) - x_j(t)\| - \Delta)}{\|x_i(t) - x_j(t)\|}(x_i(t) - x_j(t)).$$
$$(7.25)$$

It should be noted that as the agents move around, the neighborhood set will change; a collection of examples of executing this control law are seen in Figure 7.11. It is clear that we are close to achieving combinatorial coverage by producing the desired triangulations.

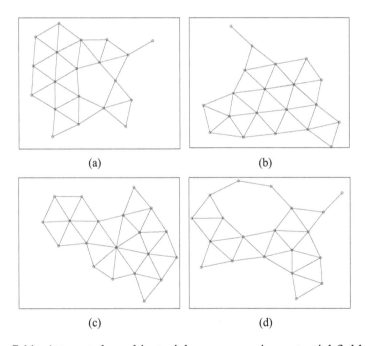

(a)   (b)

(c)   (d)

Figure 7.11: Attempted combinatorial coverage using potential field-based controllers together with Gabriel graph interaction topologies. The initial networks are given by 20 randomly placed planar nodes, as in Figure 7.10.

7.5.3 Voronoi-based Coverage Algorithms

In the previous section, we saw that by using a particular Gabriel graph structure it was possible to get close to perfect triangulations. What was needed to produce these structures was the ability to go beyond close-range interactions and incorporate longer-range interactions if needed. These longer-range interactions were based on the relative placements of the nodes. However, one can adjust this viewpoint by basing the longer-range interactions on the areas covered by the sensor nodes directly. By doing so, a formulation involving Voronoi partitions of the space emerges quite naturally.

As before, denote the (planar) area, over which the coverage task is defined, as $\Omega \subset \mathbf{R}^2$, which is assumed to be a closed, connected, and compact set. Moreover, let the agents' positions be denoted $x_1, \ldots, x_n$. We can now define a so-called tessellation of Ω as $\{W_i\}$, with sets W_i such that

$$\bigcup_{i=1}^{n} W_i = \Omega,$$
$$W_i \cap W_j = \emptyset, \ i \neq j.$$

The interpretation here is that W_i is the region in Ω that agent i is responsible for; we refer to this region as the ith agent dominance region.

If we let W denote the tessellation, and let $x = (x_1^T, \ldots, x_n^T)^T$, we can define a locational cost function

$$H(x, W) = \sum_{i=1}^{n} \int_{W_i} \|q - x_i\|^2 dq.$$

The interpretation of this function is that it divides the space Ω into the W_i regions, and then parameterizes how well these regions are covered by the agents, with the coverage quality (how well x_i can sense the point q) degrading quadratically as a function of $\|q - x_i\|$.

Intuitively, it makes sense to simplify the problem of minimizing H over x and W if we, instead, assume that W is the *Voronoi partition* of Ω, that is, $W = \mathcal{V}(x) = \{\mathcal{V}_i(x)\}$, where

$$\mathcal{V}_i(x) = \{q \in \Omega \mid \|q - x_i\| \leq \|q - x_j\| \ \text{for all } j \neq i\}.$$

In this formulation, we get the locational optimization problem in the form of minimizing

$$H_{\mathcal{V}}(x) = H(x, \mathcal{V}(x)) = \int_{\Omega} \min_{i \in \{1, \ldots, n\}} \|q - x_i\|^2 dq.$$

The idea now is to use a gradient descent algorithm for moving the agents, that is, to let

$$\dot{x}_i(t) = -\frac{\partial H_V(x)}{\partial x_i} = -2 \int_{\mathcal{V}_i} (x_i(t) - q)dq.$$

This can be further improved upon if we allow time-varying weights in the gradient descent algorithm. In particular, if we set

$$\dot{x}_i(t) = -\frac{1}{2\int_{\mathcal{V}_i} dq} \frac{\partial H_V(x)}{\partial x_i},$$

we get that

$$\dot{x}_i(t) = \rho_i(x(t)) - x_i(t),$$

where $\rho_i(x(t))$ is the center of mass of the Voronoi cell i at time t.

A few things should be pointed out about this (seemingly) simple algorithm. The first is that, for its computation, the Voronoi region $\mathcal{V}_i(x)$ must be computed. For this, agent i needs not only to be able to do a certain amount of geometry, but also to know the relative location of all agents whose Voronoi cells are adjacent to $\mathcal{V}_i$. This is where the (potentially) long-range interactions are needed since there are no guarantees that, for example, these agents are within a certain distance of each other. An example of using this approach is shown in Figures 7.12 and 7.13. In Figure 7.12(a), the initial positions of the agents are shown together with the corresponding Voronoi region. In Figure 7.12(b), the interaction graph is shown. This type of proximity graph is called a Voronoi graph, and the adjacency relationship in the graph corresponds to Voronoi cells being adjacent. In Figure 7.13 the final configuration is shown after running the gradient descent algorithm as previously described, together with the Voronoi graph.

From Figure 7.13(a), we can also make the second observation about this particular gradient descent algorithm. The final placement of the agents corresponds to them being at the centroids of their particular Voronoi cells, achieving a so-called *central Voronoi tessellation*. In fact, this way of moving the agents is very similar to an algorithm known as Lloyd's algorithm for obtaining such tessellations.

Even though the coverage in Figures 7.12 - 7.13 is perfect in the sense of being a triangulation, no such guarantees can generally be given. However, this construction constitutes another approach in which the underlying graph structure combines short and long-range interactions in a natural way in order to tackle the coverage problem.

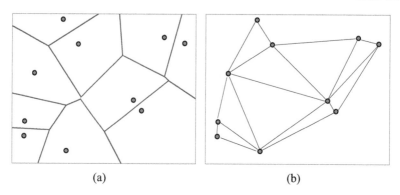

Figure 7.12: The initial placement of the agents together with the corresponding Voronoi partition (a) and the resulting Voronoi proximity graph (b)

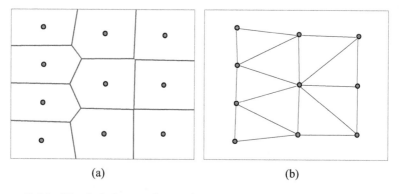

Figure 7.13: The left figure shows the position of the agents after running the gradient descent method over the locational cost function. As can be seen, what is obtained is a central Voronoi tessellation. The corresponding Voronoi proximity graph is given in the right figure.

As a final remark, it should be noted that this construction can be generalized in a number of ways. For example, instead of letting $\|q - x_i\|^2$ denote the degradation of the sensing performance, any $f(\|q - x_i\|)$ with f being nondecreasing and differentiable would do the trick. Also, one can associate a density function over the mission space $\phi : \Omega \to \mathbf{R}_+$ that captures the "event density" across the space. Using this notation, the locational optimization problem becomes that of minimizing

$$H(x, W) = \sum_{i=1}^{n} \int_{W_i} f(\|q - x_i\|)\phi(q)dq.$$

We note that the same gradient descent method as before is applicable in this case.

SUMMARY

In this chapter, the underlying sensing geometry became an important aspect of the control and coordination algorithms. In particular, we showed how the introduction of nonlinear weighted agreement protocols could be employed to ensure that the network does not become disconnected when each node is a mobile robot with a limited effective sensing range. This was done for both the rendezvous problem and the formation control problem. Extensions to the mobile sensor coverage problem were then discussed in the context of triangulations based on Gabriel graphs and central Voronoi tessellations.

NOTES AND REFERENCES

In this chapter, nonlinear weights were introduced in the agreement protocol, based mainly on the work of Ji and Egerstedt [126]. Linear time-varying weights were used by Fax and Murray [86], Lin, Broucke, and Francis [147] (for continuous time), and Jadbabaie, Lin, and Morse [124], Ren and Beard [202] (for discrete time). Nonlinear weights were also proposed by Olfati-Saber and Murray [181] and Tanner, Jadbabaie, and Pappas [230]. In addition, a robust (in the sense of disturbance rejection) rendezvous algorithm is presented by Cortés, Martínez, and Bullo in [55].

The formation control problem for limited sensing-range mobile robots has also been extensively studied in the literature. Generally speaking, there are two kinds of formation control approaches, the leader-follower approach and the leaderless approach. In the leader-follower approach, either an agent or a virtual leader is chosen as the leader, whose movement is constrained by a predefined trajectory; the remaining agents simply track the leader while obeying some coordination rules to keep the formation. A representative set of works in this direction include those by Desai, Ostrowski, and Kumar [66], Egerstedt, Hu, and Stotsky [73], Ögren, Egerstedt, and Hu [180], and Leonard and Fiorelli [145]. The other approach to formation control is the leaderless approach; see for example the works by Balch and Arkin [14] and Beard, Lawton, and Hadaegh [17]. Here the controller is typically given by a mixture of formation-maintenance, obstacle-avoidance, and trajectory-following terms.

Since few mobile networks have a static network topology due to both the movements of the individual nodes and temporal variations in the available communication channels, interest in networks with changing topologies has been growing rapidly. For example, Mesbahi proposed a dynamic extension of the static graph in [155],[156],[157] to address network problems with time-varying topologies that are induced by the dynamic states of the agents. Ren and Beard [202] find that under a dynamically changing interaction topology, if the union of the interaction graph across some time interval contains a spanning tree at a sufficient frequency as the system evolves, an information consensus is still achievable. An average consensus problem is solved for switching topology networks by Olfati-Saber and Murray [182], where a common Lyapunov function is obtained for directed balanced graphs, based on a so-called disagreement function.

The terminology used in the latter parts of this chapter "just because two nodes are neighbors it doesn't follow that they are friends" appeared in a paper by McNew and Klavins [152].

SUGGESTED READING

The use of Gabriel graphs is well explained in [217], while connectedness-preserving formation control can be found in [126]. The section on Voronoi-based coverage algorithms is taken in large part from the excellent paper [151]. For a representative sample of multirobot systems using graph theory see for example the pioneering works of Ando, Oasa, Suzuki, and Yamashita [9], Fax and Murray [85], and Lin, Broucke, and Francis [147].

EXERCISES

Exercise 7.1. This chapter mainly dealt with Δ-disk graphs, that is, proximity graph (V, E) such that $\{v_i, v_j\} \in E$ if and only if $\|x_i - x_j\| \le \Delta$, where $x_i \in \mathbf{R}^p$, $i = 1, \ldots, n$, is the state of robot i. In this exercise, we will be exploring another type of proximity graph, namely the *wedge graph*.

Assume that instead of single integrator dynamics, the agents' dynamics are defined as unicycle robots, that is,

$$\dot{x}_i(t) = v_i(t) \cos \phi_i(t),$$
$$\dot{y}_i(t) = v_i(t) \sin \phi_i(t),$$
$$\dot{\phi}_i(t) = \omega_i(t).$$

Here $[x_i, y_i]^T$ is the position of robot i, while ϕ_i denotes its orientation. Moreover, v_i and ω_i are the translational and rotational velocities, which are the controlled inputs.

Now, assume that such a robot is equipped with a rigidly mounted camera, facing in the forward direction. This gives rise to a directed *wedge graph*, as seen in the figure. For such a setup, if robot j is visible from robot i, the available information is $d_{ij} = \|[x_i, y_j]^T - [x_j, y_j]^T\|$ (distance between agents) and $\delta\phi_{ij}$ (relative interagent angle) as per the figure below. Explain how you would solve the rendezvous (agreement) problem for such a system.

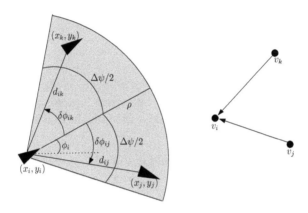

Exercise 7.2. Show that if a δ-disk proximity graph with 4 agents starts connected, then, using the linear agreement protocol, it stays connected for all times.

Exercise 7.3. Consider n agents and m leaders (all with scalar dynamics) placed in the same initial position, the origin. Assume that there is a goal located at x_g, known only to the leaders. Let the dynamics of each leader be

$$\dot{x}_{i_\ell}(t) = - \sum_{j \in \mathcal{N}(i_\ell)} (x_{i_\ell}(t) - x_j(t)) + c(x_g - x_{i_\ell}(t)),$$

for some positive gain $c > 0$. Moreover, assume that the followers are executing

$$\dot{x}_{i_f}(t) = - \sum_{j \in \mathcal{N}(i_f)} (x_{i_f}(t) - x_j(t)).$$

Now, under the assumption that the interaction graph is a Δ-disk proximity graph, find conditions on n, m, c, x_g, Δ such that the network stays connected during the transport from the origin to x_g.

Exercise 7.4. In Theorem 7.3, the edge tension energy $w_k(\Delta, \|\ell_k\|) \to \infty$ as $\|\ell_k\|$ approaches Δ (from below). Explain why this is necessary if the tension energy is not allowed to depend on n, that is, the number of agents. Also, explain how this can be avoided if w_k is allowed to depend on n.

Exercise 7.5. Show that in a planar triangulation with equidistant edge lengths, the maximum degree is 6.

Exercise 7.6. What is the maximum degree in a planar Gabriel graph?

Exercise 7.7. Verify the identity (7.17).

Exercise 7.8. Assume that we have weighted edges in the graph, that is, we have a weighted graph Laplacian L_W. Show that, as long as the weights are nonzero, the null space is not affected by the introduction of weights.

Exercise 7.9. How should the control laws in §7.3 be modified if the robots have different sensing ranges, that is, $\Delta_i \neq \Delta_j$, $i \neq j$.

Exercise 7.10. In order to facilitate a transition from rendezvous to formation control, a broadcast scheme was employed in §7.3. Is it possible to achieve such a synchronous transition in a decentralized manner if no communications capabilities are present, that is, using sensing only.

Exercise 7.11. One way of achieving translationally invariant formations is to let the desired position for agent i be ξ_i, and to run the control protocol

$$\dot{x}_i(t) = - \sum_{j \in N(i)} ((x_i(t) - x_j(t)) - (\xi_i - \xi_j)).$$

Now, consider two connected agents on the line. Assume that there is some confusion about where the target positions really are. In particular, let agent 1 run the above protocol with $\xi_1 = 0$ and $\xi_2 = 1$. At the same time, agent 2 runs the protocol with $\xi_1 = 0$ and $\xi_2 = 2$. What happens to $x_1(t)$, $x_2(t)$, and $x_1(t) - x_2(t)$, as $t \to \infty$?

Exercise 7.12. Explain why no a priori bound can be given on the edge distances in Gabriel or Voronoi graphs.

Chapter Eight

Distributed Estimation

"It is not certain that everything is uncertain."
— Blaise Pascal

In this chapter, we present two complementary areas of distributed estimation, namely, distributed linear least squares and distributed Kalman filtering over sensor networks. For the former case, the recursive least squares algorithm is adapted for sensor networks modeled as undirected graphs; in this venue, we provide necessary and sufficient conditions for the convergence of the corresponding distributed algorithm. Subsequently, we extend our analysis to networks that have a "clustered" structure and consider *pulsed* intercluster updates. In this latter scenario, intercluster communications occur every β time steps, with β a positive integer greater than one, and the corresponding updates are *held* until the next update instant. Finally, we turn our attention to distributed, discrete-time Kalman filtering and expand on a few architectures of particular interest for sensor networks.

Estimation theory is a truth-seeking endeavor; it is the scientific means of designing processes by which a static or dynamic variable of interest can be uncovered by processing a noisy signal that functionally depends on it. Estimation is a rich discipline with a wide range of applications in signal processing and control. Our emphasis in this chapter is naturally on the distributed and networked aspects of certain discrete-time estimation algorithms, namely, distributed linear least squares and distributed Kalman filtering.

8.1 DISTRIBUTED LINEAR LEAST SQUARES

We start our discussion by examining how linear least squares can be viewed and analyzed in the distributed setting. Estimators that are based on least squares do not require a probabilistic assumption on the noise signal that corrupts the underlying variable or the estimated state, and are therefore

easily implementable and applicable for a broad class of estimation problems. The underlying model involves the observation of a linear function of
a variable $\theta \in \mathbf{R}^q$ that is additively corrupted by noise v,

$$z = H\theta + v,$$

where $z, v \in \mathbf{R}^p$ and $H \in \mathbf{R}^{p \times q}$ ($p > q$); we refer to each component
of the vector z as a *measurement channel* and H is the *observation matrix*
that is assumed to be of row rank q. The rank condition on H ensures that
the measurement channels are not entirely redundant. In a centralized setting and absence of information about the noise statistics, the least squares
estimation proceeds by minimizing the cost function

$$J(\theta) = (z - H\theta)^T (z - H\theta). \tag{8.1}$$

Since J in (8.1) is a differentiable and convex function of the underlying
state θ, its optimal value is found by setting its gradient to zero, and declar-
ing its optimum, that is, the least squares estimate, as

$$\widehat{\theta} = \left(H^T H\right)^{-1} H^T z. \tag{8.2}$$

It is rather a nontrivial fact that the above framework can also be adopted
for finding *optimal estimators* in other settings, such as the maximum like-
lihood estimates or minimum variance Bayes estimates, when the distributions of the additive noise or that of the state are assumed to be Gaussian.
Thus, for example, when v is a zero-mean Gaussian noise with covariance
Σ, minimizing the weighted objective functional

$$J(\theta) = (z - H\theta)^T \Sigma^{-1} (z - H\theta), \tag{8.3}$$

leads to the optimal estimate

$$\widehat{\theta} = \left(H^T \Sigma^{-1} H\right)^{-1} H^T \Sigma^{-1} z. \tag{8.4}$$

Inclusion of the weighting matrix Σ^{-1} in (8.4), induced by the noise covariance, is motivated by the desire to skew the optimal estimate toward
measurements that are less uncertain.

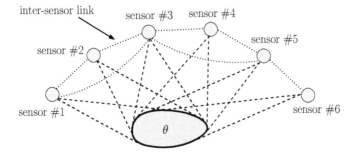

Figure 8.1: Sensor networks estimating a random vector.

8.1.1 Least Squares over Sensor Networks

We now consider *distributed least squares* for the case when $\Sigma = I$ in (8.3) and leave the extension of our discussion for more general scenarios as an exercise. In such a distributed setting, there are n sensors available for making measurements, and the observation vector for the ith sensor, $1 \leq i \leq n$, is given by $z_i \in \mathbf{R}^{p_i \times 1}$; hence

$$z_i = H_i \theta + v_i,$$

where $H_i \in \mathbf{R}^{p_i \times q}$ and $z_i = H_i \theta \in \mathbf{R}^{p_i \times 1}$; see Figure 8.1. Consider next the integration of the observation matrices H_i $(i = 1, 2, \cdots, n)$ as the equivalent *centralized* observation matrix $H \in \mathbf{R}^{p \times q}$, as

$$H = [H_1; H_2; \ldots; H_n], \tag{8.5}$$

where $p = \sum_{i=1}^{n} p_i$ and the ";" operation denotes vertical concatenation of matrices and vectors.[1] With respect to the matrix H, the centralized least squares estimator in (8.2) can now be written as

$$\widehat{\theta} = \left(\sum_{i=1}^{n} H_i^T H_i \right)^{-1} \left(\sum_{i=1}^{n} H_i^T z_i \right), \tag{8.6}$$

provided that the additive noise signals are statistically independent.

The additive nature of (8.6) suggests that if each sensor provides the raw measurement z_i to the *fusion center* which has prior knowledge of each observation matrix H_i, then the fusion center can efficiently find the estimate $\widehat{\theta}$ (8.6). However, due to scalability, modularity, and fault tolerance, it might be desirable to compute (8.6) without including a fusion center. As we will

---

[1] A Matlab notation.

see shortly, the agreement protocol of Chapter 3 provides a convenient avenue for developing a *distributed least squares* algorithm that utilizes the network as the means of computing the estimate $\widehat{\theta}$ *without a fusion center*.

Let us first consider the proposed algorithm for the simplest scenario, namely, in the context of estimating a scalar variable. Specifically, our task is to estimate a scalar variable $\theta \in \mathbf{R}$, based on the noisy observations, via a distributed algorithm operating on the measurements $z_i = \theta + v_i$, $i = 1, 2, \ldots, n$. In this case, (8.6) can be used to deduce that

$$\widehat{\theta} = \frac{1}{n} \sum_{i=1}^{n} z_i. \tag{8.7}$$

The form of (8.7) now calls for the applicability of the agreement protocol as a mechanism for computing the solution of the least squares problem in a distributed way. In this direction, consider the interconnection topology between the different sensors abstracted in terms of the graph $\mathcal{G} = (V, E)$, with V and E representing, respectively, the sensors and the ability of a sensor pair to interchange their respective intermediate estimates. Moreover, we let

$$W = \mathbf{Diag}\left([w_1, \ldots, w_m]^T\right), \tag{8.8}$$

where $w_i > 0$ is the weight on the ith edge of the graph, indexed consistently with the column ordering in the corresponding incidence matrix $D(\mathcal{G})$.[2] Next, consider the iteration for the ith sensor as

$$\widehat{\theta}_i(k+1) = \widehat{\theta}_i(k) + \Delta \sum_{j \in N(i)} w_{ij}(\widehat{\theta}_j(k) - \widehat{\theta}_i(k)), \tag{8.9}$$

where $\widehat{\theta}_i(k)$ is the estimate of the variable θ by sensor i at time instant k and $\Delta \in (0, 1)$ is the step size for the update scheme (8.9); we will have more to say on the selection of Δ. Define the weighted Laplacian of $\mathcal{G}$ as

$$L_w(\mathcal{G}) = D(\mathcal{G})WD(\mathcal{G})^T, \tag{8.10}$$

and set

$$M_w(\mathcal{G}) = I - \Delta L_w(\mathcal{G}). \tag{8.11}$$

A convenient terminology associated with the weighted Laplacian (8.10) is that of *generalized degree* of vertex i, $d_w(i)$, defined as the sum of the weights of the edges incident on i,

$$d_w(i) = [L_w(\mathcal{G})]_{ii}. \tag{8.12}$$

---

[2]Thus w_j refers to the weight on the jth edge, whereas w_{ij} refers to the weight on the edge connecting vertices i and j.

Using a matrix-vector notation, the iteration (8.9) can be written as

$$\widehat{\theta}(k+1) = M_w(\mathcal{G})\,\widehat{\theta}(k),$$

with $\widehat{\theta} = [\widehat{\theta}_1, \widehat{\theta}_2, \ldots, \widehat{\theta}_n]^T$. Hence

$$\widehat{\theta}(k) = M_w(\mathcal{G})^k\,\widehat{\theta}(0), \quad k = 1, 2, \ldots, \tag{8.13}$$

where $\widehat{\theta}(0)$ is the "prior estimate" of θ at the initialization of the estimation process. The convergence of this iterate therefore depends on the behavior of the powers of the matrix $M_w(\mathcal{G})$, which in turn depends on its spectral radius. The following lemma states conditions under which the right-hand side of (8.13) converges to a value that has statistical significance.

Lemma 8.1. *Consider the sequence (8.13), arbitrary initialized as*

$$\widehat{\theta}(0) = [\widehat{\theta}_1(0), \widehat{\theta}_2(0), \ldots, \widehat{\theta}_n(0)]^T,$$

with $\widehat{\theta}_i(k)$ denoting the estimate of sensor i of variable θ at time instance k. Then

$$\lim_{k \to \infty} \widehat{\theta}(k) = \left(\frac{1}{n}\sum_{i=1}^{n} z_i\right)\mathbf{1}$$

if and only if the underlying sensor network is connected and

$$\rho(L_w(\mathcal{G})) < \frac{2}{\Delta}, \tag{8.14}$$

where $\rho(L_w(\mathcal{G}))$ is the maximum eigenvalue of $L_w(\mathcal{G})$ in absolute value.[3]

Proof. We first observe that the spectrum of the matrix $M_w(\mathcal{G})$ (8.11) is given by the set

$$\{1 - \Delta\lambda_i(L_w(\mathcal{G})); \ i = 1, 2, \ldots, n\},$$

where $\lambda_i(L_w(\mathcal{G}))$ is the ith eigenvalue of $L_w(\mathcal{G})$. Moreover, since the smallest eigenvalue of the weighted Laplacian is zero, the maximum eigenvalue of $M(\mathcal{G})$ in absolute value is given by

$$\rho(M(\mathcal{G})) = \begin{cases} 1, & \text{if } \Delta\,\rho(L_w(\mathcal{G})) < 2 \\ |1 - \Delta\,\rho(L_w(\mathcal{G}))| & \text{if } \Delta\,\rho(L_w(\mathcal{G})) \geq 2. \end{cases} \tag{8.15}$$

---

[3] In other words, its spectral radius; note, however, that $L_w(\mathcal{G})$ is a symmetric matrix.

We now note that when $\Delta\,\rho(L_w(\mathcal{G})) > 2$, one has $\rho(M_w(\mathcal{G})) > 1$, and the sequence $M_w(\mathcal{G})^k$ becomes unbounded as $k \to \infty$. On the other hand, when $\Delta\,\rho(L_w(\mathcal{G})) = 2$, the matrix $M_w(\mathcal{G})$ has at least one eigenvalue equal to -1 and $\lim_{k\to\infty} M(\mathcal{G})^k$ fails to exist. We proceed to show that when $\Delta\rho(L_w(\mathcal{G})) < 2$, it follows that

$$\lim_{k\to\infty} M_w(\mathcal{G})^k = \frac{1}{n}\mathbf{1}\mathbf{1}^T. \tag{8.16}$$

Along this line, observe that when $\Delta\,\rho(L_w(\mathcal{G})) < 2$, the eigenvalues of $M_w(\mathcal{G})$ satisfy

$$-1 < \lambda_i(M_w(\mathcal{G})) \le 1 \quad \text{for } i = 1, 2, \ldots, n.$$

Moreover, the normalized eigenvector of $M_w(\mathcal{G})$, corresponding to its largest eigenvalue $\lambda_n(M_w(\mathcal{G})) = 1$, is $(1/\sqrt{n})\,\mathbf{1}$, which is also the eigenvector corresponding to $\lambda_1(L_w(\mathcal{G})) = 0$. This follows from the identities

$$M_w(\mathcal{G})\left(\frac{1}{\sqrt{n}}\mathbf{1}\right) = (I - \Delta L_w(\mathcal{G}))\left(\frac{1}{\sqrt{n}}\mathbf{1}\right) = \frac{1}{\sqrt{n}}\mathbf{1}.$$

Since for a connected graph $\lambda_2(L_w(\mathcal{G})) > 0$, one has

$$\lambda_1(M_w(\mathcal{G})) \le \cdots \le \lambda_{n-1}(M_w(\mathcal{G})) < 1. \tag{8.17}$$

The inequality (8.17), in conjunction with the spectral factorization of $M_w(\mathcal{G})$, leads us to the observation that when $\Delta\,\rho(L_w(\mathcal{G})) < 2$,

$$\lim_{k\to\infty} \widehat{\theta}(k) = \left(\lim_{k\to\infty} M_w(\mathcal{G})^k\right)\widehat{\theta}(0) = \frac{1}{n}\mathbf{1}\mathbf{1}^T\widehat{\theta}(0) = \left(\frac{1}{n}\sum_{i=1}^{n} z_i\right)\mathbf{1}.$$

Thereby, the identify (8.16) follows. Moreover, each sensor converges to the (centralized) linear least squares estimate (8.7). ∎

An important observation pertaining to Lemma 8.1 is that the spectral condition (8.14) does not preclude the entries of the iteration matrix $M_w(\mathcal{G})$ (8.13) from being negative. As an example, consider a three-node path graph $\mathcal{G}$ with node set $V = \{1, 2, 3\}$ and edge set $E = \{\{1, 2\}, \{2, 3\}\}$. Suppose that the weights on the edges $\{1, 2\}$ and $\{2, 3\}$ are 0.5 and 0.6, respectively. The weighted Laplacian matrix of this graph and the corresponding iteration matrix (letting $\Delta = 1$) are then

$$L_w(\mathcal{G}) = \begin{bmatrix} 0.5 & -0.5 & 0 \\ -0.5 & 1.1 & -0.6 \\ 0 & -0.6 & 0.6 \end{bmatrix} \text{ and } M_w(\mathcal{G}) = \begin{bmatrix} 0.5 & 0.5 & 0 \\ 0.5 & -0.1 & 0.6 \\ 0 & 0.6 & 0.4 \end{bmatrix}.$$

We invite the reader to verify that the maximum eigenvalue of $L_w(\mathcal{G})$ in absolute value satisfies condition (8.14) and that each entry of $M_w(\mathcal{G})^k$ approaches $\frac{1}{3}$ as $k \to \infty$.

Choosing the required weights for the distributed least squares iteration (8.9) can be approached in an optimization framework which attempts to place the spectral radius of the matrix $M_w(\mathcal{G})$ at a desired location (see Chapter 12). There are also a few convenient optimization-free means of choosing these weights, one of which we briefly expand upon.

Corollary 8.2. *Suppose that the weighting diagonal matrix in (8.8) is defined in such a way that its jth diagonal entry, representing the weight on the jth edge $e = uv$, is*

$$[W]_{jj} = (\max\{d_w(u), d_w(v)\})^{-1}. \qquad (8.18)$$

Then for any $0 \leq \Delta < 1$, one has $\Delta \, \rho(L_w(\mathcal{G})) < 2$.

Proof. Our first observation is that the maximum eigenvalue of the weighted Laplacian matrix in absolute value is bounded as

$$\rho(L_w(\mathcal{G})) \leq \max\{d_w(u) + d_w(v) \,|\, uv \in E\} \leq 2\bar{d}_w, \qquad (8.19)$$

where $\bar{d}_w = \max_{v \in V} d_w(v) = \max_i [L_w(\mathcal{G})]_{ii}$ denotes the maximum generalized vertex degree. Next, using (8.19), a sufficient condition for $\rho(L_w(\mathcal{G})) \leq 2$ to hold is that $d_w(v) \leq 1$, for all $v \in V$. Hence if the weighting matrix W (8.8) is constructed according to (8.18), one has $d_w(v) \leq 1$ for all $v \in V$. Consequently, for any $\Delta \in [0, 1)$, it follows that $\Delta \, \rho(L_w(\mathcal{G})) < 2$. ∎

A direct consequence of Corollary 8.2 is the following observation.

Corollary 8.3. *If the edge weights are constructed according to (8.18) and $\Delta \in (0, 1)$, then the distributed least squares iteration (8.9) converges to the centralized least squares estimate.*

8.1.2 Distributed Least Squares Estimation: Vector Case

In this section, we point out how the results of the previous section can be extended to the vector parameter set. In this venue, the observation matrix for the ith sensor, H_i, is allowed to be arbitrary as long as the corresponding vertical concatenation (8.5) is full row rank. This is in contrast to our discussion for the scalar case, where, in order to streamline the discussion, we assumed $H_i = 1$, for all i. Moreover, for the vector case, we consider two classes of networks, referred to as *monolithic* and *clustered* networks.

Monolithic networks are those that do not possess significant clustering or hierarchical structure–although the terminology can be more formally defined, we will appeal to the natural intuition of the readers for the distinction.[4] In the vector case, each sensor maintains two arrays, $P_i \in \mathbf{R}^{q \times q}$ and $\widehat{\theta}_i \in \mathbf{R}^{q \times 1}$, where q is the length of the parameter vector, and executes the iterations,

$$P_i(k+1) = P_i(k) + \Delta \sum_{j \in N(i)} w_{ij}(P_j(k) - P_i(k)), \qquad (8.20)$$

$$\widehat{\theta}_i(k+1) = \widehat{\theta}_i(k) + \Delta \sum_{j \in N(i)} w_{ij}(\widehat{\theta}_j(k) - \widehat{\theta}_i(k)), \qquad (8.21)$$

where $i = 1, 2, \ldots n$, is the sensor index, and the iterations are initialized as

$$P_i(0) = H_i^T H_i \quad \text{and} \quad \widehat{\theta}_i(0) = H_i^T z_i, \quad i = 1, \ldots, n, \qquad (8.22)$$

where $z_i \in \mathbf{R}^{p_i \times 1}$ denotes the observation vector for the ith sensor. By a straightforward extension of the discussion in the previous section, it can be seen that

$$\lim_{k \to \infty} P_i(k) = \frac{1}{n} \sum_{i=1}^{n} H_i^T H_i \qquad (8.23)$$

and

$$\lim_{k \to \infty} \widehat{\theta}_i(k) = \frac{1}{n} \sum_{i=1}^{n} \widehat{\theta}_i(0) = \frac{1}{n} \sum_{i=1}^{n} H_i^T z_i, \qquad (8.24)$$

when the step size Δ and the weights w_{ij} are chosen according to Lemma 8.1. Therefore, each sensor asymptotically computes the centralized linear least squares estimate according to

$$\widehat{\theta} = \lim_{k \to \infty} P_i(k)^{-1} \widehat{\theta}_i(k). \qquad (8.25)$$

---

[4]See §5.4.2.

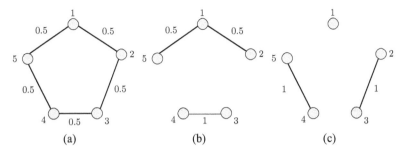

Figure 8.2: Numbers above the edges represent their weights computed according to (8.18). (a) Monolithic system. The edges represent the communication links (bidirectional). Communication updates occur at every time step, at times $t = k\Delta$, $k = 1, 2, \ldots$ (b) Distributed system with 2 clusters. Cluster 1 consists of nodes 1, 2, and 5 and cluster 2 consists of nodes 3 and 4. The edges represent the communication links over which updates occur at every time step, $t = k\Delta$, $k = 1, 2, \ldots$. (c) The edges represent the links in the distributed system over which communication updates occur every β time steps, at times $t = k\beta\Delta$, $k = 1, 2, \ldots$.

Note that the iterates $P_i(k)$ may not be invertible for all values of k; hence the local least squares estimate (8.25) at node i can only be computed once $P_i(k)$ becomes nonsingular.

8.2 PULSED INTERCLUSTER COMMUNICATION

In this section, we extend our analysis to clustered networks with *pulsed inter-cluster updates*. We assume that all intracluster updates occur at every time step, at times $t = k\Delta$, $k = 1, 2, 3, \ldots$, while all intercluster updates occur at times $t = k\beta\Delta$, $k = 1, 2, 3, \ldots$, with β as a positive integer greater than one. The set of time instants $[k\beta\Delta + \Delta, k\beta\Delta + 2\Delta, \ldots, (k + 1)\beta\Delta]$ constitutes an *update cycle*. Figure 8.2 shows inter- and intracluster communication patterns as compared with the monolithic system. Consider now a pair of nodes i and j belonging to two distinct partitions of $\mathcal{G} = (V, E)$, namely, $\mathcal{C}_1$ and $\mathcal{C}_2$, where $\mathcal{C}_i \subseteq V$ ($i = 1, 2$) and $\mathcal{C}_1 \cap \mathcal{C}_2 = \emptyset$. If i and j exchange their state information at time $t = k\beta\Delta$, we assume that node i (respectively, node j) holds node j's (respectively, node i's) state value until it receives the next update, which occurs at time $t = (k + 1)\beta\Delta$. We refer to this mechanism as pulsed intercluster updating. A node i in cluster $\mathcal{C}_j$ ($j \in \{1, 2\}$) is a *gateway node* if it connects to one or more nodes in clusters other than $\mathcal{C}_j$ (besides having neighbors in its own cluster). We do not impose any conditions on the number of gateway nodes per cluster. However, we assume that due to the distributed nature of sensors, it may well be that intercluster communications require higher transmitter power support compared to intracluster communications. From a power efficiency or network lifetime point of view, it will therefore be beneficial to limit the extent of intercluster communications, without significantly degrading the convergence time of the overall distributed estimation algorithm.

Let E_1 denote the set of edges which are activated at every time step (each of length Δ) and E_2 the set of edges which are activated every $\beta > 1$ time steps, where β is an integer. We note that $E_1 \cup E_2 = E$ and $E_1 \cap E_2 = \emptyset$, where E denotes the set of edges corresponding to the monolithic system. Furthermore, let D_1 and D_2 denote the incidence matrices defined by the edge sets E_1 and E_2, respectively. For example, D_1 is obtained from $D(\mathcal{G})$ by zeroing out the entries that correspond to edges in E_2. The corresponding weighted Laplacian matrices are denoted by $L_w(\mathcal{G}_1)$ and $L_w(\mathcal{G}_2)$. As an example, referring to Figure 8.2, the matrices D_1^T and D_2^T are

$$D_1^T = \begin{bmatrix} -1 & 1 & 0 & 0 & 0 \\ -1 & 0 & 0 & 0 & 1 \\ 0 & 0 & -1 & 1 & 0 \\ 0 & 0 & 0 & 0 & 0 \\ 0 & 0 & 0 & 0 & 0 \end{bmatrix} \tag{8.26}$$

and

$$D_2^T = \begin{bmatrix} 0 & 0 & 0 & 0 & 0 \\ 0 & 0 & 0 & 0 & 0 \\ 0 & 0 & 0 & 0 & 0 \\ 0 & -1 & 1 & 0 & 0 \\ 0 & 0 & 0 & -1 & 1 \end{bmatrix}. \tag{8.27}$$

The corresponding weighting matrices W_1 and W_2 and the weighted Laplacian matrices, $L_w(\mathcal{G}_1)$ and $L_w(\mathcal{G}_2)$, are then

$$W_1 = \begin{bmatrix} \frac{1}{2} & 0 & 0 & 0 & 0 \\ 0 & \frac{1}{2} & 0 & 0 & 0 \\ 0 & 0 & 1 & 0 & 0 \\ 0 & 0 & 0 & 0 & 0 \\ 0 & 0 & 0 & 0 & 0 \end{bmatrix}, \quad W_2 = \begin{bmatrix} 0 & 0 & 0 & 0 & 0 \\ 0 & 0 & 0 & 0 & 0 \\ 0 & 0 & 0 & 0 & 0 \\ 0 & 0 & 0 & 1 & 0 \\ 0 & 0 & 0 & 0 & 1 \end{bmatrix},$$

and

$$
L_w(\mathcal{G}_1) = \begin{bmatrix} 1 & -\frac{1}{2} & 0 & 0 & -\frac{1}{2} \\ -\frac{1}{2} & \frac{1}{2} & 0 & 0 & 0 \\ 0 & 0 & 1 & -1 & 0 \\ 0 & 0 & -1 & 1 & 0 \\ -\frac{1}{2} & 0 & 0 & 0 & \frac{1}{2} \end{bmatrix}, \tag{8.28}
$$

$$
L_w(\mathcal{G}_2) = \begin{bmatrix} 0 & 0 & 0 & 0 & 0 \\ 0 & 1 & -1 & 0 & 0 \\ 0 & -1 & 1 & 0 & 0 \\ 0 & 0 & 0 & 1 & -1 \\ 0 & 0 & 0 & -1 & 1 \end{bmatrix}.
$$

Note that $W_1 + W_2 = W$ and therefore

$$
L_w(\mathcal{G}_1) + L_w(\mathcal{G}_2) = L_w(\mathcal{G}), \tag{8.29}
$$

where W and $L_w(\mathcal{G})$ are, respectively, the weighting matrix and the weighted Laplacian matrix for the monolithic system. A consequence of (8.29) is that

$$
\max\{\lambda_n(L_w(\mathcal{G}_1)), \lambda_n(L_w(\mathcal{G}_2))\} \le \lambda_n(L_w(\mathcal{G})). \tag{8.30}
$$

Let us define

$$
M_w(\mathcal{G}_1) = I - \Delta L_w(\mathcal{G}_1) \quad \text{and} \quad M_w(\mathcal{G}_2) = \Delta L_w(\mathcal{G}_2); \tag{8.31}
$$

we note that $M_w(\mathcal{G}_1) - M_w(\mathcal{G}_2) = I - \Delta L_w(\mathcal{G})$. With the above notation, we can now express the evolution of the estimated state $\widehat{\theta}^c$ on clustered networks over an update cycle as

$$
\widehat{\theta}^c((k\beta + 1)\Delta) = (M_w(\mathcal{G}_1) - M_w(\mathcal{G}_2))\,\widehat{\theta}^c(k\beta\Delta),
$$
$$
\widehat{\theta}^c((k\beta + 2)\Delta) = M_w(\mathcal{G}_1)\,\widehat{\theta}^c((k\beta + 1)\Delta) - M_w(\mathcal{G}_2)\,\widehat{\theta}^c(k\beta\Delta),
$$
$$
\vdots
$$
$$
\widehat{\theta}^c((k+1)\beta\Delta) = M_w(\mathcal{G}_1)\,\widehat{\theta}^c(((k+1)\beta - 1)\Delta) - M_w(\mathcal{G}_2)\,\widehat{\theta}^c(k\beta\Delta).
$$

Since $M_w(\mathcal{G}_1)\mathbf{1} = \mathbf{1}$, the vector $\widehat{\theta}^c((k+1)\beta\Delta)$ can be expressed in terms of $\widehat{\theta}^c(k\beta\Delta)$ as

$$
\widehat{\theta}^c((k+1)\beta\Delta) = \left(M_w(\mathcal{G}_1)^\beta - \left(\sum_{\alpha=0}^{\beta-1} M_w(\mathcal{G}_1)^\alpha \right) M_w(\mathcal{G}_2) \right) \widehat{\theta}^c(k\beta\Delta)
$$
$$
= \widetilde{M}_w\,\widehat{\theta}^c(k\beta\Delta), \tag{8.32}
$$

where

$$\widetilde{M_w} = \left(M_w(\mathcal{G}_1)^\beta - \left(\sum_{\alpha=0}^{\beta-1} M_w(\mathcal{G}_1)^\alpha \right) M_w(\mathcal{G}_2) \right). \qquad (8.33)$$

We now use the matrix identity

$$X^k - Y^k = \sum_{\alpha=0}^{k-1} X^{k-1-\alpha}(X-Y)Y^n \qquad (8.34)$$

for any $X, Y \in \mathbf{R}^{n \times n}$, to rewrite (8.33) as

$$\widetilde{M_w} = I - \left(\left(I - M_w(\mathcal{G}_1)^\beta \right) + \left(\sum_{\alpha=0}^{\beta-1} M_w(\mathcal{G}_1)^\alpha \right) M_w(\mathcal{G}_2) \right)$$

$$= I - \left(\sum_{\alpha=0}^{\beta-1} (I - M_w(\mathcal{G}_1)) M_w(\mathcal{G}_1)^\alpha + \left(\sum_{\alpha=0}^{\beta-1} M_w(\mathcal{G}_1)^\alpha \right) M_w(\mathcal{G}_2) \right)$$

$$= I - \left(\sum_{\alpha=0}^{\beta-1} M_w(\mathcal{G}_1)^\alpha (I - M_w(\mathcal{G}_1)) + \left(\sum_{\alpha=0}^{\beta-1} M_w(\mathcal{G}_1)^\alpha \right) M_w(\mathcal{G}_2) \right)$$

$$= I - \left(\left(\sum_{\alpha=0}^{\beta-1} M_w(\mathcal{G}_1)^\alpha \right) \Delta L_w(\mathcal{G}_1) + \left(\sum_{\alpha=0}^{\beta-1} M_w(\mathcal{G}_1)^\alpha \right) \Delta L_w(\mathcal{G}_2) \right)$$

$$= I - \Delta Z L_w(\mathcal{G}), \qquad (8.35)$$

where $L_w(\mathcal{G}) = L_w(\mathcal{G}_1) + L_w(\mathcal{G}_2)$ and Z is a scaling matrix defined by

$$Z = \sum_{\alpha=0}^{\beta-1} M_w(\mathcal{G}_1)^\alpha = \sum_{\alpha=0}^{\beta-1} (I - \Delta L_w(\mathcal{G}_1))^\alpha. \qquad (8.36)$$

Note that although the matrix Z (8.36) is symmetric, the product $Z L_w(\mathcal{G})$ that appears in (8.35) is not necessarily symmetric.

Let us now make a few observations that will subsequently be used in our convergence analysis for distributed least squares estimation over a network with pulsed intercluster communications.

Lemma 8.4. *The scaling matrix Z (8.36) is positive definite if*

$$\Delta \rho(L_w(\mathcal{G}_1)) < 1,$$

where $\rho(L_w(\mathcal{G}_1))$ denotes the largest eigenvalue of $L_w(\mathcal{G}_1)$ in absolute value.

Proof. We first note that the matrix $(I - \Delta L_w(\mathcal{G}_1))^n$ is symmetric for all $n \geq 0$, and its eigenvalues are

$$(1 - \Delta \lambda_i(L_w(\mathcal{G}_1)))^n \quad \text{for } i = 1, 2, \ldots, n.$$

Thereby $\lambda_1(Z) > 0$ if $\Delta \rho(L_w(\mathcal{G}_1)) < 1$.   ∎

Before we state the next observation, we mention the following linear algebraic fact. For a pair of matrices $A, B \in \mathbf{R}^{n \times n}$, if A is positive definite and B is symmetric, then the product AB is a diagonalizable matrix, all of whose eigenvalues are real. Moreover, the matrix product AB has the same number of positive, negative, and zero eigenvalues, as B.

Lemma 8.5. *When $\Delta \rho(L_w(\mathcal{G}_1)) < 1$, the eigenvalues of the matrix product $Z L_w(\mathcal{G})$, where Z is defined as in (8.36), are real and nonnegative. Moreover, $\lambda_1(Z L_w(\mathcal{G})) = 0$ and the corresponding normalized eigenvector is $(1/\sqrt{n})\mathbf{1}$.*

Proof. The first part of the lemma follows from Lemma 8.4 and the above observation on eigenvalues of matrix products; note that $L_w(\mathcal{G})$ is symmetric and positive semidefinite and Z is positive definite if $\Delta \rho(L_w(\mathcal{G}_1)) < 1$. The second part of the lemma holds since $Z L_w(\mathcal{G}) \mathbf{1} = 0$.   ∎

Lemma 8.6. *Let the matrix Z be defined as in (8.36). If $\Delta \rho(L_w(\mathcal{G}_1)) < 1$, then $\lambda_2(Z L_w(\mathcal{G})) > 0$.*

Proof. First, note that Z is positive definite if $\Delta \rho(L_w(\mathcal{G}_1)) < 1$ (by Lemma 8.4). Since $L_w(\mathcal{G})$ is symmetric, the matrix $Z L_w(\mathcal{G})$ has the same number of positive, negative, and zero eigenvalues, as $L_w(\mathcal{G})$. As $L_w(\mathcal{G})$ has only one zero eigenvalue if the graph consisting of edges $E_1 \cup E_2$ is connected, $\lambda_2(Z L_w(\mathcal{G})) > 0$.   ∎

We now consider the following system of equations that describe the evolution of the system every $\beta \Delta$ time steps (see (8.32) and (8.35)),

$$\widehat{\theta}^c(\beta \Delta) = (I - \Delta Z L_w(\mathcal{G})) \, \widehat{\theta}^c(0),$$
$$\widehat{\theta}^c(2\beta \Delta) = (I - \Delta Z L_w(\mathcal{G})) \, \widehat{\theta}^c(\beta \Delta),$$
$$\vdots$$
$$\widehat{\theta}^c(k\beta \Delta) = (I - \Delta Z L_w(\mathcal{G})) \, \widehat{\theta}^c((k-1)\beta \Delta),$$

or alternately,

$$\widehat{\theta}^c(k\beta \Delta) = (I - \Delta Z L_w(\mathcal{G}))^k \, \widehat{\theta}^c(0), \quad \text{for } k \geq 0. \qquad (8.37)$$

For comparison, we note that the evolution of the monolithic system at time $t = k\Delta$ is

$$\widehat{\theta}(k\Delta) = (I - \Delta L_w(\mathcal{G}))^k \, \widehat{\theta}(0), \quad \text{for } k \geq 0. \tag{8.38}$$

We are ready to state the main result of this section.

Lemma 8.7. *Consider the sequence generated by (8.37), initialized as*

$$\widehat{\theta}^c(0) = [\,\widehat{\theta}_1(0), \widehat{\theta}_2(0), \dots, \widehat{\theta}_n(0)\,]^T.$$

Assuming that the network is connected and

$$\Delta\rho(L_w(\mathcal{G}_1)) < 1 \quad and \quad \Delta\rho(ZL_w(\mathcal{G})) < 2, \tag{8.39}$$

one has

$$\lim_{k\to\infty} \widehat{\theta}^c(k\beta\Delta) = \alpha\mathbf{1},$$

where

$$\alpha = \frac{1}{n}\sum_{i=1}^{n} \widehat{\theta}_i^c(0).$$

Proof. The proof is identical to that of Lemma 8.1 and can be reconstructed by noting that the spectral constraint on $L_w(\mathcal{G}_1)$ guarantees that (see Lemmas 8.5 and 8.6):

1. all eigenvalues of $ZL_w(\mathcal{G})$ are real and nonnegative,

2. $\lambda_1(ZL_w(\mathcal{G})) = 0$ and its corresponding normalized eigenvector is $(1/\sqrt{n})\,\mathbf{1}$,

3. the algebraic multiplicity of the zero eigenvalue is 1,

4. $ZL_w(\mathcal{G})$ is diagonalizable, and therefore admits a decomposition of the form $ZL_w(\mathcal{G}) = T\Lambda T^{-1}$.

We note that since the matrix product $ZL_w(\mathcal{G})$ is not necessarily symmetric, the columns of the matrix T may not be orthonormal. Nevertheless, the methodology of the proof of Lemma 8.1 can still be adopted using T^{-1} instead of T^T. ∎

Corollary 8.8. *If the weighting matrix W (8.8) is constructed according to (8.18), selecting $\beta > 1$ and $\Delta \in (0, 1/\beta)$ guarantees that the sequence (8.37) converges to the centralized least squares solution.*

Proof. Since $\rho(ZL_w(\mathcal{G})) \leq \rho(Z)\rho(L_w(\mathcal{G}))$, for the inequality

$$\Delta\rho(ZL_w(\mathcal{G})) < 2$$

to hold, it suffices to ensure that

$$\Delta\rho(Z)\rho(L_w(\mathcal{G})) < 2.$$

On the other hand, since $\rho(L_w(\mathcal{G})) \leq 2$ if the weighting matrix W is constructed according to (8.18) (see the proof of Corollary 8.2), it suffices to guarantee that $\Delta\rho(Z) < 1$. If $\Delta\rho(L_w(\mathcal{G}_1)) < 1$, we know that $\rho(Z) = \beta$. Therefore, any value of Δ in the range $(0, 1/\beta)$ is sufficient to ensure that $\Delta\rho(ZL_w(\mathcal{G})) < 2$. Furthermore, the range of acceptable values of Δ which ensures that $\Delta\rho(L_w(\mathcal{G}_1)) < 1$ is $(0, \frac{1}{2})$ since

$$\Delta\rho(L_w(\mathcal{G}_1)) \leq \Delta\rho(L_w(\mathcal{G})) \leq 2\Delta, \tag{8.40}$$

where the first inequality follows from (8.30) and the second inequality follows from the fact that $\rho(L_w(\mathcal{G})) \leq 2$ when W is constructed according to (8.18). Consequently, when $\beta > 1$, having $\Delta \in (0, 1/\beta)$ ensures the convergence of the sequence (8.37). ∎

While the upper bound $\Delta < 1/\beta$ is sufficient, it may be overly conservative. For static networks, however, it is possible to evaluate the maximum value of β offline, such that for a given value of Δ, the inequalities in (8.39) are satisfied.

8.2.1 Clustered Networks: Vector Case

For clustered networks with positive integer $\beta > 1$, gateway nodes and nongateway nodes execute distinct iterations for estimating the underlying random vector. In this direction, let V_g denote the set of gateway nodes and $\overline{V_g}$ the set of nongateway nodes; hence $V_g \cup \overline{V_g} = V(\mathcal{G})$. Moreover, let W_e denote the diagonal element of the weighting matrix W corresponding to the edge e. All nodes in $\overline{V_g}$ now execute the iterations

$$P_i(k+1) = P_i(k) + \Delta \sum_{e=ij\in E_1} W_e\left(P_j(k) - P_i(k)\right), \tag{8.41}$$

$$\widehat{\theta}_i^c(k+1) = \widehat{\theta}_i^c(k) + \Delta \sum_{e=ij\in E_1} W_e\left(\widehat{\theta}_j^c(k) - \widehat{\theta}_i^c(k)\right), \tag{8.42}$$

when initialized as (8.22); recall that E_1 denotes the set of all edges that are active at every time step. Nodes in V_g on the other hand execute another set of iterations; in particular,

if $k + 1 = 0 \pmod{\beta}$,

$$P_i(k+1) = P_i(k) + \Delta \sum_{e=ij\in E} W_e\left(P_j(k) - P_i(k)\right), \quad (8.43)$$

$$\widehat{\theta}_i^c(k+1) = \widehat{\theta}_i^c(k) + \Delta \sum_{e=ij\in E} W_e\left(\widehat{\theta}_j^c(k) - \widehat{\theta}_i^c(k)\right). \quad (8.44)$$

else

$$P_i(k+1) = P_i(k) + \Delta \sum_{e=ij\in E_1} W_e(P_j(k) - P_i(k))$$

$$+\Delta \sum_{e=ij\in E_2} W_e(P_j(\widehat{k}) - P_i(\widehat{k})), \quad (8.45)$$

$$\widehat{\theta}_i^c(k+1) = \widehat{\theta}_i^c(k) + \Delta \sum_{e=ij\in E_1} W_e(\widehat{\theta}_j^c(k) - \widehat{\theta}_i^c(k)),$$

$$+\Delta \sum_{e=ij\in E_2} W_e(\widehat{\theta}_j^c(\widehat{k}) - \widehat{\theta}_i^c(\widehat{k})), \quad (8.46)$$

where $\widehat{k}$ is the largest integer such that $\widehat{k} < k + 1$ and $\widehat{k} \bmod \beta = 0$. Note that the initializations $P_i(0)$ and $\widehat{\theta}_i^c(0)$ are as in (8.22). Gateway nodes therefore need to maintain two additional arrays to store the values of the third expression on the right-hand side of (8.45) and (8.46), corresponding to the time instances which are multiples of β. These arrays are refreshed every β time steps when updates are available from intercluster neighbors. Using the same techniques as in §8.2, it can be shown that each sensor asymptotically converges to the centralized linear least squares estimate using this adjusted protocol.

In order to demonstrate the performance of the proposed algorithm, the simulation results for a 49-node sensor network with 5 clusters are shown in Figure 8.3. The total number of edges in the network in Figure 8.3 is 163, of which 11 correspond to the intercluster network (shown as dotted lines). The average distance between any pair of intracluster nodes is 1.29 whereas the average distance between any pair of intercluster nodes is 2.72. The node degree statistics, considering both inter- and intracluster edges, are: $\min_i d(i) = 3$, $\max_i d(i) = 9$, $\text{mean}\{d(i)\} = 6.52$ and standard deviation of $d(i) = 1.46$. For all simulations, the length of the unknown parameter vector θ is 5, the observation matrix for sensor i is $H_i \in \mathbf{R}^{5\times 5}$, and each entry of H_i has been chosen from a uniform distribution on the unit interval.

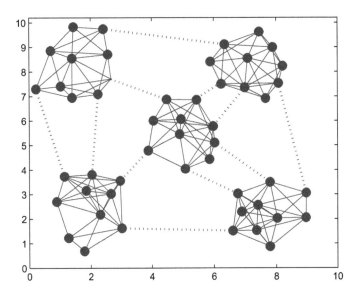

Figure 8.3: A 49-node 5-cluster network. The dotted lines represent the intercluster edges and the solid lines the intracluster edges.

Moreover, the observation vector for sensor i is $z_i \in \mathbf{R}^{5 \times 1}$, which has been generated according to the model $z_i = H_i \theta + \eta_i$, where η_i is a zero mean Gaussian noise with covariance matrix $0.1\,I$. All noise vectors are assumed to be statistically independent.

In Figure 8.4(a), we have depicted the squared error of the distributed least squares estimate $\widehat{\theta}_{LLS}^{d}$ with respect to the centralized estimate $\widehat{\theta}_{LLS}$, for the nodes with the highest and lowest node degrees. The solid lines correspond to the protocol parameters selected as $\beta = 1$ and $\Delta = 0.99$; the dashed lines correspond to $\beta = 2$ and $\Delta = 0.49$. We note that these values are such that the inequality $\Delta < 1/\beta$ holds. As remarked in the previous section, the upper bound $\Delta < 1/\beta$ may be overly conservative. This is supported by the simulations shown in Figure 8.4(b), where we plot the squared error for $\Delta = 0.99$ and $\beta \in \{1, 2\}$.[5] The curves for the two nodes are practically indistinguishable now, though, somewhat interestingly, the squared error for the highest degree node appears to be slightly smaller when $\beta = 2$ compared to $\beta = 1$, for intermediate values of k. Using the intercluster edges for only half the iteration time, as is the case for $\beta = 2$, is attractive since it reduces the associated transmission energy expenditure.

---

[5]It was verified offline that $\Delta = 0.99$ satisfies the conditions in (8.39) for $\beta = 2$.

We proceed with simulations by increasing the value of β, while keeping Δ fixed at 0.99. Figure 8.5(a) shows the squared error for the smallest degree node, when $\beta \in \{1, 8, 12, 16\}$. As can be seen from this figure, the distributed method converges even for $\beta = 8$ (in fact, it also converges for $\beta = 10$ but exhibits oscillatory behavior and ultimately diverges when $\beta \geq 11$).

We conclude this section with a comment on the performance of a *pulsed* intercluster update scheme as opposed to an *intermittent* intercluster update scheme. In the latter approach, an intracluster update received at step k is used only for computing the state values at step $k + 1$ and is not "held" for the duration of the update cycle. Figure 8.5(b) compares the squared error for the two approaches.

8.3 IMPLEMENTATION OVER WIRELESS NETWORKS

In this section, we outline how the distributed estimation algorithms discussed above may be implemented over wireless networks. Channel access mechanisms can broadly be divided into two categories: *contention-free protocols* and *contention-based protocols*. Contention-free protocols eliminate interference by proper scheduling of resources and ensure that wireless transmissions, while in process, are always successful. Contention-based protocols, on the other hand, allow users to contend for the wireless channel and prescribe mechanisms to resolve conflicts which may occur if users attempt to transmit simultaneously. Examples of contention-free protocols include *frequency division multiple access* (FDMA), *time division multiple access* (TDMA), and *code division multiple access* (CDMA). Examples of contention-based protocols are ALOHA,[6] slotted-ALOHA, and different variants of *carrier sense multiple access* (CSMA) schemes. Below, we describe the TDMA protocol in slightly more detail and explain how it may be used to implement our synchronous, distributed algorithms.

In the traditional TDMA scheme, the time axis is divided into equal sized slots and one slot is assigned to each user. Users are permitted to transmit only during their assigned slots. The slot assignments repeat periodically with each period known as a TDMA *frame*. In the simplest scheme, accommodating n users requires n slots. Obviously, users need to maintain time synchronization in a TDMA protocol. For the monolithic case ($\beta = 1$), given an undirected communication graph $\mathcal{G} = (V, E)$, one could therefore design a TDMA frame comprising $2|E|$ slots, each used to accommodate a unidirectional transmission between nodes i and j.

6

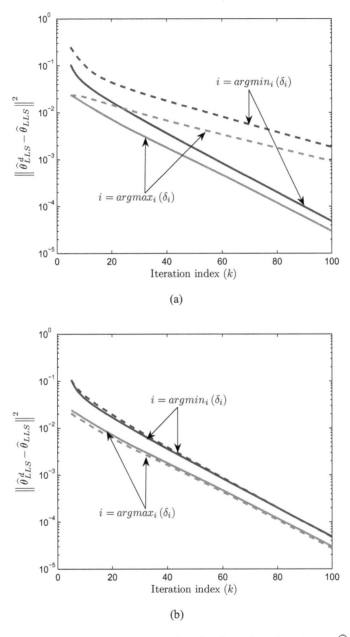

Figure 8.4: Plots of squared error of the distributed LLS estimate $(\widehat{\theta}_{\mathrm{LLS}}^{d})$ with respect to the centralized estimate $(\widehat{\theta}_{\mathrm{LLS}})$, for the nodes with the highest and lowest node degrees. (a) The solid lines correspond to $\beta = 1$ and $\Delta = 0.99$ and the dashed lines correspond to $\beta = 2$ and $\Delta = 0.49$. These values satisfy the condition $\Delta < 1/\beta$. (b) The solid lines correspond to $\beta = 1$ and the dashed lines correspond to $\beta = 2$. For both values of β, we chose $\Delta = 0.99$.

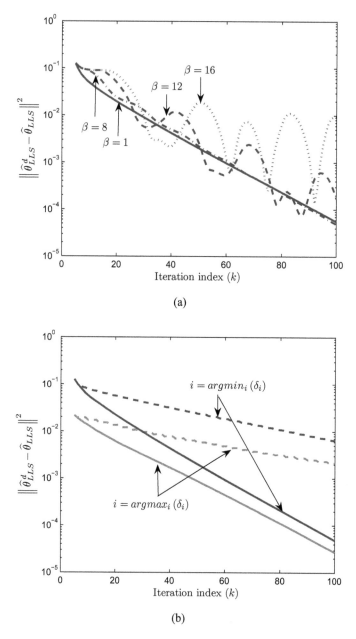

(a)

(b)

Figure 8.5: Plots of squared error of the distributed LLS estimate $(\widehat{\theta}_{\mathrm{LLS}}^{d})$ with respect to the centralized estimate $(\widehat{\theta}_{\mathrm{LLS}})$, for the nodes with the highest and lowest node degrees. (a) $\beta \in \{1, 8, 12, 16\}$ and $\Delta = 0.99$ for all values of β. (b) The solid lines represent a pulsed intracluster update scheme and the dashed lines represent an intermittent intercluster update scheme, where an intercluster update received at step k is used only for computing the state values at step $k + 1$ and is not "held" for the duration of the update cycle. For both schemes, we chose $\beta = 4$ and $\Delta = 0.99$.

We note that the factor 2 above is necessary since sensor nodes are typically equipped with a single transceiver and therefore cannot transmit and receive simultaneously. Consequently, two-way information exchange between nodes i and j must be achieved by two unidirectional transmissions between nodes i and j. Instead of a link-based scheduling approach, one can also use a node-based scheduling approach in which a TDMA frame comprises n slots ($n = |V|$) with each slot being used by a node to *broadcast* its message to its neighbors. The set of neighbors of any node i is essentially dependent on its transmit power and the physical distance between the transmitter and the receiver(s). Irrespective of which scheduling approach is used, it is important that the nodes update their states *synchronously at the end of every TDMA frame* and not within a frame immediately after receiving the state value(s) from its neighbor(s), which effectively amounts to asynchronous updating. Under a synchronous updating policy, the iteration index k is interpreted as the kth TDMA frame, with all nodes updating their state values simultaneously at the end of every frame.

For implementation purposes, it is necessary that each node maintain an one-dimensional array of length equal to the cardinality of its neighboring node set, to store the messages it receives from its neighbors during a particular TDMA frame. Of course, the contents of the array are refreshed every frame. To illustrate this, consider a 3-node path graph on the node set $V = \{1, 2, 3\}$ with $E = \{\{1, 2\}, \{2, 3\}\}$. The TDMA frame structure is such that node 1 broadcasts in the first slot, node 2 in the second slot, and node 3 in the third slot. The edge weight matrix, computed according to (8.18), is $W = \mathbf{Diag}\left([0.5, 0.5]^T\right)$, and the corresponding weighted Laplacian matrix is

$$
L_w(\mathcal{G}) = \begin{bmatrix} 0.5 & -0.5 & 0 \\ -0.5 & 1 & -0.5 \\ 0 & -0.5 & 0.5 \end{bmatrix}.
$$

Suppose that the initial state vector is $[10, 20, 30]^T$ and $\Delta = 1$. If nodes are allowed to update their own states only at the end of every frame, node 1 will broadcast its state value in slot 1 of the first frame, which is received and stored by node 2, its only neighbor. Node 2 then broadcasts its state value in slot 2, which is received and stored by both nodes 1 and 3, and finally, node 3 broadcasts its state value in slot 3, which is received and stored by node 2. All three nodes now update their state values simultaneously. It can be verified that the state vector at the end of the first cycle is $[15, 20, 25]^T$, and that it takes about 17 TDMA frames for the states of all nodes to converge (within a tolerance of 10^{-4}) to the average initial state, which is 20. On the other hand, if nodes are allowed to update their own states as soon as they receive an update from any of their neighbors,

- node 1 broadcasts its state to node 2 in slot 1, following which node 2 immediately updates its own state to $20 + 0.5(10 - 20) = 15$;

- node 2 broadcasts its current state (which is 15) to nodes 1 and 3 in slot 2, following which they update their own states to $10 + 0.5(15 - 10) = 12.5$ and $30 + 0.5(15 - 30) = 22.5$, respectively, and finally

- node 3 broadcasts its current state (which is 22.5) to node 2 in slot 3, following which node 2 updates its own state to $15 + 0.5(22.5 - 15) = 18.75$.

The state vector at the end of the first frame is therefore $[12.5, 18.75, 22.5]^T$. Figure 8.6 shows that updating the state values within a TDMA frame may not guarantee convergence of the algorithm to the average consensus. Moreover, the value to which the states converge may depend on the specific link transmission/node broadcast schedule used.

A TDMA protocol can also be used for clustered networks with $\beta > 1$, with some minor adjustments. To account for the different update rates for intra- and intercluster communications, two different frame structures should be used: an *intercluster frame structure* for frame numbers which are multiples of $\beta\Delta$ and an *intracluster frame structure* for all other frame numbers. As before, either link-based or node-based transmission schedules can be used within each frame structure.[7]

We conclude our discussion on practical implications of distributed least squares algorithms by noting that the efficiency of the TDMA schemes can be improved by exploiting *spatial diversity* and allowing more than one transmission to occur in any slot, provided (at least) that the signal-to-interference noise ratio (SINR) at the intended receivers are all above desired thresholds. Reusing time slots to accommodate multiple transmissions reduces the latency and therefore leads to faster convergence. This variant of TDMA is commonly known as *spatial TDMA* (S-TDMA).

8.4 DISTRIBUTED KALMAN FILTERING

In this section, we delve into distributed Kalman filtering, where a group of sensors make noisy observations of a dynamic state driven by a discrete time linear system, exchange information with other sensors, and collectively reach an estimate of the underlying state.

---

[7]For applications where low probability of interception/detection is essential, it may also be possible to adopt a hybrid TDMA/CDMA scheme, particularly for intercluster communications if they require high transmit power support.

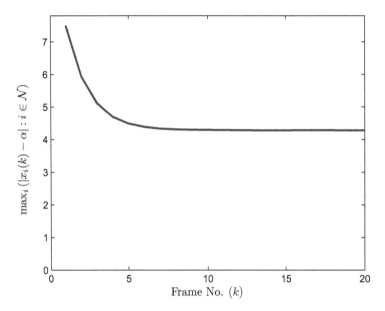

Figure 8.6: Plot of the maximum absolute divergence from the average initial state value (denoted by α) using the state equation (8.13). The topology is a 3-node path graph and the initial state vector is $[10, 20, 30]$. For this simulation, we used node broadcast scheduling in each frame (frame length = 3 slots) but allowed each node to update its own state value within a TDMA frame, as soon as it received a message from its neighboring node. Clearly, the node states do not converge to the average consensus, which is equal to 20. In fact, the final state of each node turns out to be 15.7143.

8.4.1 The Kalman Filter

We first consider the situation when the underlying variable of interest in the estimation procedure is the state of a linear dynamic system, which leads us directly to the realm of *filtering*. Before we examine distributed filtering in detail, however, let us examine the mechanism for designing an optimal linear filter for a single sensor. In this venue, consider the discrete time system

$$x(k+1) = Ax(k) + w(k), \qquad (8.47)$$
$$z(k) = H(k)x(k) + v(k), \qquad (8.48)$$

where the system matrix $A \in \mathbf{R}^{n \times n}$ is assumed to be time-invariant; the observation matrix $H(k) \in \mathbf{R}^{m \times n}$, in the meantime, is allowed to be time-varying. In (8.47) the vector $w(k)$ is a stochastic process representing the process noise and $v(k)$ in (8.48) denotes the observation noise at time instance k. Both $w(k)$ and $v(k)$ are assumed to be Gaussian stochastic processes with zero mean and covariance matrices W and V, respectively. Furthermore, the process noise and the measurement noise are assumed to be statistically independent.

We now consider a process, a *filtering algorithm*, by which one observes the vector z (8.48) and produces an estimate of the system's underlying state x. Moreover, we consider the situation when this process is *recursive*, that is, at each iteration of the algorithm, we produce an estimate that is subsequently improved upon as more measurements are obtained. In this avenue, let us denote by $\hat{x}(k|k-1)$ the estimate of the process $x(k)$ *prior* to the measurement received at time k; this is called the *prior estimate* of x at time k; similarly, the vector $\hat{x}(k|k)$ is called the *posterior estimate* of $x(k)$ *after* the measurement at time k is received and incorporated in the estimate. Following the standard argument for constructing the sought after estimation algorithm, one is led to the update rule

$$\hat{x}(k|k) = \hat{x}(k|k-1) + K(k)(z(k) - H\hat{x}(k|k-1)), \qquad (8.49)$$

with K chosen as the solution of the optimization problem

$$\min_{K(k)} \quad \textbf{trace}\, \Sigma(k|k), \qquad (8.50)$$

where $\Sigma(k|k) = \mathbf{E}\{\tilde{x}(k|k)\,\tilde{x}(k|k)^T\}$ is the covariance matrix of the error vector

$$\tilde{x}(k|k) = \hat{x}(k|k) - x(k),$$

and the constraint set for (8.50) is defined by a recursion on $\Sigma(k|k)$ that involves $K(k)$. The resulting filtering architecture scheme is shown in Figure 8.7 in feedback form. It can be shown that the optimal K for (8.50) is given by the Kalman gain

$$\begin{aligned} K(k) &= \Sigma(k|k-1)H(k)^T \left(H(k)\Sigma(k|k-1)H(k)^T + R \right)^{-1} \\ &= \Sigma(k|k)H(k)^T R^{-1}, \end{aligned} \qquad (8.51)$$

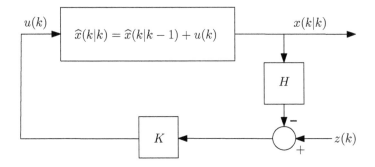

Figure 8.7: Kalman filter as a feedback mechanism

where $\Sigma(k|k-1) = \mathbf{E}\{\tilde{x}(k|k-1)\tilde{x}(k|k-1)^T\}$. Substituting the expression of the Kalman gain in the expression for the covariance update, we obtain

$$
\begin{aligned}
\Sigma(k|k) &= \Sigma(k|k-1) \\
&\quad -\Sigma(k|k-1)H(k)^T(H(k)\Sigma(k|k-1)H(k)^T + R)^{-1}H(k)\Sigma(k|k-1) \\
&= (I - K(k)H(k))\Sigma(k|k-1),
\end{aligned}
$$

which can also be expressed as

$$
I - K(k)H(k) = \Sigma(k|k)\Sigma(k|k-1)^{-1};
$$

pseudo-inverses can be used if the required inverses do not exist.

However, as we know the underlying model via which the state evolves, we can actually do a predictive step and improve on the posterior estimate until the next measurement is available. Thus

$$
\Sigma(k+1|k) = A\Sigma(k|k)A^T + W(k) \quad \text{and} \quad \hat{x}(k+1|k) = A\hat{x}(k|k),
$$

where $\Sigma(k+1|k)$ denotes the error covariance prior to incorporating the measurement at time $k+1$ into the estimate.

The Kalman filter and its recursion can be represented in an alternative form, which proves to be advantageous when we take up distributed Kalman filtering in the next section. This equivalent form, referred to as the *information filter*, is obtained by representing the Kalman filter in terms of the information matrix $\mathcal{I}(k)$ defined as

$$
\mathcal{I}(k) = \Sigma(k)^{-1}.
$$

It can then be verified that by letting

$$Y(k) = H(k)^T V^{-1} H(k) \quad \text{and} \quad y(k) = H(k)^T V^{-1} z(k),$$

with $\hat{y}(k|k) = \mathcal{I}(k|k)\,\hat{x}(k|k)$ and $\hat{y}(k|k-1) = \mathcal{I}(k|k-1)\,\hat{x}(k|k-1)$, the recursion for the information filter assumes the form

$$\mathcal{I}(k|k) = \mathcal{I}(k|k-1) + Y(k) \quad \text{and} \quad \hat{y}(k|k) = \hat{y}(k|k-1) + y(k). \quad (8.52)$$

The Kalman gain can be represented using the information matrix as

$$\begin{aligned} K(k) &= A(\mathcal{I}(k|k-1) + H(k)^T V^{-1} H(k) + V)^{-1} H(k) V^{-1} \\ &= A\mathcal{I}(k|k) H(k)^T V^{-1}. \end{aligned}$$

Lastly, the prediction step in the realm of the information matrix assumes the form

$$\begin{aligned} \mathcal{I}(k+1|k) &= L(k)M(k)L(k)^T + C(k)W^{-1}C(k)^T, \\ y(k+1|k) &= L(k)A^{-T}\hat{y}(k|k), \end{aligned}$$

where $M(k) = A^{-T}\mathcal{I}(k|k)A^{-1}$, $C(k) = M(k)(M(k) + W^{-1})^{-1}$, and $L(k) = I - C(k)$; once again, pseudo-inverses can be used if the required inverses do not exist.

8.4.2 Kalman Filtering over Sensor Networks

We now take up the notion of *distributed* Kalman filtering and, in particular, Kalman filtering over a sensor network. Our setup consists of the discrete time linear system

$$x(k+1) = Ax(k) + w(k) \qquad (8.53)$$

that is observed by n sensors

$$z_i(k) = H_i(k)x(k) + v_i(k) \quad i = 1, 2, \ldots, n, \qquad (8.54)$$

each with its own time-varying observation matrix H_i that is corrupted by a zero-mean Gaussian noise v_i with covariance V_i; once again, see Figure 8.1. One natural way to approach Kalman filtering using multiple distributed sensors is to fuse the raw sensor measurements at a *fusion center*. That is, we gather all measurements z_i in terms of one vector

$$z(k) = [\,z_1(k)^T, z_2(k)^T, \ldots, z_n(k)^T\,]^T,$$

which can be processed by a *centralized Kalman filter* for the system

$$x(k+1) = Ax(k) + w(k), \quad z(k) = H(k)x(k) + v(k), \qquad (8.55)$$

where

$$H = \begin{bmatrix} H_1(k) \\ H_2(k) \\ \vdots \\ H_n(k) \end{bmatrix} \quad \text{and} \quad v(k) = \begin{bmatrix} v_1(k) \\ v_2(k) \\ \vdots \\ v_n(k) \end{bmatrix};$$

this can be accomplished in an optimal way via the Kalman filter update (8.49) incorporating the Kalman gain (8.51). Provided that the noise vector on each sensor is independent zero mean Gaussian, both in time and across the sensors, the covariance matrix for noise on the fused measurement assumes the form

$$V = \begin{bmatrix} V_1 & 0 & \cdots & 0 & 0 \\ 0 & V_2 & 0 & \cdots & 0 \\ \vdots & \vdots & \vdots & \vdots & \vdots \\ 0 & 0 & 0 & 0 & V_n \end{bmatrix}.$$

However, the centralized scheme has the disadvantage that all the computational work is performed at the fusion center and the sensors are only employed for gathering the measurements and relaying them to the center. A closer examination of the information filter (8.52) reveals an interesting additive property for the information matrix of the form

$$\mathcal{I}(k|k) = \mathcal{I}(k|k-1) + \sum_{i=1}^{n} Y_i(k), \qquad (8.56)$$

$$\hat{y}(k|k) = \hat{y}(k|k-1) + \sum_{i=1}^{n} y_i(k), \qquad (8.57)$$

where

$$Y_i(k) = H_i(k)^T V_i^{-1} H_i(k) \quad \text{and} \quad y_i(k) = H_i(k)^T V_i^{-1} z_i(k).$$

The summation formulas (8.56) - (8.57) suggest a direct method for making the recursive steps of the Kalman filter over the sensor network more distributed. This is done by letting each sensor keep a local copy of the information matrix $\mathcal{I}(k|k-1)$ and the information vector $\widehat{y}(k|k-1)$. Then, when each sensor performs a local Kalman filter based on its local measurements, one has

$$H_i(k)^T V_i^{-1} H_i(k) = \mathcal{I}_i(k|k) - \mathcal{I}_i(k|k-1),$$

and therefore the information matrix can be updated at each node by receiving the difference $\mathcal{I}_i(k|k) - \mathcal{I}_i(k|k-1)$, summing them up across all sensors, and then adding them to obtain $\mathcal{I}(k|k-1)$. Similarly, the information vector can be updated by summing up the received $y_i(k)$ from each sensor, which is also the difference

$$\widehat{y}_i(k|k) - \widehat{y}_i(k|k-1),$$

with $y(k|k-1)$ as in (8.57). The prediction step can now be executed at each node in its original form or in the information filter form.

The above scheme can also be considered in terms of the state and covariance update by including a coordinator. In this setting, the global update assumes the form

$$\begin{aligned}\widehat{x}(k|k) &= \widehat{x}(k|k-1) + K(k)(z(k) - H(k)\widehat{x}(k|k-1))\\ &= (I - K(k)H(k))\widehat{x}(k|k-1) + K(k)z(k).\end{aligned}$$

However

$$K(k)z(k) = \Sigma(k|k)H(k)^T V^{-1} z(k) = \Sigma(k|k)\sum_i H_i(k)^T V_i^{-1} z_i(k)$$

and

$$I - K(k)H(k) = \Sigma(k|k)\Sigma(k|k-1)^{-1},$$

and therefore

$$\widehat{x}(k|k) = \Sigma(k|k)\Sigma(k|k-1)^{-1}\widehat{x}(k|k-1) + \Sigma(k|k)\sum_i H_i(k)^T V_i^{-1} z_i(k).$$

In the meantime,

$$H_i(k)^T V_i^{-1} z_i = \mathcal{I}_i(k|k)\widehat{x}_i(k|k) - \mathcal{I}_i(k|k)(I - k_i H_i(k))\widehat{x}_i(k|k-1),$$

and hence

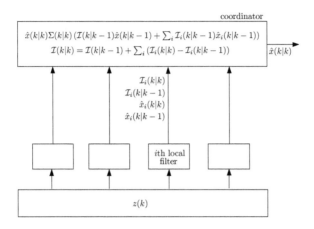

Figure 8.8: Distributed Kalman filtering with a coordinator

$$\widehat{x}(k|k) = \Sigma(k|k) \left[\mathcal{I}(k|k)\widehat{x}(k|k-1) \right.$$
$$\left. + \sum_i \mathcal{I}_i(k|k)\widehat{x}_i(k|k) - \mathcal{I}_i(k|k-1)\widehat{x}_i(k|k-1) \right].$$

The information matrix can then be updated as

$$\mathcal{I}(k|k) = \mathcal{I}(k|k-1) + \sum_i H_i(k)^T V_i H_i(k)$$
$$= \mathcal{I}(k|k-1) + \sum_i \mathcal{I}_i(k|k) - \mathcal{I}_i(k|k-1).$$

We note that the time update at the coordinator site can be implemented by receiving $\widehat{x}_i(k|k-1), \widehat{x}_i(k|k), \mathcal{I}_i(k-1|k)$, and $\mathcal{I}_i(k|k)$ from each sensor as depicted in Figure 8.8. We also note that in the above model, the communication is from the sensors to the coordinator but not vice versa.

8.4.3 Relaxing the Communication Requirement

The proposed distributed Kalman filter that relies on the compact and esthetically pleasing summations (8.56) - (8.57) allows for utilizing the local computational capability of the sensors, upgrading their status from merely

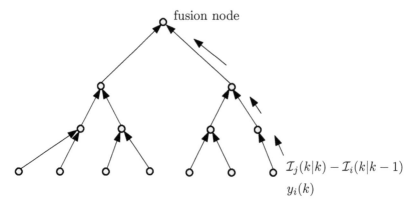

$$\mathcal{I}_j(k|k) - \mathcal{I}_i(k|k-1)$$
$$y_i(k)$$

Figure 8.9: In-branching facilitates either a routing architecture or staged summing for decentralized Kalman filter

relaying raw measurement to the fusion center. In the meantime, the proposed architecture implicitly requires that each sensor can receive the required information from *all other sensors*; in other words, the information-exchange graph is assumed to be complete. This requirement can be relaxed in multiple ways. First, if it is desired that one of the sensors, or equivalently a fusion center, calculates the estimate, it is necessary that this sensor receives information from all other local Kalman filters. This can be achieved, for example, by organizing the sensor network as an *in-branching*, which facilitates routing the required summands to the fusion center or allows the summations (8.56) - (8.57) to be done in stages; this is shown in Figure 8.9. This distributed implementation of the filters over sensor networks, assuming synchronous operation among the sensors, is completely equivalent to the original Kalman filter, with its guaranteed optimality and stability. An approximate sum can also be used at each sensor node, possibly, obtained by running a discrete time agreement protocol on the variables $\Delta \mathcal{I}_i = \mathcal{I}_i(k|k) - \mathcal{I}_i(k|k-1)$ and $y(k)$, and then multiplying the resulting consensus value at each sensor by the number of sensors in the network.

Another approach for accomplishing the summation task (8.56) - (8.57) in a distributed way, while each sensor is allowed to operate on its local measurement, is to combine the update equation for the local Kalman filters with an appropriately weighted consensus-type algorithm that implicitly determines the sum in a distributed way. For example, this adjustment can be accomplished via the update rule

$$\widehat{x}_i(k|k) = \widehat{x}_i(k|k-1) + K_i^o(z_i(k) - H_i(k)\widehat{x}_i(k|k-1))$$
$$+ \sum_{j \in N(i)} K_{ij}^c(\widehat{x}_j(k|k-1) - \widehat{x}_i(k|k-1)), \qquad (8.58)$$

where K_i^o is the gain by sensor i on the update term obtained from its observation, possibly a local Kalman filter, and K_{ij}^c is the gain used by sensor i on the update terms obtained from communication with its neighbor j. Although this scheme seems to resolve the inherent assumption on the synchronous operation of the sensor network in terms of information-exchange, it does have certain drawbacks, most notably, the lack of optimality guarantees; see notes and references for pointers to the existing literature on this and similar fully distributed filtering schemes.

SUMMARY

In this chapter, we presented two pillars of distributed estimation, namely, distributed least squares and distributed Kalman filters, of particular relevance in areas such as sensor networks, data fusion, and distributed signal processing. The emphasis has been on the structure of information-exchange in relation to the particular distributed character of the estimation tasks.

NOTES AND REFERENCES

Estimation theory, like a number of other disciplines in systems sciences, was motivated by studying astronomical phenomena, that is, the motion of comets and planets. This is the case both for the least squares approach as invented by Gauss and Legendre, as well as *dynamic* estimation as it arises, for example, in satellite tracking. Distributed estimation, as motivated in the chapter, on the other hand, is mainly motivated by the area of distributed robotics and sensor networks. Our treatment of the distributed least squares parallels the works of Xiao, Boyd, and Lall in [253] and Das and Mesbahi in [62]. §8.4 and our treatment of Kalman filtering is based on the work of Rao and Durrant-Whyte [201] who stressed the utility of the information filter in the context of distributed estimation. We also relied on the paper by Hashemipour, Roy, and Laub [112] for their discussion on the parallel implementation of the Kalman filter. Earlier references in the area of distributed Kalman filtering include those by Speyer [223] and Willsky, Bello, Castanon, Levy, and Verghese [246]. Among the papers devoted to relaxing the all-to-all communication using the consensus algorithm we refer to those by Spanos, Olfati-Saber, and Murray [221], Olfati-Saber and Sandell [186], Stanković, Stanković, and Stipanović [225], Nabi, Mesbahi, Fathpour, and Hadaegh [171], Khan and Moura [132], and Kirti and Scaglione [136].

SUGGESTED READING

There are a number of excellent books devoted to Kalman filtering and estimation theory. We particularly recommend the two classics by Gelb [96] and Anderson and Moore [8], as well the more recent book by Crassidis and Junkins [59].

EXERCISES

Exercise 8.1. Let H_i, $i = 1, 2, 3$, be the rows of the 3×3 identity matrix in the observation scheme $z_i = H_i x + v_i$ for a three-node sensor network, observing state $x \in \mathbf{R}^3$. It is assumed that the nodes form a path graph and that v_i is a zero-mean, unit variance, Gaussian noise. Choose the weighting matrix W (8.8) and the step size Δ in (8.20) - (8.21), conforming to the condition (8.14). Experiment with the selection of the weights for a given value of Δ and their effect on the convergence properties of the distributed least square estimation (8.20) - (8.21).

Exercise 8.2. Show that the iterations (8.20) - (8.21) converge to the centralized least squares solution.

Exercise 8.3. Suppose that $w(k)$ and $v(k)$ in (8.47) - (8.48) are replaced by $Bw(k)$ and $Dv(k)$, for matrices B and D of appropriate dimensions. Find the corresponding update equation for the Kalman gain analogous to (8.51).

Exercise 8.4. The probability density function for a Gaussian distribution with mean μ and variance σ is defined as

$$f(x; \mu, \sigma^2) = \frac{K}{\sqrt{2\pi\sigma^2}} \, e^{-(x-\mu)^2/(2\sigma^2)},$$

where K is a normalization constant independent of x.

Now, consider two Gaussian distributions $f_1(x; \mu_1, \sigma_1^2)$ with K_1 and $f_2(x; \mu_2, \sigma_2^2)$ with K_2. Show that the product of these two Gaussian distributions is also a Gaussian distribution $f_3(x; \mu_3, \sigma_3^2)$ with

$$\mu_3 = \frac{\sigma_2^2}{\sigma_1^2 + \sigma_2^2}\mu_1 + \frac{\sigma_1^2}{\sigma_1^2 + \sigma_2^2}\mu_2, \quad \sigma_3^2 = \frac{\sigma_1^2\sigma_2^2}{\sigma_1^2 + \sigma_2^2}.$$

What is the corresponding value of K?

Exercise 8.5. Verify the iteration (8.32).

Exercise 8.6. Bayes Theorem is central to the Kalman/Information Filter. It states that given a sequence of observations $\{z_1, z_2, \ldots, z_k\} = Z^k$, then

$$f(x|Z^k) = \frac{f(Z^k|x)f(x)}{f(Z^k)}$$

where $f(x|Z^k)$ is the probability density function of x given observations Z^k (the posterior distribution), $f(x)$ is the probability density function of x, and $f(Z^k)$ is the joint probability density function of $\{z_1, z_2, \ldots, z_k\}$. With the assumption that information about x obtained at time $k-1$ is independent of the information about x at time k for all k, then $f(Z^k|x) = f(z_1|x)f(z_2|x)\ldots f(z_k|x)$.

(a) Show that the posterior distribution is

$$f(x|Z^k) = \frac{f(z_k|x)f(x|Z^{k-1})}{f(z_k|Z^{k-1})}.$$

Hint: $f(z_1|\{z_1, z_2\}) = 1$ and $f(\{z_1, z_2\}|z_1) = f(z_2|z_1)$.

(b) The Kalman/ Information Filter models $f(x|Z^k)$ as a Gaussian, that is, as defined by a mean and variance. Under what assumptions is $f(x|Z^k)$ Gaussian for all k. (*Hint:* Use Exercise 8.4.)

(c) A set of bearing only sensors produce a sequence of observations Z^k of an object's location $x \in \mathbf{R}^2$. Can the Kalman/Information Filter be used to calculate $f(x|Z^k)$? (A bearing only sensor is a sensor that indicates with a certain uncertainty which direction the object's location x is from the sensor's location).

Exercise 8.7. Mutual independence of sensor measurements can not always be guaranteed. An alternative sensor fusion method to the Kalman filter is the Covariance Intersection algorithm, where no assumptions are made about data correlation between sensor 1 with mean $\hat{a}$ and covariance A and sensor 2 with mean $\hat{b}$ and covariance B. The covariance intersection algorithm with $\omega \in [0, 1]$, produces the ellipsoid

$$\{x \,|\, (x - \hat{c})^T C(x - \hat{c})^T \leq 1\},$$

with

$$C^{-1} = wA^{-1} + (1-w)B^{-1}, \ \hat{c} = C\left(wA^{-1}\hat{a} + (1-w)B^{-1}\hat{b}\right).$$

(a) Substitute the information matrix and vector into this algorithm. What is the interpretation of the parameter w?

(b) Plot the covariance ellipses defined by $(\hat{a}, A)$, $(\hat{b}, B)$ and $(\hat{c}, C)$ where

$$\hat{a} = \begin{bmatrix} 10 \\ 4.9 \end{bmatrix}, A = \begin{bmatrix} 9 & 3 \\ 3 & 4 \end{bmatrix}, \hat{b} = \begin{bmatrix} 10 \\ 5 \end{bmatrix}, B = \begin{bmatrix} 5 & -1 \\ -1 & 7 \end{bmatrix}.$$

Now compare to the Kalman update data fusion $(\hat{d}, D)$, is the Covariance Intersection algorithm conservative? Why or Why not?

(c) Show that the choice of w that minimizes the area of the update covariance C is equivalent to minimizing the determinant of C^{-1} for the case where C is a 2×2 matrix. Find this w for the $(\hat{a}, A)$ and $(\hat{b}, B)$ of part (b).

Exercise 8.8. Consider the linear system (8.53) with $A = -.02I_{n\times n}$, where $I_{n\times n}$ is the $n \times n$ identity matrix. Let H_i, $i = 1, 2, \cdots, n$ be the ith row of $I_{n\times n}$ in (8.54). Implement the information form of the Kalman filter on the complete graph when $n = 3$, assuming that the updates (8.56) - (8.57) can be executed on all nodes between the updates. Now, assume that the graph is a path graph and that the nodes can run an agreement protocol (on an undirected and unweighted graph; see Chapter 3) in order to calculate (8.56) - (8.57). Implement this scheme for $n = 2$ and $n = 3$. Experiment and comment on how the quality of this distributed Kalman filter is affected as n increases.

Exercise 8.9. Verify the matrix identity (8.34).

Exercise 8.10. Consider the proposed update scheme for the distributed Kalman filter (8.58). Given the local Kalman filter gain, propose an optimization problem for choosing the weights K_{ij} such that the covariance of the estimated error is minimized at each iteration. Can this optimization problem be formulated with a convex objective function and a convex constraint set?

Exercise 8.11. Using an appropriate Lyapunov function, show that the Kalman filter is stable.

Exercise 8.12. Show that the filtering scheme (8.58) can be made to be stable by appropriately choosing the gains K_{ij}^c for each pair of sensors i and j, when K_i^o is the local Kalman gain for sensor i and the underlying information graph is connected.

Exercise 8.13. Show that the Kalman filter is unbiased, that is, if the noisy measurement is a zero-mean, the filter output is zero-mean as well. Does this property remain valid for the distributed filter discussed in §8.4.2?

Exercise 8.14. Implement the coordinated distributed Kalman filter shown in Figure 8.8 for the case when the linear system (8.53) is specified with $A = -.05I_{3\times3}$, where $I_{3\times3}$ is the 3×3 identity matrix and H_i ($i = 1, 2, 3$) is the ith row of $I_{3\times3}$ in (8.54).

Exercise 8.15. What is the primary bottleneck in removing the coordinator from the coordinated Kalman filter from Figure 8.8?

Exercise 8.16. Given a network of sensor nodes, each of which is measuring the value $\tau_i(t) = r(t) + v_i(t)$ at time t, where $r(t)$ is the true value, and $v_i(t)$ is noise. A so-called consensus filter is given by

$$\dot{\xi}_i(t) = - \sum_{j \in N(i)} (\xi_i(t) - \xi_j(t)) - \sum_{j \in N(i)} (\xi_j(t) - \tau_j(t)) - (\xi_i(t) - \tau_i(t)),$$

where $\xi_i(t)$ is agent i's estimate of what r is at time t.
    Show that the consensus filter is a low-pass filter in that

$$\lim_{s \to \infty} G(s) = 0,$$

where the transfer function is given by $G(s) = \widehat{\xi}(s)/\widehat{\tau}(s)$, where

$$\widehat{\xi}(s) = \int_0^\infty e^{-st}\xi(t)\,dt \quad \text{and} \quad \widehat{\tau}(s) = \int_0^\infty e^{-st}\tau(t)\,dt$$

are the Laplace transforms of the respective signals. Moreover, show that the filter is bounded-in, bounded-out, which implies that $\lim_{t\to\infty} \|\xi(t) - 1r(t)\| \le \epsilon$ if $\|\dot{r}(t)\| \le \delta$ for some $\epsilon, \delta > 0$.

Chapter Nine

Social Networks, Epidemics, and Games

> "Consistency is the last refuge of the unimaginative."
> — Oscar Wilde

Social networks, epidemics, and games offer rich areas for a network-centric inquiry, requiring a blend of ideas from dynamical systems and graph theory for their study. In this chapter, we delve into a representative set of problems in these areas: diffusion on social networks, analysis of epidemic models using Lyapunov techniques, and chip firing games. This is done in order to give the reader a glimpse of a graph theoretic perspective on dynamic systems that are traditionally considered outside engineering. The notes and references at the end of the chapter provide pointers to references for each of these disciplines. Our focus will be on scenarios where the underlying interconnection is assumed to be static; the protocols running on these static networks can have a probabilistic component. As interaction models for certain social interactions often have a strong stochastic character, to this end, we refer the reader to Chapter 5 for an introduction to random networks and processes that evolve over them.

9.1 DIFFUSION ON SOCIAL NETWORKS–THE MAX PROTOCOL

Our first example, which has a strong resemblance to the agreement problem, is inspired by considering a social network of friends, viewed as nodes in an undirected graph $\mathcal{G}$, that adopt a certain level of "fashionability," measured in terms of a real number on the unit interval $[0, 1]$. Thus $x_i(k) = 1$ and $x_j(k) = 0$ refer to the scenario where nodes i and j are, respectively, the most and least fashionable a member of the social group can be at time index k. Let us initialize the group fashionability state at $k = 0$ by choosing $0 \leq x_i(0) \leq 1$, $i = 1, 2, \ldots, n$, which can be done, for example, using a normalized random number generator. We then let the social group evolve according to the following rule: at every time step, a randomly selected person in the network influences one of his/her neighbors–chosen according to the uniform distribution–to be at least as fashionable as him/her. Note that this protocol does not require that a given node have information about his/her standing in the fashionability spectrum of the social network.

The proposed update rule for this social network can be expressed by

$$x_i(k+1) = \max\{x_i(k), x_j(k)\}, \quad k = 0, 1, 2, \ldots, \tag{9.1}$$

where node j is a randomly chosen node from the set of neighbors of i, $N(i)$. We refer to the update rule (9.1) as the *max-protocol*. Intuitively, when the underlying network is connected, we expect that the max-protocol steer the group toward the value

$$M = \max_i x_i(0), \tag{9.2}$$

that is, connectedness of the network with the (monotonic) updating scheme (9.1), steers the social group toward being highly fashionable across the board. In this section, we provide an analysis for this observation in terms of the properties of the underlying network, provided that none of the nodes in the network are socially isolated.

Consider the probability $p_i(k)$ that node i possesses the maximum fashionability index M (9.2) after applying the protocol (9.1) k times. It is straightforward to verify that this probability admits a recursive representation as

$$p_i(k) = p_i(k-1) + \frac{1}{d(i)}(1 - p_i(k-1)) \sum_{j \in N(i)} p_j(k-1). \tag{9.3}$$

In turn, the recursion (9.3) admits the compact representation

$$p(k) = p(k-1) + \mathbf{Diag}(1 - p(k-1)) \Delta(\mathcal{G})^{-1} A(\mathcal{G}) p(k-1), \tag{9.4}$$

where

$$p(k) = [p_1(k), p_2(k), \ldots, p_n(k)]^T$$

denotes the "fashionability state" of the social group, and $\Delta(\mathcal{G})$ and $A(\mathcal{G})$ denote, respectively, the degree matrix and the adjacency matrix of the graph $\mathcal{G}$. Note the nonlinear form of (9.4) as compared with recursions we encountered when studying Markov chains or the agreement protocol in Chapter 3.

The intuitive observation of this section is as follows.

Proposition 9.9. *Given a connected network and a nonzero initial probability vector $p(0)$, under the max-protocol (9.1), every node asymptotically assumes the maximum fashionability level (9.2) in probability.*

Proof. We prove the proposition using the Lyapunov method. First note that (9.4) can be rewritten as

$$x(k) = x(k-1) - \mathbf{Diag}\left(x(k-1)\right)\Delta(\mathcal{G})^{-1}A(\mathcal{G})\left(\mathbf{1} - x(k-1)\right),$$

where

$$x(k) = \mathbf{1} - p(k).$$

In view of the equality

$$\mathbf{Diag}\left(x(k-1)\right)\Delta(\mathcal{G})^{-1}A(\mathcal{G})\,\mathbf{1} = x(k-1)$$

we have

$$x(k) = \mathbf{Diag}\left(x(k-1)\right)\Delta(\mathcal{G})^{-1}A(\mathcal{G})\,x(k-1).$$

It now suffices to show that the sequence $\{x(k)\}_{k\geq 1}$ converges to the origin. For this purpose, consider the Lyapunov function

$$V(x(k)) = \mathbf{diag}\left(\Delta(\mathcal{G})\right)^{T}x(k), \tag{9.5}$$

where $\Delta(\mathcal{G})$ is the diagonal matrix of node degrees. Note that $V(x(k)) > 0$ for any nonzero $x(k)$. Furthermore,

$$\begin{aligned}
\Delta V(k) &= V(x(k+1)) - V(x(k)) \\
&= \mathbf{diag}\left(\Delta(\mathcal{G})\right)^{T}\mathbf{Diag}\left(x(k)\right)\Delta(\mathcal{G})^{-1}A(\mathcal{G})\,x(k) \\
&\quad - \mathbf{diag}\left(\Delta(\mathcal{G})\right)^{T}x(k) \\
&= x(k)^{T}\Delta(\mathcal{G})\Delta(\mathcal{G})^{-1}A(\mathcal{G})\,x(k) - \mathbf{diag}\left(\Delta(\mathcal{G})\right)^{T}x(k) \\
&= x(k)^{T}A(\mathcal{G})x(k) - \mathbf{diag}\left(\Delta(\mathcal{G})\right)^{T}x(k).
\end{aligned}$$

First notice that when $x(k) = \alpha\mathbf{1}$ for $\alpha \in (0,1)$,

$$\Delta V(k) = (\alpha^{2} - \alpha)\,\mathbf{diag}\left(\Delta(\mathcal{G})\right)^{T}\mathbf{1} < 0 \tag{9.6}$$

and the statement of the proposition follows immediately by viewing $V(x(k))$ (9.5) as a Lyapunov function for the probabilistic model of (9.1). Otherwise, we can write

$$x(k) = \alpha\mathbf{1} + x_{\perp}(k),$$

such that $\mathbf{1}^T x_\perp(k) = 0$ and $\alpha \in (0,1)$ is chosen appropriately. Observe that

$$\mathbf{diag}\left(\Delta(\mathcal{G})\right)^T x(k) \geq x(k)^T \Delta(\mathcal{G})\,x(k),$$

since every entry of $x(k)$ lies in the open interval $(0,1)$. Hence,

$$\begin{aligned}
\Delta V(k) &= x(k)^T A(\mathcal{G}) x(k) - \mathbf{diag}\left(\Delta(\mathcal{G})\right)^T x(k)\\
&\leq x(k)^T A(\mathcal{G}) x(k) - x(k)^T \Delta(\mathcal{G}) x(k)\\
&= -x(k)^T L(\mathcal{G}) x(k)\\
&= -(\alpha\mathbf{1} + x_\perp(k))^T L(\mathcal{G})\,(\alpha\mathbf{1} + x_\perp(k))\\
&= -x_\perp(k)^T L(\mathcal{G})\,x_\perp(k) \leq -\lambda_2(\mathcal{G})\,\|x_\perp(k)\|^2 < 0, \quad (9.7)
\end{aligned}$$

where, once again, $L(\mathcal{G})$ is the graph Laplacian and $\lambda_2(\mathcal{G})$ is its second smallest eigenvalue. The last inequality follows from the fact that $\lambda_2(\mathcal{G}) > 0$ when the underlying graph is connected. The inequality (9.7) now leads to the proof of the proposition. ∎

From (9.6), we conclude that when $x(k) = \alpha\mathbf{1}$, for $m \geq k$,

$$\begin{aligned}
V(x(m)) &= \alpha^{2^{m-k}-1}\,V(x(k)) \leq \alpha^{m-k}\,V(x(k))\\
&= e^{-(\log\frac{1}{\alpha})(m-k)}\,V(x(k)).
\end{aligned}$$

In other words, once $x(k) = \alpha\mathbf{1}$ for some α and time index k, $x(k)$ exponentially converges to the origin. In the meantime, while away from the subspace spanned by the vector $\mathbf{1}$, the sequence $x(k)$ asymptotically converges to it with its behavior governed by (9.7). For this case, a larger second smallest Laplacian eigenvalue results in a more substantial reduction $-\Delta V$ at each step, and hence, a better convergence to a homogeneous, and equally fashionable, social group.

9.2 THE THRESHOLD PROTOCOL

Our second model pertains to dynamic sociological models over graphs. In this setting, the members of the population are represented by nodes of the graph $\mathcal{G} = (V, E)$; the graph is assumed to have an infinite number of nodes while every node has only a finitely many neighbors.[1] At a given time instance k, each node has the ability to choose among two states represented by A and B; it might be convenient to think of these states as the political

---

[1] This is the only time in this book that we will be encountering infinite graphs.

affiliation of the individual node in a two-party system. Thus for a given time index k, one has

$$x_i(k) = A \quad \text{or} \quad x_i(k) = B.$$

Having initialized the nodes in $\mathcal{G}$ to assume one of the two possible states A and B at $k = 0$, we allow the population to evolve according to the following payoff scheme. We fix a parameter $q \in (0,1)$ and let the nodes update their states knowing that

- if they both choose state A they receive payoff q,

- if they both choose state B they receive payoff $1 - q$, and

- their payoff is zero if they choose opposite states.

Hence, if $q = 0$, all the nodes in the graph will choose state B right after their initialization and stay at this state subsequently.

Suppose that at time k, every node in the graph except node i has fixed its state, and node i, being greedy, faces the decision of choosing its state in order to maximize its payoff. Let us denote by $d_A(i)$ and $d_B(i)$ the number of neighbors of node i who have adopted, respectively, states A and B, right before i makes its decision. Since the payoff at time index k, $p_i(k)$, is

$$p_i(k) = qd_A(i) + (1-q)d_B(i),$$

it is judicious for node i to choose state B if

$$d_B(i)/d(i) > q;$$

to handle ties, we also allow node i to adopt B if $q = d_B(i)/d(i)$. Thus node i will adopt state B if the fraction of its neighbors that have adopted B, right before it makes its decision is greater than q. We call this policy that underlies the evolution of the state of each node the *threshold protocol*.

In this context, consider the scenario when the majority of nodes have initially adopted state A; state B represents the novel, initially unpopular, choice that has been adopted by a few. We are interested in determining under what conditions, particularly in relation to the choice of q, the initially unpopular state B can spread throughout the network. In this venue, consider $S \subseteq V(\mathcal{G})$ as the set of initial adopters of state B and denote by $h_q^k(S)$ the set of nodes that have adopted being a B node after k applications of the threshold protocol. Hence, h_q defines a map from $V(\mathcal{G})$ to itself.

Definition 9.10. *The set S is called contagious (with respect to the threshold q) if, for any finite set $M \subseteq V(\mathcal{G})$, there exists k such that $M \subseteq h_q^k(S)$.*

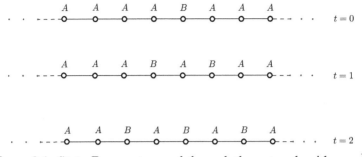

Figure 9.1: State B cannot spread through the network with $q = \frac{1}{2}$.

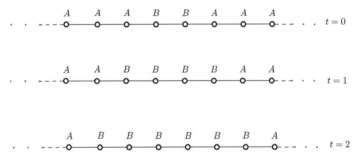

Figure 9.2: State B spreads through the network with $q = \frac{1}{2}$.

Figures 9.1 - 9.2 illustrate this definition, exemplifying how being B can spread throughout the network depending on the set of initial adopters and the threshold value q.

Consider next a node, initially in state A, which has converted to state B as the result of a threshold protocol. This node can conceivably reinstate its A-state as a result of the state of its neighbors and the threshold value. If we *do not* allow the node to reinstate its A state after switching to state B, we call the resulting protocol a *progressive threshold protocol*. We denote by $\bar{h}_q^k(S)$ the set of nodes in state B after k applications of the progressive threshold protocol, with threshold value q, to the set S whose nodes are in state B; naturally, $\bar{h}_q$ defines yet another map from $V(\mathcal{G})$ to itself. Progressive threshold protocols are conceptually easier to analyze, as members of the population are not allowed to "oscillate" between the two states. It is therefore insightful that, as far as being contagious, the two protocols are in fact equivalent.

Theorem 9.11. *For any graph, there exists a finite contagious set with respect to h_q if and only if there exists one with respect to $\bar{h}_q$.*

Proof. Clearly, if S is contagious with respect to h_q it is also contagious with respect to $\bar{h}_q$; in other words, the progressive threshold protocol is "more contagious" than its nonprogressive version. Thus it suffices to show that if S is contagious with respect to the progressive threshold map $\bar{h}_q$, there exists a set $U \subseteq V(\mathcal{G})$ such that h_q is contagious with respect to U.

The construction of the set U proceeds as follows. Let **cl** S be the *closure* of the set S, that is, the set of vertices that not only contains S as a subset but also all vertices that are neighbors of vertices in S. Since S is contagious with respect to $\bar{h}_q$, there exists some positive integer u such that

$$\textbf{cl } S \subseteq \bar{h}_q^u(S);$$

set $U = \bar{h}_q^u(S)$. We now proceed to show that the set $U \subseteq V(\mathcal{G})$ is contagious with respect to the nonprogressive threshold protocol defined by h_q. Our strategy involves showing that repeated applications of the nonprogressive threshold on U can "match" the application of the progressive threshold on S. First, we observe that by induction, for all $W \subseteq V(\mathcal{G})$ and $j \geq 1$,

$$\bar{h}_q^j(W) = W \cup h_q(\bar{h}_q^{j-1}(W)). \tag{9.8}$$

Suppose that $v > u$. By (9.8) one has

$$\bar{h}_q^v(S) = S \cup h_q(\bar{h}_q^{v-1}(S)).$$

But since $\bar{h}_q^{v-1}(S)$ includes U, and hence the closure of S, one has $S \subseteq h_q(\bar{h}_q^{v-1}(S))$ and thus

$$\bar{h}_q^v(S) = h_q(\bar{h}_q^{v-1}(S)).$$

By induction, it then follows that

$$h_q^{v-u}(U) = h_q^{v-u}(\bar{h}_q^u(S)) = \bar{h}_q^v(S),$$

and thereby U is contagious with respect to h_q. ∎

Progressive threshold protocols are easier to analyze as they make the progression of state B in the network monotonic. Progression of state B in the network is a function of the network geometry and the threshold value. However, the threshold value can have a dominant effect on the progression of state B. For example, if $q = 1$, then state B does not have much of a chance to diffuse over the network. We call the *contagion threshold* the minimum value of q where any finite initial subset of vertices in B state can eventually diffuse their state to the entire network.

Theorem 9.12. *The contagion threshold for any graph is at most $\frac{1}{2}$.*

Proof. In light of Theorem 9.11, it suffices to consider the progressive version of the threshold protocol. In this case, for $S \subseteq V(\mathcal{G})$, $S_{j-1} \subseteq S_j$ where $S_j = \bar{h}_q(S)$ for all $j \geq 1$. Assume that $q > \frac{1}{2}$ and for any subset of vertices W, let $e(W)$ denote the number of edges with one end in W and the other node outside of W, that is, in $V(\mathcal{G})\backslash W$. We now claim that when $q > \frac{1}{2}$, if S_{j-1} is strictly contained in S_j, then $e(S_j) < e(S_{j-1})$. To see this, consider a node v in $S_j \backslash S_{j-1}$. Since v at time j has decided to adopt a B state, and given that $q > \frac{1}{2}$, v must have strictly more neighbors in S_{j-1} than in $V(\mathcal{G})\backslash S_j$. Summing over all nodes in $S_i \backslash S_{j-1}$, we can conclude that $e(S_j) < e(S_{j-1})$. Subsequently, $e(S_k)$ is a strictly decreasing sequence bounded by zero. Hence, there has to be some value of k such that

$$S_k = S_{k+1} = S_{k+2} = \cdots,$$

which implies that state B will not spread unboundedly throughout the network, and thus cannot be contagious.   ∎

9.3 EPIDEMICS

Mathematical epidemiology is another rich source of problems in dynamic multiagent systems. There are a number of models in mathematical epidemiology that have been used to shed light on the spread of diseases in a population. In this section, we consider one such model, the single population and multipopulation SEIR model. The SEIR acronym refers to the fact that this model considers the interaction between the *susceptible*, *exposed*, *infective*, and *recovered* (or removed) in a given population. One of the critical issues often examined using models such as SEIR is the stability of various equilibria. Our choice of the model in this section has been influenced by the ease by which this model lends itself to analysis via Lyapunov theory, with a crucial part of the analysis being graph theoretic.

9.3.1 Single Population

In the single population SEIR model, the population is divided into different compartments:

- susceptible group S; individuals that can potentially get infected;

- exposed group E; individuals that have been infected but they are not considered infectious;

- infective group I; individuals that are infectious; and

- recovered (or removed) group R; individuals that have either recovered or removed from the population. We note that in this model it is assumed that members of group R *do not* go back to the susceptible group.

Let us denote by $y(t)$, $z(t)$, $w(t)$, and $v(t)$ the fractions of the population that are in groups S, E, I, and R, respectively, at a given time t. Thus

$$y(t) + z(t) + w(t) + v(t) = 1 \quad \text{for all } t \geq 0.$$

In the SEIR model, the population is considered to be of a *fixed size*. The part of the removed population R that accounts for recovered, natural death, or death caused by the disease is exactly compensated for by birth in the society. The set of differential equations that governs the evolution of the various compartments of the population thereby assumes the form,

$$
\begin{aligned}
\textbf{S}: \quad & \dot{y}(t) = -\delta y(t) - \beta y(t)w(t) + \delta, & (9.9) \\
\textbf{E}: \quad & \dot{z}(t) = -(\delta + \epsilon)z(t) + \beta y(t)w(t), & (9.10) \\
\textbf{I}: \quad & \dot{w}(t) = -(\delta + \gamma)w(t) + \epsilon z(t), & (9.11)
\end{aligned}
$$

while the evolution of the fraction of removed population R is dictated by

$$\textbf{R}: \quad \dot{v}(t) = \gamma w(t) - \delta v(t); \qquad (9.12)$$

in this model, δ denotes the birth rate as well as the natural death rate in the various compartments,[2] ϵ is the rate of becoming infectious after a latent period, γ is the recovery rate of infectious individuals, and β is the constant of the bilinear term, indicating the rate by which the disease is transmitted from the infective to the susceptible group upon contact. The model scheme is shown in Figure 9.3. Note that since the population is fixed, the dynamics of the recovered group does not have to be examined explicitly for stability analysis. Denote the state of the population, the fraction of the population in the SEI categories at time t, by $x(t) = [y(t), z(t), w(t)]$. Then, it can be shown that there are exactly two equilibrium points for the system (9.9) - (9.11): the infection-free state $x_o = [1, 0, 0]^T$ and the potential "endemic" state $x^* = [y^*, z^*, w^*]^T$.

---

[2]Note that $\delta y(t) + \delta z(t) + \delta w(t) + \delta v(t) = \delta$ for all $t \geq 0$.

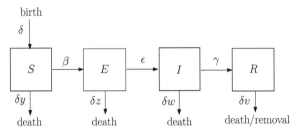

Figure 9.3: The SEIR model

The endemic equilibrium point is defined by

$$
\begin{aligned}
y^* &= \frac{1}{R_o}, \\
z^* &= \frac{\delta}{\delta + \epsilon}\left(1 - \frac{1}{R_o}\right), \\
w^* &= \frac{\delta \epsilon}{(\delta + \epsilon)(\delta + \gamma)}\left(1 - \frac{1}{R_o}\right),
\end{aligned}
\tag{9.13}
$$

where

$$
R_o = \frac{\beta \epsilon}{(\delta + \epsilon)(\delta + \gamma)}
\tag{9.14}
$$

is called the *reproduction number* for the system (9.9) - (9.11); a *positive* endemic equilibrium state x^* exists when $R_o > 1$. The reproduction number has a critical role in specifying which equilibrium state is globally asymptotically stable for the system (9.9) - (9.11). Let us start with an observation whose proof is left as an exercise.

Proposition 9.13. *The positive orthant $\mathbf{R}_+^3$ is positively invariant with respect to (9.9) - (9.11), that is, if $x(\bar{t}) \in \mathbf{R}_+^3$ for some $\bar{t}$ then $x(t) \in \mathbf{R}_+^3$ for all $t \geq \bar{t}$.*

As a result of Proposition 9.13, it is natural to examine whether $\mathbf{R}_+^3$ contains a globally stable equilibrium.

Theorem 9.14. *When the reproduction number $R_o > 1$, the positive endemic equilibrium state $x^* \in \mathbf{R}_+^3$ exists and is globally asymptotically stable with respect to $\mathbf{R}_+^3 \backslash \{x_o\}$.*

Proof. Consider the Lyapunov function

$$V(x(t)) = (y(t) - y^* \ln y(t)) + \frac{\delta + \epsilon}{\epsilon}(z(t) - z^* \ln z(t))$$
$$+ \frac{\delta + \epsilon}{\epsilon}(w(t) - w^* \ln w(t)), \tag{9.15}$$

which is built around the equilibrium point (9.13). Note that this Lyapunov function is continuous for all $x > 0$ and diverges to infinity at the boundary of the positive orthant; its time derivative is also equal to

$$\begin{bmatrix} 1 - (y^*/y(t)) \\ 1 - (z^*/z(t)) \\ ((\delta + \epsilon)/\epsilon)(1 - (w^*/w(t))) \end{bmatrix}^T \begin{bmatrix} -\delta y(t) - \beta y(t)w(t) + \delta \\ -(\delta + \epsilon)z(t) + \beta y(t)w(t) \\ -(\delta + \gamma)w(t) + \epsilon z(t) \end{bmatrix},$$

which in light of (9.13) - (9.14), can be simplified as[3]

$$\dot{V}(x(t)) = (\delta - \delta y^*) \left(-\frac{y^*}{y(t)} - \frac{z(t)w^*}{z^*w(t)} - \frac{y(t)z^*w(t)}{y^*z(t)w^*} + 3 \right)$$
$$- \delta y^* \left(\frac{y^*}{y(t)} + \frac{y(t)}{y^*} - 2 \right). \tag{9.16}$$

Now let

$$\mathbf{y}(t) = \frac{y(t)}{y^*}, \quad \mathbf{u}(t) = \frac{z^*w(t)}{z(t)w^*}, \quad \mathbf{a} = \delta y^*, \quad \text{and} \quad \mathbf{c} = \delta - \delta y^*.$$

Then (9.16) can be written as

$$\dot{V}(x(t)) = -\mathbf{a}\left(\mathbf{y}(t) + \frac{1}{\mathbf{y}(t)} - 2\right) - \mathbf{c}\left(\mathbf{y}(t)\mathbf{u}(t) + \frac{1}{\mathbf{y}(t)} + \frac{1}{\mathbf{u}(t)} - 3\right).$$

Moreover, as long as the trajectories of the SEIR dynamics remain in $\mathbf{R}_+^3$, one has $\mathbf{a}, \mathbf{c} \geq 0$. Furthermore, from the arithmetic mean geometric mean inequality,[4] it follows that for all $t \geq 0$,

$$\mathbf{y}(t) + \frac{1}{\mathbf{y}(t)} - 2 \geq 0$$

---

[3] The simplification requires some work.
[4] The arithmetic mean geometric mean (AM-GM) inequality states that for any sequence of nonnegative numbers, their arithmetic mean is an upper bound for their geometric mean.

and

$$\frac{1}{\mathbf{y}(t)} + \frac{1}{\mathbf{u}(t)} + \mathbf{y}(t)\mathbf{u}(t) - 3 \geq 0.$$

Hence, from LaSalle's invariance principle, it follows that the trajectory of the system (9.9) - (9.11) converges to the set

$$\{x(t) \mid \dot{V}(x(t)) = 0\}.$$

This condition requires that $\mathbf{y}(t) = \mathbf{u}(t) = 1$, which in turn implies the convergence of the system (9.9) - (9.11) to the set

$$M = \{(y(t), z(t), w(t)) \mid y(t) = y^*, z(t)w^* = z^*w(t)\}. \quad (9.17)$$

As $x^* = [y^*, z^*, w^*]^T$ is the only equilibrium in M (9.17), the global asymptotic stability of the positive endemic equilibrium x^* follows. ∎

Theorem 9.14 highlights the pivotal role of the reproduction number R_o (9.14) in studying the epidemic outbreak in the SEIR model. In fact, this is particularly crucial since when the reproduction number $R_o \leq 1$, there exists no positive endemic equilibrium state for (9.9) - (9.11) and the infection-free equilibrium state $x_o = [1, 0, 0]^T$ turns out to be globally asymptotically stable.

9.3.2 Multipopulation

In this section, we delve into the multipopulation extension of the SEIR epidemic model of §9.3.1. The model is as follows. We allow each population to have three different compartments, corresponding to susceptible, exposed, and infective groups, as in the single population model. However, in addition to interactions between the distinct compartments in each population, we allow for interactions among the compartments in n *different populations.* Hence, the model (9.9) - (9.11) is extended as

$$\mathbf{S}: \quad \dot{y}_i(t) = -\delta_i^y y_i(t) - \sum_{j=1}^{n} \beta_{ij} y_i(t) w_j(t) + \delta_i, \quad (9.18)$$

$$\mathbf{E}: \quad \dot{z}_i(t) = -(\delta_i^z + \epsilon_i) z_i(t) + \sum_{j=1}^{n} \beta_{ij} y_i(t) w_j(t), \quad (9.19)$$

$$\mathbf{I}: \quad \dot{w}_i(t) = -(\delta_i^w + \gamma_i) w_i + \epsilon_i z_i(t), \quad (9.20)$$

where $i = 1, 2, \ldots, n$, and the parameters for each group are direct extensions of the single population parameters; for example, δ_i^w is the rate of natural death, removed, or recovered, in the infective group for the ith population and δ_i denotes the rate of population influx or the birth rate in the ith population. We once again adopt the notation $x_i(t) = [y_i(t), z_i(t), w_i(t)]^T$ and $x(t) = [x_1(t)^T, x_2(t)^T, \ldots, x_n(t)^T]^T$ for all t. Our standing assumption for the rest of this section is that for each i,

$$\epsilon_i > 0 \quad \text{and} \quad \delta_i^* = \min\{\delta_i^y, \delta_i^z, \delta_i^w + \gamma_i\} > 0.$$

In order to follow an analysis analogous to §9.3.1, we define the set

$$\Gamma = \left\{ x \in \mathbf{R}_+^{3n} \mid y_i \leq \frac{\delta_i}{\delta_i^y}, y_i + z_i + w_i \leq \frac{\delta_i}{\delta_i^*}, i = 1, 2, \ldots, n \right\}, \quad (9.21)$$

and observe the following invariance property, whose proof is left as an exercise.

Proposition 9.15. *The set Γ (9.21) is positively invariant for the multipopulation SEIR model (9.18) - (9.20).*

The reproduction parameter for the multipopulation model assumes a matrix theoretic flavor, expressed in terms of the spectral radius of an appropriately defined matrix for the interaction model.

Definition 9.16. *The multipopulation reproduction parameter R_o is defined as the spectral radius of the matrix M_o with entries*

$$[M_o]_{ij} = \frac{\beta_{ij}\epsilon_i \, \delta_i/\delta_i^y}{(\delta_i^z + \epsilon_i)(\delta_i^w + \gamma_i)}$$

for $i, j = 1, 2, \ldots, n$, that is, $R_o = \rho(M_o)$.

The significance of the reproduction number R_o in the multipopulation SEIR model is analogous to that for the single population model: as we will see shortly, when $R_o > 1$ and the interpopulation interaction network is a strongly connected digraph, then the positive endemic equilibrium for the multipopulation SEIR model is globally stable. This observation will be substantiated via a reasoning that nicely blends a Lyapunov-type argument with graph theoretic constructs. The *existence* of such an equilibrium point follows from the fact that the trajectories of (9.18) - (9.20) remain in the interior of the bounded set Γ (9.21); see notes and references. In addition, the disease-free state turns out to be unstable when $R_o > 1$.[5]

---

[5]Moreover, the disease-free state is globally stable in Γ (9.21) when $R_o \leq 1$.

We now gather the main ingredients for the proof of global stability of the endemic equilibrium in the interior of Γ; we denote this interior by $\mathbf{int}\,\Gamma$. The proof also provides a motivation for introducing *out-degree Laplacian* as opposed to the *in-degree* Laplacian for digraphs that has been used throughout the book. In this direction, let $\mathcal{D}$ denote the weighted digraph associated with the interaction between the multiple populations, that is, we let $\mathcal{D} = (V, E)$, where $V = \{1, 2, \ldots, n\}$ and there is a weighted directed edge from node i to j if $\beta_{ij} > 0$; in this case, we set the weight on the directed edge from i to j as β_{ij}. The corresponding weighted *out-degree Laplacian* assumes the form

$$
L_o(\mathcal{D}) = \begin{bmatrix} \sum_{k=1,k\neq 1}^n \beta_{1k} & -\beta_{12} & \cdots & -\beta_{1n} \\ -\beta_{21} & \sum_{k=1,k\neq 2}^n \beta_{2k} & \cdots & -\beta_{2n} \\ \vdots & \vdots & \cdots & \vdots \\ -\beta_{n1} & \cdots & -\beta_{n,n-1} & \sum_{k=1,k\neq n}^n \beta_{nk} \end{bmatrix}.
$$

Parallel to our discussion on (in-degree) Laplacians for digraphs in Chapter 2, when $\mathcal{D}$ is strongly connected, the null space of $L_o(\mathcal{D})$ is characterized by $\mathbf{span}\{\mathbf{1}\}$ and the null space of $L_o(\mathcal{D})^T$ is parameterized by the left eigenvector of $L_o(\mathcal{D})$ associated with an eigenvalue of zero. In fact, this left eigenvector $v = [v_1, v_2, \ldots, v_n]^T$ has positive entries and can be specified by letting

$$
v_i = \sum_{T \in \mathcal{T}_i} \prod_{(r,m)\in E(T)} \beta_{rm},
$$

where $\mathcal{T}_i$ is the set of all spanning *in-branchings* of $\mathcal{D}$ that are rooted at vertex i, and $E(T)$ is the set of edges in the directed tree T. A rooted in-branching is the "dual" construct of a rooted out-branching discussed in Chapter 2, in the sense that in the former case, all directed edges are oriented *toward* the root. When a directed edge is added away from the root of a rooted in-branching toward another vertex in the digraph, we call the resulting digraph *unicyclic*. With this definition in mind, we state an auxiliary lemma which proves to be crucial in the proof of the main theorem of this section.

Lemma 9.17. *Let x^* be an arbitrary point in the interior of the set Γ (9.21) and define*

$$
H_n(x) = \sum_{i=1}^n \sum_{j=1}^n v_i \widehat{\beta}_{ij} \left(3 - \frac{y_i^*}{y_i} - \frac{y_i w_j z_i^*}{y_i^* w_j^* z_i} - \frac{z_i w_i^*}{z_i^* w_i} \right), \tag{9.22}
$$

where $\widehat{\beta}_{ij} = \beta_{ij} y_i^* w_j^*$ and v is the left eigenvector of the out-degree Laplacian

$$
L_o(\widehat{\mathcal{D}}) = \begin{bmatrix} \sum_{k=1,k\neq 1}^{n} \widehat{\beta}_{1k} & -\widehat{\beta}_{12} & \cdots & -\widehat{\beta}_{1n} \\ -\widehat{\beta}_{21} & \sum_{k=1,k\neq 2}^{n} \widehat{\beta}_{2k} & \cdots & -\widehat{\beta}_{2n} \\ \vdots & \vdots & \cdots & \vdots \\ -\widehat{\beta}_{n1} & \cdots & -\widehat{\beta}_{n,n-1} & \sum_{k=1,k\neq n}^{n} \widehat{\beta}_{nk} \end{bmatrix}
$$

*for a strongly connected weighted digraph $\widehat{\mathcal{D}}$. Then $H_n(x) \le 0$ for $x \in$ **int** Γ and $H_n(x) = 0$ implies that $x = x^*$.*

Proof. The components of the left eigenvector of the out-degree Laplacian v corresponding to its zero eigenvalue can be parameterized as

$$
v_k = \sum_{T \in \mathcal{T}_k} \prod_{(r,m) \in E(T)} \widehat{\beta}_{rm},
$$

where $\mathcal{T}_k$ is the set of all spanning in-branchings of $\widehat{\mathcal{D}}$ that are rooted at vertex k, and $E(T)$ is the set of edges in the directed tree T. Thus the products of the form $v_i \widehat{\beta}_{ij}$ in the expression of H_n (9.22) can be viewed as the product of the weights on the edges of a unicyclic digraph Q, obtained by adding a directed edge from i to j in the (spanning) in-branching rooted at vertex i. In fact, the double sum defining the expression for H_n (9.22) can be viewed as the sum of the product of the weights of the edges in the unique cycles of all unicyclic subgraphs Q of $\widehat{\mathcal{D}}$. Hence

$$
H_n(x) = \sum_{Q} H_{n,Q},
$$

where Q ranges over all unicyclic subgraphs of $\widehat{\mathcal{D}}$,

$$
H_{n,Q} = \prod_{(r,m)} \widehat{\beta}_{rm} \sum_{(i,j) \in E(Q_c)} \left(3 - \frac{y_i^*}{y_i} - \frac{y_i w_j z_i^*}{y_i^* w_j^* z_i} - \frac{z_i w_i^*}{z_i^* w_i} \right)
$$

$$
= \prod_{(r,m) \in E(Q)} \widehat{\beta}_{rm} \left(3q - \sum_{(i,j) \in E(Q_c)} \frac{y_i^*}{y_i} - \frac{y_i w_j z_i^*}{y_i^* w_j^* z_i} - \frac{z_i w_i^*}{z_i^* w_i} \right),
$$

the parameter q denotes the number of directed edges in Q, and Q_c is the unique directed cycle in the unicyclic subgraph Q. However,

$$
\prod_{(i,j) \in E(Q_c)} \left(\frac{y_i^*}{y_i} \frac{y_i w_j z_i^*}{y_i^* w_j^* z_i} \frac{z_i w_i^*}{z_i^* w_i} \right) = \prod_{(i,j) \in E(Q_c)} \frac{w_j w_i^*}{w_j^* w_i} = 1
$$

for each unicyclic graph Q. Therefore from the arithmetic mean geometric mean inequality, it follows that

$$\sum_{(i,j)\in E(Q_c)} \left(\frac{y_i^*}{y_i} + \frac{y_i w_j z_i^*}{y_i^* w_j^* z_i} + \frac{z_i w_i^*}{z_i^* w_i} \right) \geq 3q,$$

and $H_{n,Q} \leq 0$ for each Q. Moreover, when $H_{n,Q} = 0$, one has that for each $(i,j) \in E(Q_c)$,

$$\frac{y_i^*}{y_i} = \frac{y_i w_j z_i^*}{y_i^* w_j^* z_i} = \frac{z_i w_i^*}{z_i^* w_i}. \tag{9.23}$$

From the above discussion, it then follows that $H_n = 0$ if $y_i = y_i^*$ for all i. We claim that $H_n(x) = 0$ also implies that for some $\alpha > 0$,

$$z_i = \alpha z_i^* \quad \text{and} \quad w_i = \alpha w_i^*, \tag{9.24}$$

for all $i = 1, 2, \ldots, n$.

We now show that in fact $\alpha = 1$, thus completing the proof of the lemma. Let us first observe that (9.23) implies that

$$\frac{w_i}{w_i^*} = \frac{z_i}{z_i^*} = \frac{w_j}{w_j^*} \tag{9.25}$$

for every directed edge (i,j) that belongs to the (directed) cycle of some unicycle subgraph Q of $\widehat{D}$. Since $\widehat{D}$ is assumed to be strongly connected, every directed edge (i,j) belongs to the cycle of at least one such subgraph. Thus the identity (9.25) holds for every directed edge in $\widehat{D}$. As the digraph is strongly connected, the identity (9.25) can thus be extended to all pairs of vertices in $\widehat{D}$ and hence (9.24) follows. Substituting $y_i = y_i^*$, $z_i = \alpha z_i^*$, and $w_i = \alpha w_i^*$ in (9.18) results in the identity

$$\delta_i - \delta_i^y y_i^* - \alpha \sum_j \beta_{ij} y_i^* w_j^* = 0,$$

which in view of (9.18) implies that $\alpha = 1$. Hence x^* is the unique root of $H_n(x)$ (9.22).   ∎

We are now in the position to prove the main result of this section.

Theorem 9.18. *Assume that the multipopulation interaction graph is strongly connected. Then when the reproduction number $R_o > 1$, the multigroup model has a unique endemic equilibrium which is globally stable in the interior of set Γ (9.21).*

Proof. Denote by x^* the endemic equilibrium in the interior of Γ denoted by Γ^o whose existence is guaranteed when $R_o > 1$. This follows from the fact that the trajectory of the system (9.18) - (9.20) is persistent in the interior of a compact set; see Proposition 9.15 and notes and references. Let $\widehat{\beta}_{ij} = y_i^* w_j^* \beta_{ij}$ be the weights on the edges of the strongly connected interaction digraph, and let $v = [v_1, v_2, \ldots, v_n]^T$ be the left eigenvector of the corresponding out-degree Laplacian associated with its zero eigenvalue. Construct the Lyapunov function $V(x)$ as

$$\sum_i v_i \left((y_i + y_i^* \ln y_i) + (z_i + z_i^* \ln z_i) + \frac{\delta_i^z + \epsilon_i}{\epsilon_i} (w_i + w_i^* \ln w_i) \right).$$

Then

$$\dot{V} = \sum_i v_i \left[\delta_i - \delta_i^y y_i - \sum_j \beta_{ij} y_i w_j - \delta_i \frac{y_i^*}{y_i} + \delta_i^y y_i^* \right.$$

$$+ \sum_j \beta_{ij} y_i^* w_j + \sum_j \beta_{ij} y_i w_j - (\delta_i^z + \epsilon_i) z_i - \sum_j \beta_{ij} \frac{w_i^* y_i w_j}{w_i}$$

$$+ (\delta_i^z + \epsilon_i) z_i^* + (\delta_i^z + \epsilon_i) z_i - \frac{(\delta_i^z + \epsilon_i)(\delta_i^w + \gamma_i)}{\epsilon_i} w_i$$

$$\left. - (\delta_i^z + \epsilon_i) \frac{w_i^* z_i}{w_i} + \frac{(\delta_i^z + \epsilon_i)(\delta_i^w + \gamma_i)}{\epsilon_i} w_i^* \right]$$

$$= \sum_i v_i \left[(\delta_i^y y_i^* (2 - \frac{y_i^*}{y_i} - \frac{y_i}{y_i^*}) + (\sum_j \beta_{ij} y_i^* w_j - \frac{(\delta_i^z + \epsilon_i)(\delta_i^w + \gamma_i)}{\epsilon_i} w_i) \right.$$

$$\left. + (3 \sum_j \beta_{ij} y_i^* w_j^* - \sum_j \beta_{ij} w_j^* \frac{(y_i^*)^2}{y_i} - \sum_j \beta_{ij} y_i w_j \frac{z_i^*}{z_i} - (\delta_i^z + \epsilon_i) z_i \frac{w_i^*}{w_i}) \right]$$

$$\leq \sum_i v_i \left[(\sum_j \beta_{ij} y_i^* w_j - \frac{(\delta_i^z + \epsilon_i)(\delta_i^w + \gamma_i)}{\epsilon_i} w_i) \right.$$

$$\left. + (3 \sum_j \beta_{ij} y_i^* w_j^* - \sum_j \beta_{ij} w_j^* \frac{(y_i^*)^2}{y_i} - \sum_j \beta_{ij} y_i w_j \frac{z_i^*}{z_i} - (\delta_i^z + \epsilon_i) z_i \frac{w_i^*}{w_i}) \right]$$

since

$$\frac{y_i^*}{y_i} + \frac{y_i}{y_i^*} \geq 2$$

with equality holding if and only if $y_i = y_i^*$. However, as v is the left eigenvector of the weighted out-degree Laplacian defined in Lemma 9.17

corresponding to its zero eigenvalue, it follows that

$$\dot{V} \leq \sum_i v_i \left(3 \sum_j \widehat{\beta}_{ij} - \sum_j \widehat{\beta}_{ij} \frac{y_i^*}{y_i} - \sum_j \widehat{\beta}_{ij} \frac{y_i w_j z_i^*}{y_i^* w_j^* z_i} - (\delta_i^z + \epsilon_i) z_i \frac{w_i^*}{w_i} \right)$$

$$= \sum_{i=1}^n \sum_{j=1}^n v_i \widehat{\beta}_{ij} \left(3 - \frac{y_i^*}{y_i} - \frac{y_i w_j z_i^*}{y_i^* w_j^* z_i} - \frac{z_i w_i^*}{z_i^* w_i} \right)$$

$$= H_n(x).$$

From LaSalle's invariance principle and Lemma 9.17, the statement of the theorem now follows. ∎

We conclude this chapter with a turn toward yet another vista, namely, the chip firing games over graphs.

9.4 THE CHIP FIRING GAME

Let N chips be distributed among n vertices on a connected graph $\mathcal{G} = (V, E)$ with m edges. The number of chips on vertex v at time t, denoted by $x_v(t)$, will be its state. Thus

$$\mathbf{1}^T x_v(t) = N \quad \text{for all } t \geq 0.$$

We denote by $x(t)$ the "configuration" of the game at time t (the number of chips on all vertices). In the chip firing game, one chooses a vertex in the graph that has as many chips as its degree. Subsequently, one chip from this selected vertex is moved to each of its neighbors–this is referred to as the firing of that vertex. If a vertex does not have as many chips as its degree, it is spared of being fired; see Figure 9.4.

In addition to the number of chips on a given vertex, another state that we will associate with that vertex is the number of times it has fired up to time t. This will be denoted by $f_v(t)$. The game terminates when there exists no vertex that can be fired, that is, for some t,

$$x_v(t) < d(v) \quad \text{for all } v \in V.$$

In this section, we focus on examining graph theoretic conditions that govern the termination of a chip firing game. Let us warm up with a lemma on chip firing games that never terminate; we refer to them as infinite games.

Lemma 9.19. *In an infinite chip firing game, every vertex is fired infinitely often.*

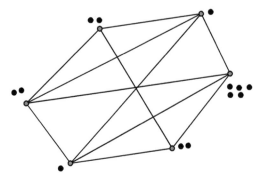

Figure 9.4: Chip firing over graphs; in this round, the node with five chips can fire whereas all other nodes cannot.

Proof. Since the game is infinite, some configuration of chips, say $\widehat{x}$, appears infinitely often. Let us consider the firing sequence between two subsequent appearances of state $\widehat{x}$, say at times t_1 and t_2. Suppose that there exists a vertex that has not fired during the interval $[t_1, t_2]$. Since the state of this vertex remains the same, none of its neighbors could fire during this interval, and thus

$$x_j(t_1) = x_j(t_2) = x_j(t) \quad \text{for all } t \in [t_1, t_2], \ j \in N(i) \cup \{i\}.$$

However, $\mathcal{G}$ is connected; thus, by extending the above argument across the network, during the interval $[t_1, t_2]$, either everybody fired or nobody did. This is, however, in contradiction with the assumption that the game is infinite; hence, every vertex in the graph has to fire infinitely often.   ∎

Now we give the dual version of Lemma 9.19.

Lemma 9.20. *If each vertex has already been fired at least once, then the game will never terminate.*

Proof. Assume that each vertex has already been fired at least once, yet the game has terminated at time T. Furthermore, assume that $\widehat{v}$ is the vertex that fired last in the game and $\widetilde{v}$ is the vertex that had been idle the longest by the termination time. Thus, all other vertices have fired after $\widetilde{v}$ last fired. Among these vertices, some are the neighbors of $\widetilde{v}$; thus $x_{\widetilde{v}}(T) \geq d(\widetilde{v})$, that is, $\widetilde{v}$ can fire after $\widehat{v}$, contradicting the assumption that the game had terminated at T by $\widehat{v}$.   ∎

The next observation further formalizes the intuition that a chip firing game with "too many" chips will never terminate.

Proposition 9.21. *Consider a terminating chip firing game on $\mathcal{G} = (V, E)$ with n vertices, m edges, and N chips. Then $N \leq 2m - n$.*

Proof. If the game is terminating, then for some finite time T,

$$x_v(T) \leq d(v) - 1$$

for all v and hence

$$N = \sum_v x_v(T) \leq \sum_v (d(v) - 1) \leq 2m - n.$$

∎

We now further examine the role of two graph parameters in the termination of a chip firing game on $\mathcal{G}$. First, a useful lemma.

Lemma 9.22. *Suppose that $uv \in E$ for a chip firing game with N chips on $\mathcal{G} = (V, E)$. Then for any t,*

$$|f_u(t) - f_v(t)| \leq N.$$

Proof. Given t, let $f_u(t) = a$ and $f_v(t) = b$ with $a < b$ without loss of generality. Consider the subgraph $\mathcal{H}$ of $\mathcal{G}$ whose nodes have not fired more than a times up to time t. Then the edge uv has contributed $b - a$ chips to the transfer of chips from $\overline{\mathcal{H}}$ to $\mathcal{H}$. Since the total number of chips in $\mathcal{H}$ cannot be more than N, it follows that $b - a \leq N$. ∎

Theorem 9.23. *Let $\mathcal{G} = (V, E)$ be a connected graph on n vertices and m edges, with $\mathrm{diam}(\mathcal{G})$ denoting its diameter, that is, the maximum (edge) distance between two vertices of $\mathcal{G}$. Then a finite chip firing game on $\mathcal{G}$ terminates within $2mn\,\mathrm{diam}(\mathcal{G})$ firings.*

Proof. Since the game is finite, by Lemma 9.20, there exists a vertex v that has never been fired, that is, $f_v(t) = 0$ for all t. Thus, by Lemma 9.22, for a vertex u at a distance d from v, one has $f_u(t) \leq dN$, that is, for any $u \in V$ one has $f_u(t) \leq \mathrm{diam}(\mathcal{G})N$. Consequently, there were at most a total of $\mathrm{diam}(\mathcal{G})nN$ firings during the game. As the game is finite, we can bound the total number of firings as

$$\mathrm{diam}(\mathcal{G})nN \leq n(2m - n)\,\mathrm{diam}(\mathcal{G}) \leq 2nm\,\mathrm{diam}(\mathcal{G}) - n^2\,\mathrm{diam}(\mathcal{G})$$
$$< 2mn\,\mathrm{diam}(\mathcal{G}).$$

∎

The above result highlights the role of the diameter of the graph on the termination of a terminating chip firing game. Our final bound–as expected– is in terms of the second smallest eigenvalue of the graph Laplacian. Again, we need an auxiliary observation first.

Proposition 9.24. *Let $\mathcal{G} = (V, E)$ be a connected graph with $V = [n]$. If $y = L(\mathcal{G})x$ and $x_n = 0$, then*

$$|\mathbf{1}^T x| \leq \frac{n}{\lambda_2(\mathcal{G})} \|y\|.$$

Proof. Define

$$L(\mathcal{G})^\dagger = \sum_{i=2}^{n} \frac{1}{\lambda_i(\mathcal{G})} u_i u_i^T,$$

with u_i as the normalized eigenvector of $L(\mathcal{G})$ associated with eigenvalue $\lambda_i(\mathcal{G})$. Then

$$L(\mathcal{G})^\dagger L(\mathcal{G}) = \left(I - \frac{1}{n} \mathbf{1} \mathbf{1}^T \right),$$

and $y = L(\mathcal{G})x$ implies that $L(\mathcal{G})^\dagger L(\mathcal{G})x = L(\mathcal{G})^\dagger y$. Now let

$$e_n = [0, 0, \dots, 1]^T$$

and observe that

$$e_n^T (I - \frac{1}{n} \mathbf{1} \mathbf{1}^T)x = \frac{-1}{n} \mathbf{1}^T x = e_n^T L(\mathcal{G})^\dagger y,$$

and

$$|\mathbf{1}^T x| = n |e_n^T L(\mathcal{G})^\dagger y| \leq \frac{n}{\lambda_2(\mathcal{G})} \|e_n\| \|y\| = \frac{n}{\lambda_2(\mathcal{G})} \|y\|.$$

∎

Let σ be a sequence of firing in the chip firing game during the time interval $[t_o, t_f]$. Then the counter function of the sequence, $f(\sigma)$, is such that $[f(\sigma)]_i$ denotes the number of times vertex i has fired in the sequence σ. Hence,

$$x(t_f) - x(t_o) = L(\mathcal{G})f(\sigma).$$

Corollary 9.25. *Let $\mathcal{G} = (V, E)$ be a connected graph with $V = [n]$. Then a terminating chip firing game on $\mathcal{G}$ with N chips terminates in at most $\sqrt{2}nN/\lambda_2(\mathcal{G})$ firings.*

Proof. Note that as $\|x(t_o)\|, \|x(t_f)\| \leq N$ one has $\|x(t_o) - x(t_f)\| \leq \sqrt{2}N$. Since the game is finite, there exists a vertex, say v_n, that has not fired. Thus $L(\mathcal{G})f(\sigma) = x(t_o) - x(t_f)$ and $f_{v_n}(\sigma) = 0$. Applying Proposition 9.24 results in

$$|\mathbf{1}^T f(\sigma)| \leq \frac{n}{\lambda_2(\mathcal{G})}\|x(t_o) - x(t_f)\| \leq \sqrt{2}Nn/\lambda_2(\mathcal{G}). \qquad (9.26)$$

∎

Hence, a terminating chip firing game terminates faster on the graph with larger algebraic connectivity, a result which is reminiscent of the agreement protocol discussed in Chapter 3.

SUMMARY

The purpose of this chapter has been to provide a glimpse into the vast area of graph theoretic inquiries in sociology, epidemiology, and games over graphs. Along the way, we examined dynamic models over networks for capturing how fashions or infections diffuse over a population or multipopulations, as well as the termination properties of chip firing games.

NOTES AND REFERENCES

The section on the max-protocol is from an unpublished work of Kim and Mesbahi, expanding on how insights obtained from the agreement protocol can be extended to nonlinear protocols evolving over lattices (where taking maximum or minimum of elements is well defined). As the reader will quickly realize, analyzing this rather intuitive scenario is streamlined by framing the problem setup in terms of the underlying probability space. Our exposition of the threshold model for the spreading of fashions, ideas, and so on, in §9.2 parallels Kleinberg's article in [176].

The example in §9.3 reinforcing the utility of a graph theoretic approach to multiagent systems in the context of epidemics is from the paper of Guo, Li, and Shuai [109], which is based on Lyapunov-type arguments. For an alternative venue for studying epidemiology over populations that are not fully mixed using generating functions, see the work of Newman [174]. Our discussion on chip firing games, also referred to as Abelian sandpiles in theoretical physics, parallels [101].

There are a number of other disciplines in sociology, biology, and games that blend notions from graph theory (in particular, the degree sequence) in the dynamic analysis of the corresponding networked system. Among these, we point out the area of evolutionary games on graphs [178], population dynamics [119], chemical reaction networks [87, 235], social learning over networks [105], referral systems like Google PageRank [142], opinion dynamics [25], and pulsed biological oscillators [160].

SUGGESTED READING

For more on social networks we refer the reader to [122]. For epidemiology, the two volume book by Murray [170] has been the classic reference. The edited volume by Nisan, Roughgarden, Tardos, and Vazirani [176] is the source of many interrelated research inquires on algorithmics, pricing, games, and networks.

EXERCISES

Exercise 9.1. Verify that the update equation (9.4) encodes the max-protocol (9.1).

Exercise 9.2. Consider a modification of the max-protocol,

$$x_i(k+1) = \max\{x_i(k), \beta x_j(k)\}$$

for $0 \leq \beta \leq 1$ and $j \in N(i)$. In this case, β reflects the deficiency of vertex i to gauge the fashionability of its neighbors. Discuss how the convergence of this protocol is influenced by the choice of β.

Exercise 9.3. Consider another modification of the max-protocol,

$$x_i(k+1) = \max\{x_i(k), 1 - \beta(1 - x_j(k))\}$$

for $0 \leq \beta \leq 1$. In which situation might this model be applicable? Discuss how the convergence of this protocol is influenced by the choice of β.

Exercise 9.4. Consider the threshold protocol in the configuration shown in Figure 9.1 at $t = 0$. Discuss how the behavior of the protocol will be altered if the infinite path graph is changed to a cycle on a large number of vertices.

Exercise 9.5. Under what conditions the threshold protocol is contagious on a finite cycle graph with $q \in [0, \frac{1}{2}]$?

Exercise 9.6. Does adding edges to a connected graph on an infinite number of vertices improves the chances that a given set is contagious?

Exercise 9.7. Using computer simulations, examine whether the states of the vertices of a finite graph, under the action of the threshold protocol, is periodic.

Exercise 9.8. In the SIS model of infectious diseases, the population consists of the susceptible, denoted by S, and infective, denoted by I, whose evolution are governed by the coupled differential equations

$$\dot{S}(t) = -\beta I(t)S(t) + \gamma I, \quad \dot{I} = \beta S(t)I(t) - \gamma I,$$

where β is the pairwise infectious contact rate and γ is the recovery rate. Define the reproduction ratio for the SIS model as $R_o = \beta N / \gamma$, where $N = S(t) + I(t)$ for all t. Show that for $R_o < 1$ the disease will die out and for $R_o > 1$ it remains endemic in the population. What is the interpretation of the case when $R_o = 1$ and comment on whether the disease remains endemic in the population in this case.

Exercise 9.9. Show that when the reproduction parameter $R_o \leq 1$ for the single population SEIR model (9.9) - (9.11) the infection-free equilibrium state $x_o = [1, 0, 0]^T$ is globally asymptotically stable.

Exercise 9.10. Verify the statement of Proposition 9.13.

Exercise 9.11. Show that there are exactly two equilibrium points for the single population SEIR dynamics (9.9) - (9.11).

Exercise 9.12. Verify the simplified expression for $\dot{V}(t)$ in (9.16).

Exercise 9.13. Let $L_o(\mathcal{D})$ be the out-degree Laplacian for the weighted strongly connected digraph where the weight on the edge (i, j) is denoted by β_{ij}. Use Theorem 2.12 to deduce that the entries of the left eigenvector of $L_o(\mathcal{D})$ corresponding to its zero eigenvalue are parameterized as

$$v_i = \sum_{T \in \mathcal{T}_i} \prod_{(r,m) \in E(T)} \beta_{rm},$$

where $\mathcal{T}_i$ denotes the set of all spanning in-branchings of $\mathcal{D}$ rooted at i and $E(T)$ denotes the set of edges in the directed tree T.

Exercise 9.14. Let δ_i^y, δ_i^z, and δ_i^w denote the natural death or removal rates in the susceptible, exposed, and infective groups in population i, respectively. Moreover, let γ_i and δ_i denote, respectively, the rate of recovery for infectious individuals and the rate of population influx or birth rate in this population. Set $\delta_i^* = \min\{\delta_i^y, \delta_i^z, \delta_i^w + \gamma_i\}$. Show that

$$\Gamma = \left\{ x \in \mathbf{R}_+^3 \mid y_i \le \frac{\delta_i}{\delta_i^y}, y_i + z_i + w_i \le \frac{\delta_i}{\delta_i^*}, i = 1, 2, \ldots, n \right\}$$

is positively invariant for the multipopulation SEIR model (9.18) - (9.20).

Exercise 9.15. In the multipopulation SEIR model (9.18) - (9.20), assume that the underlying interaction digraph is strongly connected. Show that when $R_o \le 1$ (R_o is the reproduction number defined in Definition 9.16), the unique equilibrium for the system is a disease-free state. Moreover, show that in this case, this equilibrium state is globally asymptotically stable in $\Gamma(9.21)$.

Exercise 9.16. Let $\bar{h}_q$ denote the map of the progressive threshold protocol with threshold value q on $\mathcal{G} = (V, E)$. Show that for a given $W \subseteq V$ and all $j \ge 1$,

$$\bar{h}_q^j(W) = W \cup h_q(\bar{h}_q^{j-1}(W)).$$

Exercise 9.17. Show that for a chip firing game on a connected graph with n nodes, m edges, and N chips, when $N < m$ the game is always finite.

Exercise 9.18. Show that for a chip firing game on a connected graph with n nodes, m edges, and N chips, if $m \le N \le 2m - n$, then the game can be finite or infinite depending on the initial configuration of the game. Give an example for both situations.

PART 3

NETWORKS AS SYSTEMS

Chapter Ten

Agreement with Inputs and Outputs

> "Fundamental progress has to do with
> the reinterpretation of basic ideas."
> — Alfred North Whitehead

In this chapter, we consider the input-output linear systems obtained when a collection of nodes in the network assume control and sensing roles, while the remaining nodes execute a local, agreement-like protocol. Our aim is to identify graph theoretic implications for the system theoretic properties of such systems. In particular, we show how the symmetry structure of a network with a single control/sensing node, characterized in terms of its automorphism group, directly relates to the controllability and observability of the corresponding input-output system. Moreover, we introduce network equitable partitions as means by which such controllability and observability characterizations can be extended to networks with multiple inputs and outputs.

10.1 THE BASIC INPUT-OUTPUT SETUP

The agreement protocol, as introduced in Chapter 3, provides the ambient setting for the evolution of a set of dynamic agents. Just as a stabilizing controller is typically a first step in the control design phase, the agreement protocol will provide the underlying cohesion of the network. In this chapter, we consider situations where the agreement protocol over a fixed network is also influenced by external inputs, injected at particular nodes. We also consider the case where the corresponding *linear system* can be observed. Although, in principle, one can designate network inputs and outputs at distinct nodes, we will be primarily concerned with the situation when the *input and output nodes* are identical. Hence, we postulate a scenario involving nodes in the network that are capable of influencing the network and observing their neighbors' responses as their injected signals propagate through–and are reflected back by–the network. We refer to the complements of the input and output nodes in the network as the *floating nodes*.

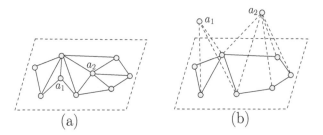

Figure 10.1: A network with input nodes as $\{a_1, a_2\}$: (a) before partitioning the nodes among input/output and floating nodes; (b) after partitioning the nodes among input/output and floating nodes; the dashed edges determine the entries of the matrix B_f in (10.3) - (10.4).

10.1.1 The Network Partition

Our initial setup involves designating some of the nodes in the agreement protocol (3.2) over a fixed network $\mathcal{G}$ as inputs and outputs. The other agents in the network, the floating nodes, continue to abide by the ambient agreement protocol. Let us use the subscripts i and f to denote attributes related to input/output nodes and floating nodes, respectively. For example, a *floating graph* $\mathcal{G}_f$ is the subgraph induced by the floating node set $V(\mathcal{G}_f) \subseteq V(\mathcal{G})$ after removing the input/output nodes as well as the edges between input/output nodes and between input/output nodes and floating nodes. An example of this is shown in Figure 10.1.

The input/output designation thus induces a partition of the incidence matrix $D(\mathcal{G})$ as

$$D(\mathcal{G}) = \left[\begin{array}{c} D_f \\ D_i \end{array} \right], \tag{10.1}$$

where $D_f \in \mathbf{R}^{n_f \times m}$ and $D_i \in \mathbf{R}^{n_i \times m}$; here n_f and n_i are the cardinalities of the sets of floating and input/output nodes, respectively, and m is the number of edges in the graph $\mathcal{G}$. The underlying assumption for this partition, without loss of generality, is that input/output nodes are indexed last in the original graph $\mathcal{G}$.

Since $L(\mathcal{G}) = D(\mathcal{G})D(\mathcal{G})^T$, the partitioning (10.1) implies that

$$L(\mathcal{G}) = \left[\begin{array}{cc} A_f & B_f \\ B_f^T & A_i \end{array} \right], \tag{10.2}$$

where

$$A_f = D_f D_f^T, \quad A_i = D_i D_i^T, \quad \text{and} \quad B_f = D_f D_i^T.$$

As an example, Figure 10.2 depicts an agreement protocol endowed with inputs and outputs with

$$V(\mathcal{G}_i) = \{5, 6\} \quad \text{and} \quad V(\mathcal{G}_f) = \{1, 2, 3, 4\}.$$

Such an input/output and floating node grouping also partitions the incidence matrix[1] of the original network as

$$D_f = \begin{bmatrix} -1 & 0 & 0 & 1 & 0 & 0 & -1 & 0 \\ 0 & 0 & 1 & -1 & -1 & 0 & 0 & 0 \\ 1 & -1 & 0 & 0 & 0 & 0 & 0 & 1 \\ 0 & 1 & -1 & 0 & 0 & 1 & 0 & 0 \end{bmatrix}$$

and

$$D_i = \begin{bmatrix} 0 & 0 & 0 & 0 & 1 & -1 & 0 & 0 \\ 0 & 0 & 0 & 0 & 0 & 0 & 1 & -1 \end{bmatrix},$$

where the columns of the matrices D_f and D_i correspond to the edges e_1 - e_8 in Figure 10.2. Hence,

$$A_f = \begin{bmatrix} 3 & -1 & -1 & 0 \\ -1 & 3 & 0 & -1 \\ -1 & 0 & 3 & -1 \\ 0 & -1 & -1 & 3 \end{bmatrix} \quad \text{and} \quad B_f = \begin{bmatrix} 0 & -1 \\ -1 & 0 \\ 0 & -1 \\ -1 & 0 \end{bmatrix}.$$

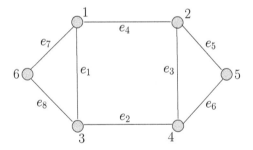

Figure 10.2: Input/output agreement network with $V(\mathcal{G}_f) = \{1, 2, 3, 4\}$ and $V(\mathcal{G}_i) = \{5, 6\}$

10.1.2 Input-Output Agreement

Based on the partitioning of the node set into input/output and floating nodes, the resulting system is a standard linear time-invariant system. We

---

[1] With an arbitrary orientation on the edges.

thereby proceed to study the *input-output agreement* in this context when the floating nodes evolve as

$$\dot{x}_f(t) = -A_f\, x_f(t) - B_f\, u(t), \qquad (10.3)$$
$$y(t) = -B_f^T\, x_f(t), \qquad (10.4)$$

where u denotes the exogenous "control" signal injected at the input nodes. Moreover, as (10.3) - (10.4) suggest, we allow the input nodes to also function as output nodes, consistent with the geometry by which they influence the floating nodes. In this sense, we are considering a collocated control structure imposed on the agreement protocol.

It is important to note that the system matrices in (10.3) - (10.4) are functions of the underlying graph $\mathcal{G}$ and the scheme by which its vertices have been partitioned among inputs and outputs. In fact, let us provide more insight into the role of the network and its partition on the system matrices A_f and B_f in (10.3) - (10.4) before proceeding to consider certain system theoretic aspects of the resulting controlled agreement protocol. A convenient tool for achieving this is the *input/output indicator vectors*.

Definition 10.1. *Let v_i be an input node in $\mathcal{G}$, that is, $v_i \in V(\mathcal{G}_i)$. The input indicator vector with respect to node i,*

$$\delta_i : V(\mathcal{G}_f) \rightarrow \{0,1\}^{n_f},$$

is such that

$$\delta_i(v_j) = \begin{cases} 1 & \text{if } \ v_j \sim v_i, \\ 0 & \text{otherwise.} \end{cases}$$

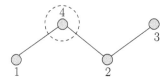

Figure 10.3: Path graph with node 4 designated as the input

For instance, the indicator vector for the node set $V(\mathcal{G}_f) = \{1,2,3\}$ in the graph shown in Figure 10.3 with respect to the input $\{4\}$ is $\delta_4 = [\,1,\ 1,\ 0\,]^T$. We now realize that since

$$[B_f]_{nm} = \sum_k [D_f]_{nk}[D_i]_{mk},$$

the nmth entry of B_f is negative if and only if v_j is an input node that is connected to the floating node v_i in the original graph $\mathcal{G}$; otherwise, this entry is zero. Hence each column of B_f in (10.3) - (10.4) is an indicator vector with respect to one of the inputs to the floating network, that is,

$$B_f = -[\delta_{n_f+1} \ \delta_{n_f+2} \ \cdots \ \delta_n] \in \mathbf{R}^{n_f \times n_i}. \tag{10.5}$$

Another useful construct for relating the system matrices in (10.3) - (10.4) to the structure of the network $\mathcal{G}$ is the *input-to-state degree matrix*. This matrix is defined in relation to the input nodes as follows: for the floating node $v_j \in V(\mathcal{G}_f)$, let $d_i(j)$ denote the number of input nodes that are adjacent to v_j. Then the input-to-state degree matrix is

$$\Delta_f = \mathbf{Diag}([\,d_i(1), d_i(2), \ldots, d_i(n_f)\,]^T) \in \mathbf{R}^{n_f \times n_f}. \tag{10.6}$$

For example, with reference to Figure 10.2 and the selection of nodes 5 and 6 as inputs, one has

$$\Delta_f = I.$$

A moment's reflection on the construction of the matrix A_f now reveals that

$$A_f = L(\mathcal{G}_f) + \Delta_f, \tag{10.7}$$

where $L(\mathcal{G}_f)$ is the Laplacian matrix of the floating graph $\mathcal{G}_f$. This follows from the observation that one can partition the matrix D_f as

$$D_f = [\,D_{ff} \mid D_{fi}\,],$$

where the columns of the submatrix D_{ff} correspond to edges incident between floating nodes, and the columns of D_{fi} correspond to edges between input/output nodes and floating nodes. We also note that the columns of D_{fi}, by construction, have one nonzero entry in each column. As a result, we have

$$A_f = D_f D_f^T = D_{ff} D_{ff}^T + D_{fi} D_{fi}^T = L(\mathcal{G}_f) + \Delta_f. \tag{10.8}$$

Since the row sum of the Laplacian matrix is zero, the sum of the jth row of A_f and that of $B_f(\mathcal{G})$ are both equal to $d_i(j)$ as

$$A_f \mathbf{1} = \Delta_f \mathbf{1} = [\,d_j(1)\, d_j(2) \cdots d_j(n_f)\,]^T. \tag{10.9}$$

Hence

$$A_f \mathbf{1} = -B_f \mathbf{1}. \tag{10.10}$$

Note that the **1** vectors on the left- and right-hand sides of (10.10) belong, respectively, to $\mathbf{R}^{n_f}$ and $\mathbf{R}^{n_i}$. For example, if there is only one input in the network, that is, $V(\mathcal{G}_i) = \{n\}$, we have

$$B_f = -\delta_n \quad \text{and} \quad \Delta_f = \mathbf{Diag}([\, d_n(1), d_n(2), \ldots, d_n(n-1)\,]^T).$$

10.1.3 Controllability and Observability of Input-Output Networks

Having a linear system induced by the agreement protocol, exemplified by (10.3) - (10.4), it is natural to proceed by considering its system theoretic properties. Due to the structure of (10.3) - (10.4), it is only necessary to consider either the controllability or observability properties of the system, as one implies the other. (Recall our standing assumption that the input and output nodes are the same and interact with the rest of the network identically.)

We note that controllability of the system (10.3) - (10.4) allows the input nodes to be used as a steering mechanism for the states of the floating nodes by locally injecting continuous signals into the network. Similarly, observability at the output nodes of the network would allow a mechanism by which a node can observe the state of the entire network by locally observing the states of its neighbors. However, before we are ready to put on our graph theoretic "shades," let us explore what the more traditional matrix theoretic point-of-view would offer in regards to the controllability and observability of (10.3) - (10.4).

As (10.3) - (10.4) is a linear, time-invariant system, its controllability and observability can be inferred via the Popov-Belevitch-Hautus test (see Appendix A.3). Specifically, (10.3) is uncontrollable and unobservable if and only if there exists a left eigenvector of A_f that is orthogonal to all columns of B_f, that is, if the system of linear equations

$$\nu^T A_f = \lambda \nu^T \quad \nu^T B_f = 0,$$

in the variables λ and ν, is feasible.[2] Since the system matrix A_f is symmetric, its left and right eigenvectors are the same. Hence, the necessary and sufficient condition for controllability and observability of (10.3) - (10.4) is that none of the eigenvectors of A_f should be simultaneously orthogonal to all columns of B_f, and we state this fact as a proposition.

Proposition 10.2. *Consider the input-output agreement protocol whose evolution is described by (10.3) - (10.4). This system is controllable and observable if and only if none of the eigenvectors of A_f are simultaneously orthogonal to all columns of B_f.*

---

[2] We note that controllability and observability of the pair (A_f, B_f) is equivalent to that of the pair $(-A_f, -B_f)$.

One useful consequence of Proposition 10.2 pertains to the relationship between the multiplicity of the eigenvalues of the matrix A_f and the network controllability in the SISO case. Specifically, suppose that one of the eigenvalues of A_f is not simple, that is, it has a geometric multiplicity greater than one. Since A_f is symmetric, this is also equivalent to A_f not having a set of distinct eigenvalues. For example, assume that ν_1 and ν_2 are two eigenvectors of A_f that correspond to the same eigenvalue with geometric multiplicity greater than one; moreover, assume that none of these eigenvectors are orthogonal to B_f. Then $\nu = \nu_1 + c\nu_2$ is also an eigenvector to A_f. In particular, by choosing $c = -\nu_1^T B_f / \nu_2^T B_f$, we get

$$\nu^T B_f = 0.$$

In other words, we are always able to find an eigenvector to A_f that is orthogonal to B_f when an eigenvalue has geometric multiplicity greater than one. Hence, we arrive at the following observation.

Proposition 10.3. *Consider the agreement protocol with a single input whose evolution is described by (10.3) - (10.4). If A_f has an eigenvalue with geometric multiplicity greater than one then the system is uncontrollable (and unobservable).*

Another matrix theoretic result pertaining to the controllability of (10.3), which holds in the SISO as well as the MIMO case, is as follows.

Lemma 10.4. *Given a connected graph, the system (10.3) is controllable if and only if $L(\mathcal{G})$ and A_f do not share an eigenvalue.*

Proof. We can reformulate the lemma as stating that the system is uncontrollable if and only if there exists at least one common eigenvalue between $L(\mathcal{G})$ and A_f.

Suppose that the system is uncontrollable. Then by Proposition 10.3 there exists a vector $\nu_i \in \mathbf{R}^{n_f}$ such that $A_f \nu_i = \lambda \nu_i$ for some $\lambda \in \mathbf{R}$, with

$$B_f^T \nu_i = 0.$$

Now, since

$$
\begin{bmatrix} A_f & B_f \\ B_f^T & A_i \end{bmatrix}
\begin{bmatrix} \nu_i \\ 0 \end{bmatrix}
=
\begin{bmatrix} A_f \nu_i \\ B_f^T \nu_i \end{bmatrix}
= \lambda
\begin{bmatrix} \nu_i \\ 0 \end{bmatrix},
$$

λ is also an eigenvalue to A_f, with eigenvector $[\nu_i^T, 0]^T$. The necessary condition thus follows.

It suffices to show that if $L(\mathcal{G})$ and A_f share an eigenvalue, then the system (A_f, B_f) is not controllable. Since A_f is a principal submatrix of $L(\mathcal{G})$, it can be represented as

$$A_f = P_f^T L(\mathcal{G}) P_f,$$

where $P_f = [I_{n_f}, 0]^T$ is the $n \times n_f$ matrix. Now, if A_f and $L(\mathcal{G})$ share a common eigenvalue, say λ, then the corresponding eigenvector satisfies

$$\nu = P_f \nu_f = \begin{bmatrix} \nu_f \\ 0 \end{bmatrix},$$

where ν and ν_f are, respectively, the eigenvectors of $L(\mathcal{G})$ and A_f corresponding to eigenvalue λ. Moreover, we know that

$$L(\mathcal{G})\nu = \begin{bmatrix} A_f & B_f \\ B_f^T & A_i \end{bmatrix} \begin{bmatrix} \nu_f \\ 0 \end{bmatrix} = \lambda \begin{bmatrix} \nu_f \\ 0 \end{bmatrix},$$

which gives us $B_f^T \nu_f = 0$; thus the system is uncontrollable.   ∎

10.2 GRAPH THEORETIC CONTROLLABILITY: THE SISO CASE

Our goal in this section is to make connections between the controllability and observability of the SISO agreement protocol and the structural properties of the underlying network. This is done by making a few observations and then proceeding to make tighter connections between graph theoretic and system theoretic facets of such networks. Our analysis will be provided in the context of the controllability of the agreement protocol with a single input; however, we will state the direct ramifications of this analysis in terms of the observability of (10.3) - (10.4).

First, we note that in view of the form of the input matrix B_f (10.5), the original Laplacian $L(\mathcal{G})$ is related to the Laplacian of the floating graph $L(\mathcal{G}_f)$ via

$$L(\mathcal{G}) = \begin{bmatrix} L(\mathcal{G}_f) + \Delta_f & -\delta_n \\ -\delta_n^T & d_n \end{bmatrix}, \tag{10.11}$$

where d_n denotes the degree of the input node v_n, Δ_f is the input-to-state degree matrix, and δ_n is the indicator vector for the floating graph. Since controllability for linear systems is essentially a linear algebraic statement, we proceed to build the necessary linear algebraic means of reasoning about the structure of the graph. The following observations are all part of this overall agenda.

Proposition 10.5. *If the original network $\mathcal{G}$ is connected then the system matrix A_f for the single-input network (10.3) is full rank.*

Proof. See Lemma 10.36.   ∎

Corollary 10.6. *The controlled agreement protocol (10.3) is controllable if and only if none of the eigenvectors of A_f are orthogonal to $\mathbf{1}$.*

Proof. Since, according to (10.31), $A_f\mathbf{1} = B_f\mathbf{1}$ in the single-input case, the elements of B_f correspond to the negation of the row sums of A_f, that is, $B_f = -A_f\mathbf{1}$. Thus,

$$\nu^T B_f = -\nu^T A_f\mathbf{1} = -\lambda(\nu^T\mathbf{1}).$$

By Proposition 10.5, one has $\lambda \neq 0$. Therefore, $\nu^T B_f = 0$ if and only if $\mathbf{1}^T\nu = 0$.   ∎

Proposition 10.7. *If the single-input network (10.3) is uncontrollable, then there exists an eigenvector ν of A_f such that*

$$\sum_{i\sim n}\nu(i) = 0.$$

Proof. From Corollary 10.6, when the system is uncontrollable, there exists an eigenvector ν orthogonal to $\mathbf{1}$. As

$$A_f\nu = \lambda\nu,$$

taking the inner product of both sides with $\mathbf{1}$, we obtain

$$\mathbf{1}^T(A_f\nu) = 0.$$

This is equivalent to

$$\nu^T\{L(\mathcal{G}_f) + \Delta_f\}\mathbf{1} = 0.$$

But $L(\mathcal{G}_f)\mathbf{1} = 0$ and so

$$\nu^T\Delta_f\mathbf{1} = \nu^T\delta_n = \sum_{i\sim n}\nu(i) = 0.$$

                                                                                ∎

Proposition 10.8. *Suppose that the single-input network (10.3) is uncontrollable. Then there exists an eigenvector of $L(\mathcal{G})$ that has a zero component on the index that corresponds to the input.*

Proof. Let ν be an eigenvector of A_f that is orthogonal to **1** (by Corollary 10.6 such an eigenvector exists). Attach a zero to ν; using the partitioning (10.11), we then have

$$L(\mathcal{G}) \begin{bmatrix} \nu \\ 0 \end{bmatrix} = \begin{bmatrix} A & -\delta_n \\ -\delta_n^T & d_n \end{bmatrix} \begin{bmatrix} \nu \\ 0 \end{bmatrix} = \begin{bmatrix} \lambda\nu \\ -\delta_n^T\nu \end{bmatrix},$$

where δ_n is the indicator vector for the floating nodes. From Proposition 10.7 we know that $\delta_n^T\nu = 0$. Thus

$$L(\mathcal{G}) \begin{bmatrix} \nu \\ 0 \end{bmatrix} = \lambda \begin{bmatrix} \nu \\ 0 \end{bmatrix}.$$

In the other words, $L(\mathcal{G})$ has an eigenvector with a zero on the index that corresponds to the input.   ∎

A direct consequence of Proposition 10.8 is the following:

Corollary 10.9. *Suppose that none of the eigenvectors of $L(\mathcal{G})$ have a zero component. Then the single-input network (10.3) is controllable for any choice of input node.*

10.2.1 Controllability and Graph Symmetry

The controllability of the single-input agreement protocol depends not only on the topology of the information exchange network, but also on the position of the input with respect to the graph topology. In this section, we will show that there is an intricate relationship between the controllability of (10.3) and the symmetry structure of the graph as captured by its *automorphism group*. We first need to introduce a few useful constructs.

Definition 10.10. *A permutation matrix is a $\{0,1\}$-matrix with a single nonzero element in each row and column. The permutation matrix J is called an involution if $J^2 = I$.*

A particular class of permutations, which will play a crucial role shortly, are those that characterize symmetries.

Definition 10.11. *The system (10.3) is input symmetric with respect to the input node if there exists a nonidentity permutation J such that*

$$JA_f = A_f J. \tag{10.12}$$

We call the system asymmetric if it does not admit such a permutation for any choice of input node.

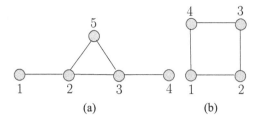

Figure 10.4: Network topologies that are input symmetric: (a) only with respect to input $\{5\}$; and (b) with respect to an input at any node

As an example, the graph in Figure 10.4(a) is input symmetric only with respect to $\{5\}$ as the input node. The graph in Figure 10.4(b) is input symmetric with respect to any single, arbitrarily chosen input node.

10.2.2 Input-symmetry via Graph Automorphisms

Before we demonstrate the utility of input symmetry in the context of network controllability, let us further refine the connection between input symmetry and graph automorphisms. Recall from Definition 10.11 that input symmetry for (10.3) - (10.4) corresponds to having

$$JA_f = A_f J,$$

where J is a nonidentity permutation. However, we know that

$$A_f = -(L(\mathcal{G}_f) + \mathcal{D}_{fl}(\mathcal{G})).$$

Thus, using the identity $L(\mathcal{G}_f) = \mathcal{D}(\mathcal{G}_f) - \mathcal{A}(\mathcal{G}_f)$, one has

$$J\left(\mathcal{D}(\mathcal{G}_f) - \mathcal{A}(\mathcal{G}_f) + \mathcal{D}_{fl}(\mathcal{G})\right) = \left(\mathcal{D}(\mathcal{G}_f) - \mathcal{A}(\mathcal{G}_f) + \mathcal{D}_{fl}(\mathcal{G})\right) J. \quad (10.13)$$

Pre- and postmultiplication by (a permutation matrix) J does not change the structure of diagonal matrices. Also, we know that all diagonal elements of $\mathcal{A}(\mathcal{G})$ are zero. As a consequence, we can rewrite (10.13) as two separate conditions,

$$J\mathcal{D}_f(\mathcal{G}) = \mathcal{D}_f(\mathcal{G})J \quad \text{and} \quad J\mathcal{A}(\mathcal{G}_f) = \mathcal{A}(\mathcal{G}_f)J, \quad (10.14)$$

with $\mathcal{D}_f(\mathcal{G}) = \mathcal{D}(\mathcal{G}_f) + \mathcal{D}_{fl}(\mathcal{G})$. The second equality in (10.14) states that J in (10.12) is in fact an automorphism of $\mathcal{G}_f$.

Proposition 10.12. *Let Ψ be the matrix associated with permutation ψ. Then*

$$\Psi \, \mathcal{D}_f(\mathcal{G}) = \mathcal{D}_f(\mathcal{G}) \, \Psi$$

if and only if, for all i,

$$d(i) + \delta_n(i) = d(\psi(i)) + \delta_n(\psi(i)).$$

In the case where ψ is an automorphism of $\mathcal{G}_f$, this condition simplifies to

$$\delta_n(i) = \delta_n(\psi(i)) \quad \text{for all } i.$$

Proof. Using the properties of permutation matrices, one has that

$$[\Psi \mathcal{D}_f(\mathcal{G})]_{ik} = \sum_t \Psi_{it} \mathcal{D}_{tk} = \begin{cases} d(k) + \delta_n(k) & \text{if } i \to k, \\ 0 & \text{otherwise,} \end{cases}$$

and

$$[\mathcal{D}_f(\mathcal{G})\Psi]_{ik} = \sum_t \mathcal{D}_{it} \Psi_{tk} = \begin{cases} d(i) + \delta_n(i) & \text{if } i \to k, \\ 0 & \text{otherwise.} \end{cases}$$

For these matrices to be equal elementwise, one needs to have $d(i) + \delta_n(i) = d(k) + \delta_n(k)$ when $\psi(i) = k$. The second statement in the proposition follows from the fact that the degree of a node remains invariant under the action of the automorphism group. ∎

The next two results follow immediately from the above discussion.

Proposition 10.13. *The networked system (10.3) is input symmetric if and only if there is a nonidentity automorphism for $\mathcal{G}_f$ such that the input indicator vector remains invariant under its action.*

Corollary 10.14. *The networked system (10.3) is input asymmetric if the automorphism group of the floating graph only contains the trivial (identity) permutation.*

10.2.3 Controllability Revisited

Although input symmetries and graph automorphisms are quite fascinating in their own rights, they are also highly relevant to the system theoretic concept of controllability. In fact, this connection is one of the main results of this chapter.

Theorem 10.15. *The system (10.3) is uncontrollable if it is input symmetric. Equivalently, the system (10.3) is uncontrollable if the floating graph admits a nonidentity automorphism for which the input indicator vector remains invariant under its action.*

Proof. If the system is input symmetric then there is a nonidentity permutation J such that

$$JA_f = A_f J. \tag{10.15}$$

Recall that, by Proposition 10.3, if the eigenvalues of A_f are not distinct then (10.3) is not controllable. We thus consider the case where all eigenvalues λ are distinct and satisfy $A_f \nu = \lambda \nu$; therefore, for all eigenvalue-eigenvector pair (λ, ν), one has

$$JA_f \nu = J(\lambda\nu).$$

Using (10.15) however,

$$A_f (J\nu) = \lambda (J\nu),$$

and as a result $J\nu$ is also an eigenvector of A_f corresponding to the eigenvalue λ. Given that λ is distinct and A_f admits a set of orthonormal eigenvectors, we conclude that for one such eigenvector ν, $\nu - J\nu$ is also an eigenvector of A_f. Moreover, $JB_f = J^T B_f = B_f$, as the elements of B_f correspond to the row sums of the matrix A_f, that is, $B_f = -A_f \mathbf{1}$. Therefore,

$$(\nu - J\nu)^T B_f = \nu^T B_f - \nu^T J^T B_f = \nu^T B_f - \nu^T B_f = 0. \tag{10.16}$$

This, on the other hand, translates to having one of the eigenvectors of A_f, namely, $\nu - J\nu$, be orthogonal to B_f. Proposition 10.3 now implies that the system (10.3) is uncontrollable, and the result follows. ∎

Theorem 10.15 states that input symmetry is a sufficient condition for uncontrollability of the system. It is instructive to examine whether the lack of such symmetry automatically leads to a controllable system.

Proposition 10.16. *Input symmetry is not a necessary condition for system uncontrollability.*

Proof. In Figure 10.5, the subgraph shown by solid lines, $\mathcal{G}_f$, is the smallest asymmetric graph in the sense that it does not admit a nonidentity automorphism. Let us augment this graph with the input node a and connect it to all vertices of $\mathcal{G}_f$. Constructing the corresponding system matrix A_f (that is, setting it equal to $-L_f(\mathcal{G})$), we have

$$-A_f = L(\mathcal{G}_f) + \mathcal{D}_{fl}(\mathcal{G}) = L(\mathcal{G}_f) + I,$$

where I is the identity matrix of proper dimensions. Consequently, A_f has the same set of eigenvectors as $L(\mathcal{G}_f)$. Since $L(\mathcal{G}_f)$ has an eigenvector

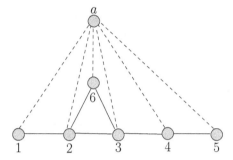

Figure 10.5: Asymmetric information topology with respect to the input node $\{a\}$. The subgraph shown by solid lines is the smallest asymmetric graph.

orthogonal to **1**, A_f also has an eigenvector that is orthogonal to **1**. Hence, the network is not controllable. Yet the system is not symmetric with respect to a.   ■

In order to demonstrate the controllability notion for the single-input agreement protocol (10.3), consider a path-like information network as shown in Figure 10.6. In this figure, the last node is chosen as the input. By Proposition 10.19, this system is controllable. The system matrices in (10.3) assume the form

$$
A_f = \begin{bmatrix} -1 & 1 & 0 \\ 1 & -2 & 1 \\ 0 & 1 & -2 \end{bmatrix} \text{ and } B_f = \begin{bmatrix} 0 \\ 0 \\ 1 \end{bmatrix}.
$$

Using (10.33), one can find a controller that drives this network from any

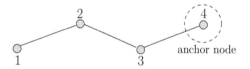

Figure 10.6: A path-like information exchange network.

initial state to an arbitrary final state. For this purpose, we chose to re-orient the planar triangle on the node set $\{1, 2, 3\}$. The maneuver time is set to be five seconds. Figure 10.7 shows the initial and final positions of the floating nodes along with their respective trajectories.

Figure 10.8 depicts the input node state trajectory as needed to perform the required maneuver. This trajectory corresponds to the speed of node

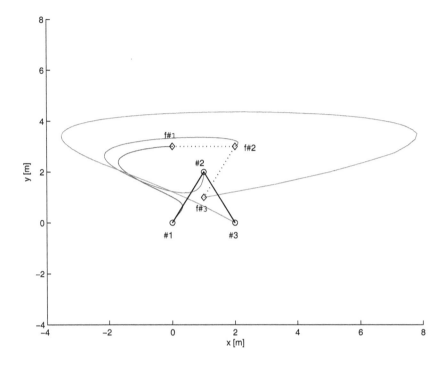

Figure 10.7: Initial and final positions of dynamic units and their respective state trajectories. The final positions are labeled f.

4 in the x, y-plane. We note that, as there are no restrictions on the input node's state trajectory, the actual implementation of this control law can become infeasible if the input node must physically assume the state that it communicates to its neighbors–particularly when the maneuver time is arbitrarily short. This observation is apparent in the previous example. In this scenario, the speed of node 4 changes rather rapidly between 20 and -50 m/s.

10.2.4 Controllability of Special Graphs

In this section, we investigate the controllability of ring and path graphs.

Proposition 10.17. *A ring graph with only one input node is never controllable.*

Proof. With only one input node in the ring graph, the floating graph $\mathcal{G}_f$ becomes the path graph with one nontrivial automorphism, its mirror image. Without loss of generality, choose the first node as the input and index

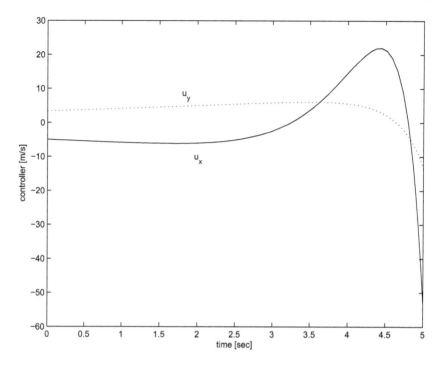

Figure 10.8: The input node's velocity acts as the control signal for a net-worked system

the remaining floating nodes by a clockwise traversing of the ring. Then the permutation $i \rightarrow n - i + 2$ for $i = 2, \ldots, n$ is an automorphism of $\mathcal{G}_f$. Moreover, the input node is connected to both node 2 and node n; hence $\delta_n = [1, 0, \ldots, 0, 1]^T$ remains invariant under the permutation. Using Proposition 10.13, we conclude that the corresponding system (10.3) is input symmetric and thus uncontrollable.   ∎

Proposition 10.18. *A path graph with odd number of vertices is uncontrollable from its center.*

Proof. Suppose that the path graph is of odd order. Then choose the middle node $(n + 1)/2$ as the input node. Note that $\psi : k \rightarrow n - k + 1$ is an automorphism for the floating subgraph. Moreover, the input node is connected to nodes $\{(n + 1)/2\} - 1$ and $\{(n + 1)/2\} + 1$, and

$$\psi\left(\frac{n + 1}{2} - 1\right) = \frac{n + 1}{2} + 1.$$

Thus

$$\delta_n = [0, \ldots, 0, 1, 1, 0, \ldots 0]^T$$

remains invariant under the permutation ψ and the system is uncontrollable.

∎

Hence, although in general the notion of input symmetry is a sufficient–yet not necessary–condition for uncontrollability of (10.3), it is necessary and sufficient for uncontrollability of the path graph.

Corollary 10.19. *A path graph endowed with the agreement protocol with a single input node is controllable if and only if it is input asymmetric.*

10.2.5 Observability from a Single Observer Post

Let us briefly summarize the main result of the previous sections in terms of the agreement protocol equipped with a single output node. In this setting, consider the system

$$\dot{x}_f(t) = -A_f\, x_f(t), \tag{10.17}$$
$$y(t) = -B_f^T\, x(t). \tag{10.18}$$

Then the following observations are direct ramification of our results on the controllability of (10.3).

Proposition 10.20. *The system (10.18) - (10.18) is unobservable if it is output symmetric. Equivalently, the system (10.18) - (10.18) is unobservable if the floating graph admits a nonidentity automorphism for which the output indicator vector remains invariant under its action.*

10.3 GRAPH THEORETIC CONTROLLABILITY: THE MIMO CASE

In this section, we examine the graph theoretic connection between network topology and controllability for the agreement protocol equipped with multiple inputs and outputs. As our subsequent discussion will show, in this case one needs an additional set of graph theoretic tools–namely, the machinery of equitable partitions–to analyze the network controllability.

The way we approach establishing this correspondence is through linear algebra, the common ground between linear system theory and equitable partitions. In particular, our approach starts from Lemma 10.4, which holds for both the SISO and the MIMO case. This is followed by showing that the matrices $L(\mathcal{G})$ and A_f are both similar to some particular block diagonal matrices. Furthermore, we show that under certain assumptions the diagonal block matrices obtained from the diagonalization of $L(\mathcal{G})$ and A_f have common diagonal block(s).

Lemma 10.21. *If a graph $\mathcal{G}$ has a nontrivial equitable partition (NEP) π with characteristic matrix P, then the corresponding adjacency matrix $A(\mathcal{G})$ is similar to a block diagonal matrix*

$$\bar{A} = \left[\begin{array}{cc} \mathcal{A}_P & 0 \\ 0 & \mathcal{A}_Q \end{array} \right],$$

where $\mathcal{A}_P$ is similar to the adjacency matrix $\hat{A} = A(\mathcal{G}/\pi)$ of the quotient graph.

Proof. Let the matrix $T = [\bar{P} \mid \bar{Q}]$ be the orthonormal matrix with respect to π, and let

$$\bar{A} = T^T A T = \left[\begin{array}{cc} \bar{P}^T A(\mathcal{G})\bar{P} & \bar{P}^T A(\mathcal{G})\bar{Q} \\ \bar{Q}^T A(\mathcal{G})\bar{P} & \bar{Q}^T A(\mathcal{G})\bar{Q} \end{array} \right]. \qquad (10.19)$$

Since $\bar{P}$ and $\bar{Q}$ have the same column spaces as P and Q, respectively, they inherit their $A(\mathcal{G})$-invariance property, that is, there exist matrices B and C such that

$$A(\mathcal{G})\bar{P} = \bar{P}B \quad \text{and} \quad A(\mathcal{G})\bar{Q} = \bar{Q}C.$$

Moreover, since the column spaces of $\bar{P}$ and $\bar{Q}$ are orthogonal complements of each other, one has

$$\bar{P}^T A(\mathcal{G})\bar{Q} = \bar{P}^T \bar{Q}C = 0$$

and

$$\bar{Q}^T A(\mathcal{G})\bar{P} = \bar{Q}^T \bar{P}B = 0.$$

In addition, by letting $D_p^2 = P^T P$, we obtain

$$\begin{aligned} \bar{P}^T A(\mathcal{G})\bar{P} &= D_P^{-1} P^T A(\mathcal{G}) P D_P^{-1} \\ &= D_P (D_P^{-2} P^T A(\mathcal{G}) P) D_P^{-1} \qquad (10.20) \\ &= D_P \hat{A} D_P^{-1}, \end{aligned}$$

and therefore the first diagonal block is similar to $\hat{A}$.   ∎

Lemma 10.22. *Let P be the characteristic matrix of an NEP in $\mathcal{G}$. Then $\mathcal{R}(P)$ is K-invariant, where K is any diagonal block matrix of the form*

$$K = \mathbf{Diag}(\underbrace{[k_1, \ldots, k_1}_{n_1}, \underbrace{k_2, \ldots, k_2}_{n_2}, \ldots, \underbrace{k_r, \ldots, k_r]^T}_{n_r}) = \mathbf{Diag}([k_i 1_{n_i}]_{i=1}^r),$$

where $k_i \in \mathbf{R}$, $n_i = card(C_i)$ is the cardinality of the ith cell and $r = |\pi|$ is the cardinality of the partition. Consequently,

$$\bar{Q}^T K \bar{P} = 0,$$

where $\bar{P} = P(P^T P)^{-\frac{1}{2}}$ and $\bar{Q}$ is chosen in such a way that $T = [\bar{P} \mid \bar{Q}]$ is an orthonormal matrix.

Proof. We note that

$$P = \begin{bmatrix} P_1 \\ P_2 \\ \vdots \\ P_r \end{bmatrix} = \begin{bmatrix} p_1 & p_2 & \cdots & p_r \end{bmatrix},$$

where $P_i \in \mathbf{R}^{n_i \times r}$ is a row block which has 1s in column i and 0s elsewhere. On the other hand, p_i is a characteristic vector representing C_i, which has 1s in the positions associated with C_i and 0s elsewhere.

Recalling the example given in (2.20), with

$$P = \left[\begin{array}{c|c|cc} 1 & 0 & 0 & 0 \\ \hline 0 & 1 & 0 & 0 \\ 0 & 1 & 0 & 0 \\ \hline 0 & 0 & 1 & 0 \\ 0 & 0 & 0 & 1 \end{array} \right], \tag{10.21}$$

we can then find

$$P_2 = \begin{bmatrix} 0 & 1 & 0 & 0 \\ 0 & 1 & 0 & 0 \end{bmatrix},$$

while $p_2 = [0, 1, 1, 0, 0]^T$. A little algebra reveals that

$$KP = \begin{bmatrix} k_1 P_1 \\ k_2 P_2 \\ \vdots \\ k_r P_r \end{bmatrix} = \begin{bmatrix} k_1 p_1 & k_2 p_2 & \cdots & k_r p_r \end{bmatrix} = P\hat{K},$$

where $\hat{K} = \mathbf{Diag}([k_1, k_2, \ldots, k_r]^T) = \mathbf{Diag}([k_i]_{i=1}^r)$; hence $\mathcal{R}(P)$ is K-invariant. Since $\mathcal{R}(\bar{Q}) = \mathcal{R}(P)^\perp$, it is K-invariant as well, and

$$\bar{Q}^T K \bar{P} = \bar{Q}^T \bar{P} \hat{K} = 0.$$

■

By the definition of equitable partitions, the subgraph induced by a cell is regular and every node in the same cell has the same number of neighbors outside the cell. Therefore, the nodes belonging to the same cell have the same degree, and thus by Lemma 10.22, $\mathcal{R}(\bar{Q})$ and $\mathcal{R}(P)$ are Δ-invariant, where Δ is the degree matrix given by

$$\Delta = \mathbf{Diag}([d_i \mathbf{1}_{n_i}]_{i=1}^r),$$

with $d_i \in \mathbf{R}$ denoting the degree of the nodes in cell i.

Since the graph Laplacian satisfies $L(\mathcal{G}) = \Delta(\mathcal{G}) - A(\mathcal{G})$, Lemmas 10.21 and 10.22 imply that $\mathcal{R}(\bar{Q})$ and $\mathcal{R}(P)$ are $L(\mathcal{G})$-invariant. We have thus obtained the following corollary.

Corollary 10.23. *Given the same condition as in Lemma 10.21, $L(\mathcal{G})$ is similar to a diagonal block matrix*

$$T^T L(\mathcal{G}) T = \begin{bmatrix} L_P & 0 \\ 0 & L_Q \end{bmatrix}, \tag{10.22}$$

where $L_P = \bar{P}^T L(\mathcal{G}) \bar{P}$ and $L_Q = \bar{Q}^T L(\mathcal{G}) \bar{Q}$, and $T = [\bar{P} \mid \bar{Q}]$ defines an orthonormal basis for $\mathbf{R}^n$ with respect to π.

As (10.22) defines a similarity transformation, it follows that L_P and L_Q carry all the spectral information of $L(\mathcal{G})$, that is, they share the same eigenvalues as $L(\mathcal{G})$. And, as the input-output designation in the agreement protocol partitions the graph Laplacian as

$$L(\mathcal{G}) = \begin{bmatrix} A_f & B_f \\ B_f^T & A_i \end{bmatrix},$$

transformations similar to (10.22) can also be found for A_f in the presence of NEPs in the floating graph $\mathcal{G}_f$.

Corollary 10.24. *Let $\mathcal{G}_f$ be a floating graph, and let A_f be the submatrix of $L(\mathcal{G})$ corresponding to $\mathcal{G}_f$. If there is a NEP π_f in $\mathcal{G}_f$ and a π in $\mathcal{G}$, such that all the nontrivial cells in π_f are also cells in π, there exists an orthonormal matrix T_f such that*

$$\bar{A}_f = T_f^T A_f T_f = \begin{bmatrix} A_{fP} & 0 \\ 0 & A_{fQ} \end{bmatrix}. \tag{10.23}$$

Proof. Let $\bar{P}_f = P_f(P_f^T P_f)^{\frac{1}{2}}$, where P_f is the characteristic matrix for π_f. Moreover, let $\bar{Q}_f$ be defined on an orthonormal basis of $\mathcal{R}(P_f)^\perp$. In this way, we obtain an orthonormal basis for $\mathbf{R}^{n_f}$ with respect to π_f. Moreover, by (10.7), $A_f = \mathcal{D}_f^l(\mathcal{G}) + L(\mathcal{G}_f)$, where $L(\mathcal{G}_f)$ denotes the Laplacian matrix of $\mathcal{G}_f$ while $\mathcal{D}_f^l$ is the diagonal input-to-state degree matrix defined in (10.6). Since all the nontrivial cells in π_f are also cells in π, $\mathcal{D}_f$ satisfies the condition in Lemma 10.22, that is, nodes from an identical cell in π_f have the same degree. Hence, by Lemma 10.21 and Lemma 10.22, $\mathcal{R}(\bar{P}_f)$ and $\mathcal{R}(\bar{Q}_f)$ are A_f-invariant, and consequently

$$\bar{A}_f = T_f^T A_f T_f = \begin{bmatrix} A_{fP} & 0 \\ 0 & A_{fQ} \end{bmatrix}, \tag{10.24}$$

where $T_f = [\bar{P}_f \mid \bar{Q}_f]$, $A_{fP} = \bar{P}_f^T A_f \bar{P}_f$ and $A_{fQ} = \bar{Q}_f^T A_f \bar{Q}_f$. ∎

Again, the diagonal blocks of $\bar{A}_f$ contain the complete spectral information of A_f. We are now in the position to prove the main result of this section.

> **Theorem 10.25.** *Given a connected graph $\mathcal{G}$ and the induced floating graph $\mathcal{G}_f$, the system (10.3) is not controllable if there exist NEPs on $\mathcal{G}$ and $\mathcal{G}_f$, say π and π_f, such that all nontrivial cells of π are contained in π_f, that is, for all $C_i \in \pi \backslash \pi_f$, one has $\text{card}(C_i) = 1$.*

Proof. In Corollary 10.23, we saw that $L(\mathcal{G})$ and A_f are similar to some block diagonal matrices. Here we further expand on the relationship between these matrices.

Assume that $\pi \cap \pi_f = \{C_1, C_2, \ldots, C_{r_1}\}$. According to the underlying condition, one has $\text{card}(C_i) \geq 2$, $i = 1, 2, \ldots, r_1$. Without loss of generality, we can index the nodes in such a way that the nontrivial cells comprise the first n_1 nodes,[3] where

$$n_1 = \sum_{i=1}^{r_1} \text{card}(C_i) \leq n_f < n.$$

As all the nontrivial cells of π are in π_f, their characteristic matrices have

---

[3] We have introduced n_1 for notational convenience. It is easy to verify that $n_1 - r_1 = n - r = n_f - r_f$.

similar structures

$$P = \begin{bmatrix} P_1 & 0 \\ 0 & I_{n-n_1} \end{bmatrix}_{n \times r} \quad \text{and} \quad P_f = \begin{bmatrix} P_1 & 0 \\ 0 & I_{n_f-n_1} \end{bmatrix}_{n_f \times r_f},$$

where P_1 is an $n_1 \times r_1$ matrix containing the nontrivial part of the characteristic matrices. Since $\bar{P}$ and $\bar{P}_f$ are the normalizations of P and P_f, respectively, they have the same block structures. Consequently $\bar{Q}$ and $\bar{Q}_f$, the matrices containing orthonormal basis of $\mathcal{R}(P)$ and $\mathcal{R}(P_f)$, have the structures

$$\bar{Q} = \begin{bmatrix} Q_1 \\ 0 \end{bmatrix}_{n \times (n_1-r_1)} \quad \text{and} \quad \bar{Q}_f = \begin{bmatrix} Q_1 \\ 0 \end{bmatrix}_{n_f \times (n_1-r_1)}$$

where Q_1 is an $n_1 \times (n_1 - r_1)$ matrix that satisfies

$$Q_1^T P_1 = 0.$$

We observe that $\bar{Q}_f$ is different from $\bar{Q}$ only by $n - n_f$ rows of zeros. In other words, the special structures of $\bar{Q}$ and $\bar{Q}_f$ lead to the relationship

$$Q_f = R^T Q,$$

where $R = [I_{n_f}, 0]^T$. Now, recalling the definitions of L_Q and L_{fQ} from (10.22) and (10.23) leads us to

$$L_Q = \bar{Q}^T L(\mathcal{G}) \bar{Q} = \bar{Q}_f^T R^T L(\mathcal{G}) R \bar{Q}_f = \bar{Q}_f^T A_f \bar{Q}_f = L_{fQ}. \qquad (10.25)$$

Therefore L_f and $L(\mathcal{G})$ have the same eigenvalues associated with L_Q; hence by Lemma 10.4, the system is not controllable. ∎

Theorem 10.25 provides a method to identify uncontrollable multiagent systems in the presence of multiple inputs. In an uncontrollable multiagent system, vertices in the same cell of an NEP, satisfying the condition in Theorem 10.25, are *not distinguishable from the input nodes' point of view.* In other words, agents belonging to a shared cell among π and π_f, when identically initialized, remain undistinguished to the input nodes throughout the system evolution. Moreover, the controllable subspace for this multiagent system can be obtained by collapsing all the nodes in the same cell into a single "meta-agent."

Two immediate ramifications of the above theorem are as follows.

Corollary 10.26. *Given a connected graph $\mathcal{G}$ with the induced floating node graph $\mathcal{G}_f$, a necessary condition for (10.3) to be controllable is that no NEPs π and π_f, on $\mathcal{G}$ and $\mathcal{G}_f$, respectively, share a nontrivial cell.*

Corollary 10.27. *If $\mathcal{G}$ is disconnected, a necessary condition for (10.3) to be controllable is that all of its connected components are controllable.*

Example 10.28. *In Figure 2.12, if we choose node 5 as the leader, the symmetric pair (2,3) in the floating graph renders the network uncontrollable. The dimension of the controllable subspace is three, while there are four nodes in the follower group. This result can also be interpreted via Theorem 10.25, since the corresponding automorphisms introduce equitable partitions.*

Example 10.29. *We have shown in Figure 2.11 that the Peterson graph has two NEPs. One is introduced by the automorphism group and the other, π_2, by the equal distance partition. Based on π_2, if we choose node 1 as the input node, the network ends up with a controllable subspace of dimension two. Since there are four orbits in the automorphism group,[4] this dimension pertains to the two-cell equal distance partitions.[5]*

Example 10.30. *This example is a modified graph based on the Peterson graph. In Figure 10.9, we add another node (11) connected to the nodes in the set $\{3,4,7,8,9,10\}$ as the second leader in addition to node 1. In this network, there is an equal distance partition with four cells, $\{1\}$, $\{2,5,6\}$ $\{3,4,7,8,9,10\}$ and $\{11\}$. In this case, the dimension of the controllable subspace is still two, which is consistent with Example 10.29.*

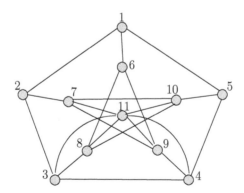

Figure 10.9: A two-leader network based on the Peterson graph

---

[4]They are $\{2,5,6\}$, $\{7,10\}$, $\{8,9\}$, $\{3,4\}$
[5]They are $\{2,5,6\}$ and $\{3,4,7,8,9,10\}$.

10.4 AGREEMENT REACHABILITY

Equipped with a collection of controllability results, we now shift our attention to the issue of whether we can steer the system to agreement subspace, for which we first need some additional notation.

10.4.1 Representation

Another way to construct the system matrices A_f and B_f in (10.3) - (10.4) is from the Laplacian of the original graph via

$$A_f = P_f^T L(\mathcal{G}) P_f \quad \text{and} \quad B_f = P_f^T L(\mathcal{G}) T_{fl}, \qquad (10.26)$$

where $P_f \in \mathbf{R}^{n \times n_f}$ is constructed by eliminating the columns of the $n \times n$ identity matrix that correspond to the input nodes, and $T_{fl} \in \mathbf{R}^{n \times n_l}$ is formed by grouping these eliminated columns in a new matrix.

For example, in Figure 10.2, these matrices assume the form

$$P_f = \begin{bmatrix} I_4 \\ 0_{2 \times 4} \end{bmatrix} \quad \text{and} \quad T_{fl} = \begin{bmatrix} 0_{4 \times 2} \\ I_2 \end{bmatrix}.$$

Proposition 10.31. *If a single node is chosen as the input node, one has*

$$T_{fl} = (I_n - \tilde{P})\mathbf{1}_n \text{ and } l_{fl} = -A_f \mathbf{1}_{n_f}$$

in (10.26), where $\tilde{P} = [P_f\ 0_{n \times n_l}]$ is the $n \times n$ square matrix obtained by expanding P_f with a zero block of proper dimensions.

Proof. The first equality follows directly from the definitions of P_f and T_{fl}. Without loss of generality, assume that the last node is the input node; then $[P_f\ T_{fl}] = I_n$. Multiplying both sides by $\mathbf{1}_n$ and noting that $\tilde{P}\,\mathbf{1}_n = P_f\mathbf{1}_{n_f}$, one has $T_{fl} = (I_n - \tilde{P})\mathbf{1}_n$.

Moreover,

$$\ell_{fl} = P_f^T L(\mathcal{G})\{(I - \tilde{P})\mathbf{1}_n\} = P_f^T L(\mathcal{G})\mathbf{1}_n - P_f^T L(\mathcal{G})P_f\mathbf{1}_{n_f}. \quad (10.27)$$

The first term on the right-hand side of the equality is zero as $\mathbf{1}$ belongs to the null space of $L(\mathcal{G})$; the second term is simply $A_f \mathbf{1}$.   ∎

Alternatively, for the case when the exogenous signal is constant, the dynamics can be rewritten as

$$\begin{bmatrix} \dot{x}_f(t) \\ \dot{u}(t) \end{bmatrix} = - \begin{bmatrix} A_f & B_f \\ 0 & 0 \end{bmatrix} \begin{bmatrix} x_f(t) \\ u(t) \end{bmatrix}. \qquad (10.28)$$

This corresponds to zeroing-out the rows of the original graph Laplacian associated with the leader. Zeroing-out a row of a matrix can be accomplished via a reduced identity matrix Q_r, with zeros at the diagonal elements that correspond to the leaders and all other diagonal elements being kept as one. In this case

$$\begin{bmatrix} A_f & B_f \\ 0 & 0 \end{bmatrix} = Q_r L(\mathcal{G}), \tag{10.29}$$

where

$$Q_r = \begin{bmatrix} I_{n_f} & 0 \\ 0 & 0 \end{bmatrix},$$

and all the zero matrices are of appropriate dimensions.

10.4.2 Steering to the Agreement Subspace

First, we examine whether we can steer the controlled agreement protocol to the agreement subspace, **span**$\{1\}$, when the exogenous signal is constant, that is, $x_i = c$, for all $i \in V_i$ and $c \in \mathbf{R}$ is a constant. As shown in (10.29), in this case the controlled agreement can be represented as

$$\dot{x}(t) = -Q_r L(\mathcal{G})\, x(t) = -L_r(\mathcal{G})\, x(t), \tag{10.30}$$

where Q_r is the reduced identity matrix and $L_r(\mathcal{G}) = Q_r L(\mathcal{G})$ is the reduced Laplacian matrix.

Let us now examine the convergence properties of (10.30) with respect to **span**$\{1\}$. Define $\zeta(t)$ as the projection of the followers' state $x_f(t)$ onto the subspace orthogonal to the agreement subspace **span**$\{1\}$. This subspace is denoted by $\mathbf{1}^\perp$ and it is sometimes referred to as the *disagreement* subspace.

One can then model the disagreement dynamics as

$$\dot{\zeta}(t) = -L_r(\mathcal{G})\, \zeta(t). \tag{10.31}$$

Choosing a standard quadratic Lyapunov function for (10.31),

$$V(\zeta(t)) = \frac{1}{2}\, \zeta(t)^T \zeta(t),$$

reveals that its time rate of change assumes the form

$$\dot{V}(\zeta(t)) = -\zeta(t)^T \overline{L}_r(\mathcal{G})\, \zeta(t),$$

where $\overline{L}_r(\mathcal{G}) = (1/2)\,[\, L_r(\mathcal{G}) + L_r(\mathcal{G})^T \,]$.

Proposition 10.32. *The agreement subspace is reachable for the controlled agreement protocol.*

Proof. Since $\dot{V}(\zeta) < 0$ for all $\zeta \neq 0$ and $Q_r L(\mathcal{G}) \mathbf{1} = 0$, for any input nodes, the agreement subspace remains a globally attractive subspace for (10.30). ∎

Proposition 10.33. *In the case of a single input node, the matrix $L_r(\mathcal{G})$ has a real spectrum and the same number of zero and positive eigenvalues as $L(\mathcal{G})$.*

Proof. Let $E = \mathbf{11}^T$ denote the matrix of all ones. Since $EL(\mathcal{G}) = 0$ and $Q_r L(\mathcal{G}) = L_r(\mathcal{G})$, we have that

$$(Q_r + E) L(\mathcal{G}) = L_r(\mathcal{G}).$$

Hence $L_r(\mathcal{G})$ is a product of a positive definite matrix, namely, $Q_r + E$, and the symmetric matrix $L(\mathcal{G})$. As a consequence, $L_r(\mathcal{G})$ is diagonalizable and has a real spectrum, it has the same number of zero and positive eigenvalues as $L(\mathcal{G})$. ∎

10.4.3 Rate of Convergence

In previous sections, we discussed the controllability properties of the controlled agreement dynamics in terms of the symmetry structure of the network. When the resulting system is controllable, the nodes can reach agreement arbitrary fast.

Proposition 10.34. *A controllable agreement dynamics can reach the agreement subspace arbitrarily fast.*

Proof. The (invertible) controllability Gramian for the controlled agreement dynamics is defined as

$$W_a(t_0, t_f) = \int_{t_0}^{t_f} e^{sA_f} B_f B_f^T e^{sA_f^T} \, ds. \qquad (10.32)$$

For any $t_f > t_0$, the input node can then transmit the signal

$$u(t) = B_f^T e^{A_f^T (t_f - t_0)} W_a(t_0, t_f)^{-1} \left(x_f - e^{A_f (t_f - t_0)} x_0 \right), \qquad (10.33)$$

to its neighbors; in (10.33) x_0 and x_f are the initial and final states for the floating nodes, and t_0 and t_f are the prespecified initial and final maneuver times. ∎

Next, let us examine the convergence properties of the input network with an input node that transmits a constant signal (10.30). In this venue, define the quantity

$$\mu_2(L_r(\mathcal{G})) = \min_{\substack{\zeta \neq 0 \\ \zeta \perp \mathbf{1}}} \frac{\zeta^T \overline{L}_r(\mathcal{G}) \zeta}{\zeta^T \zeta}. \qquad (10.34)$$

Proposition 10.35. *The rate of convergence of the disagreement dynamics (10.31) is bounded by $\mu_2(L_r(\mathcal{G}))$ and $\lambda_2(L(\mathcal{G}))$, when the input node transmits a constant signal.*

Proof. Using the variational characterization of the second smallest eigenvalue of graph Laplacian, we have

$$\lambda_2(L(\mathcal{G})) = \min_{\substack{\zeta \neq 0 \\ \zeta \perp \mathbf{1}}} \frac{\zeta^T L(\mathcal{G}) \zeta}{\zeta^T \zeta}$$

$$\leq \min_{\substack{\zeta \neq 0 \\ \zeta \perp \mathbf{1} \\ \zeta = Q\beta}} \frac{\zeta^T L(\mathcal{G}) \zeta}{\zeta^T \zeta}$$

$$= \min_{\substack{Q\beta \neq 0 \\ Q\beta \perp \mathbf{1}}} \frac{\beta^T Q L(\mathcal{G}) Q\beta}{\beta^T Q\beta}$$

$$= \min_{\substack{Q\beta \neq 0 \\ Q\beta \perp \mathbf{1}}} \frac{\beta^T Q \left\{ \frac{1}{2}(Q L(\mathcal{G}) + L(\mathcal{G})Q) \right\} Q\beta}{\beta^T Q\beta}$$

$$= \min_{\substack{Q\beta \neq 0 \\ Q\beta \perp \mathbf{1}}} \frac{\beta^T Q \left(\frac{1}{2}(L_r(\mathcal{G}) + L_r(\mathcal{G})^T) \right) Q\beta}{\beta^T Q\beta}$$

$$= \min_{\substack{\zeta \neq 0 \\ \zeta \perp \mathbf{1}}} \frac{\zeta^T \overline{L}_r(\mathcal{G}) \zeta}{\zeta^T \zeta} = \mu_2(\overline{L}_r(\mathcal{G})),$$

where β is an arbitrary vector with the appropriate dimension, Q is the matrix introduced in (10.29), and $Q^2 = Q$. In the last variational statement, we observe that ζ should have a special structure, that is, $\zeta = Q\beta$ (a zero at the row corresponding to the leader). An examination of the error dynamics suggests that such a structure always exists. As the input node does not update its value, the difference between the input node's state and the agreement value is always zero. Thus, with respect to the disagreement dynamics (10.31),

$$\dot{V}(\zeta) = -\zeta^T \overline{L}_r(\mathcal{G})\, \zeta \le -\mu_2(L_r(\mathcal{G}))\zeta^T\zeta \le -\lambda_2(L(\mathcal{G}))\, \zeta^T\zeta.$$

■

It is intuitive that a highly connected input node (or anchor) will result in faster convergence to the agreement subspace. However, a highly connected anchor also increases the chances that a symmetric graph (with respect to the anchor) emerges. A limiting case for this latter scenario is the complete graph. In such a graph, $n-1$ anchors are needed to make the corresponding system controllable. This requirement is of course not generally desirable as it means that the anchor group includes all nodes except one! The complete graph is in fact the the "worst case" configuration for its controllability properties.

Generally, at most $n-1$ anchors are needed to make any information exchange network controllable. In the meantime, a path graph with an anchor at one end is controllable. Thus it is possible to make a complete graph controllable by keeping the links on the longest path between an anchor and all other nodes, deleting the unnecessary information exchange links to break its inherent symmetry. This procedure is not always feasible; for example, a star graph is not amenable to such graphical alterations.

10.5 NETWORK FEEDBACK

Once one starts thinking of networks as dynamical systems, by viewing individual nodes as inputs, it becomes imperative to investigate how this point of view can be used to make the network perform useful things.

Loosely speaking, one can think of the problems under investigation in this chapter as variants of the "autonomous sheep-herding" problem. In other words, *how should the herding dogs move in order to maneuver the herd in the desired way?* Based on the previous sections, we can select the leaders (herding dogs)[6] as inputs to the network. Once such a set of leaders is selected, we will apply optimal control techniques for driving the system between specified positions. In fact, it will be shown that this problem is equivalent to the problem of driving an invertible linear system between

---

[6]In this section, we use "leaders" for the input nodes and "followers" for the floating nodes, due to the historic robotics context in which this notation arose.

quasi-static equilibrium points.[7]

As before, we will let the state of an individual agent be described by a vector in $\mathbf{R}^n$. Moreover, under the linear agreement protocol, the dynamics along each dimension can be decoupled, which allows us to analyze the performance of our proposed control methods along a single dimension.[8] In other words, let $x_i \in \mathbf{R}$, $i = 1, 2, \ldots, n$, be the state of the ith agent, and let $x(t) = [x_1(t), x_2(t), \ldots, x_n(t)]^T$ be the state vector of the group of agents, where n is the total number of agents. As we have seen, the agreement protocol will solve the rendezvous problem (drive all agents to a common point) as long as the network is connected. We will use this as the basic coordination scheme executed by the follower agents. The reason for this is not that we are interested in solving the rendezvous problem per se, but rather that it provides some cohesion among the follower agents.

Given the partition of the network into leaders and followers, as specified in (10.2), we have the following refinement of Propositions 10.5 and 10.33.

Lemma 10.36. *If the graph is connected, then the matrix A_f is positive definite.*

Proof. We know that $L(\mathcal{G})$ is positive semidefinite. In addition, if the graph $\mathcal{G}$ is connected, we have that $\mathcal{N}(L(\mathcal{G})) = \mathbf{span}\{\mathbf{1}\}$. Moreover, since

$$x_f^T A_f x_f = [x_f^T \ 0] L(\mathcal{G}) \begin{bmatrix} x_f \\ 0 \end{bmatrix}$$

and $[x_f^T \ 0]^T \notin \mathcal{N}(L(\mathcal{G}))$, when $x_f \neq 0$, we have that

$$[x_f^T \ 0] L(\mathcal{G}) \begin{bmatrix} x_f \\ 0 \end{bmatrix} > 0 \quad \text{for all nonzero } x_f \in \mathbf{R}^{n_f},$$

and the statement of the lemma follows.   ∎

As we have seen repeatedly throughout this book, the agreement protocol works because it averages the contribution from all neighbors in a distributed way. As such, it seems like a natural starting point when determining the movements of the followers, that is, by letting

$$\dot{x}_f(t) = -A_f x_f(t) - B_f x_l(t). \tag{10.35}$$

---

[7]A process is called quasi-static when it follows a succession of equilibrium states. In such a process, a sufficiently slow transition of a thermodynamic system from one equilibrium state to another occurs such that at every moment in time the state of the system is close to an equilibrium state.

[8]See Chapter 3.

Theorem 10.37. *Given fixed leader positions x_l, the quasi-static equilibrium point under the follower dynamics in (10.35) is*

$$x_f = -A_f^{-1}B_f x_l, \tag{10.36}$$

which is globally asymptotically stable.

Proof. From the previous lemma, we know that L_f is invertible and hence (10.36) is well defined. Hence the equilibrium point is unique. Moreover, since L_f is positive definite, this equilibrium point is in fact globally asymptotically stable. ∎

10.6 OPTIMAL CONTROL

Since we will be using the leader positions as the inputs to the network, for the sake of notational convenience (and to harmonize with the standard notation in the controls literature), we will equate x_f with x and x_l with u throughout this section. Moreover, we will identify A with $-A_f$ and B with $-B_f$. Using this notation, the leader-follower system can be rewritten as

$$\dot{x}(t) = Ax(t) + Bu(t). \tag{10.37}$$

Moreover, since the leaders are unconstrained in their motion, we let

$$\dot{u}(t) = v(t),$$

where $v(t)$ is the control input.

For a fixed u, the quasi-static equilibrium to (10.37) is given by

$$x^* = -A^{-1}Bu. \tag{10.38}$$

An example of letting a single leader agent drive three followers close (in the least squares sense) to a number of intermediary targets is shown in Figure 10.10.

The problem under consideration here is the quasi-static equilibrium process problem, that is, the problem of transferring (x, u) from an initial point satisfying (10.38) to a final point also satisfying (10.38). Moreover, we want to achieve this in a finite amount of time, and we define our performance function as follows:

$$J = \frac{1}{2} \int_0^T \left(\dot{x}(t)^T P \dot{x}(t) + \dot{u}(t)^T Q \dot{u}(t) \right) dt, \tag{10.39}$$

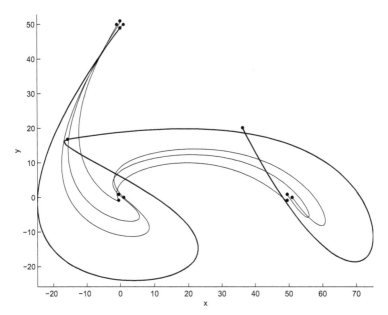

Figure 10.10: Starting at a formation close to $(0, 50)$ at time $t = 0$, the leader (thick curve) maneuvers the followers to the new positions close to $(0, 0)$ at $t = 5$, and close to $(50, 0)$ at $t = 10$. This is done while expending the smallest possible control energy.

where P and Q are assumed to be positive semidefinite and positive definite, respectively. The optimal control problem can now be formulated as

$$\min_{v} J, \qquad (10.40)$$

under the constraints that $\dot{x}(t) = Ax(t) + Bu(t)$, with the boundary conditions $x(0) = -A^{-1}Bu_0$ and $x(T) = -A^{-1}Bu_T$, given u_0 and u_T.

It should be noted that we do not, in fact, need for the network to be controllable in this particular case, even though, for the general point-to-point transfer problem, we do need controllability. To see this assume, without loss of generality, that we have a (partial) Kalman decomposition,

$$\dot{x} = \begin{bmatrix} \dot{x}_c(t) \\ \dot{x}_u(t) \end{bmatrix} = \begin{bmatrix} A_{11} & A_{12} \\ 0 & A_{22} \end{bmatrix} \begin{bmatrix} x_c(t) \\ x_u(t) \end{bmatrix} + \begin{bmatrix} B_1 \\ 0 \end{bmatrix} u(t),$$

where x_c is controllable and x_u is uncontrollable.[9] Now, given a fixed u^e, where the superscript e denotes equilibrium, the quasi-static equilibrium is

---

[9] We referred to this as partial Kalman decomposition since there is no observation matrix involved.

given by

$$0 = \begin{bmatrix} A_{11}x_c^e + A_{12}x_u^e + B_1 u^e \\ A_{22}x_u^e \end{bmatrix}.$$

Since A is invertible (and hence also A_{22}), this means that $x_u^e = 0$. Hence the quasi-static process will simply drive $x_u(0) = 0$ to $x_u(T) = 0$ and we can restrict our attention to the nontrivial part of the system, namely,

$$\dot{x}_c(t) = A_{11}x_c(t) + A_{12}x_u(t) + B_1 u(t).$$

But, since $x_u(t) = 0$ on the interval $[0, T]$, we only have

$$\dot{x}_c(t) = A_{11}x_c(t) + B_1 u(t),$$

and since (A_{11}, B_1) is a controllable pair, point-to-point transfer is always possible.

Now, in order to solve the optimal control problem,[10] we first form the Hamiltonian

$$
\begin{aligned}
\mathcal{H} &= \frac{1}{2}(\dot{x}(t)^T P \dot{x}(t) + \dot{u}(t)^T Q \dot{u}(t)) + \lambda(t)^T (Ax(t) + Bu(t)) + \mu(t)^T v(t) \\
&= \frac{1}{2}\Big[x(t)^T A^T P A x(t) + 2x(t)^T A^T P B u(t) + u(t)^T B^T P B u(t) \\
&\quad + v(t)^T Q v(t) \Big] + \lambda(t)^T (Ax(t) + Bu(t)) + \mu(t)^T v(t),
\end{aligned}
$$

$$(10.41)$$

where λ and μ are the co-states. The first-order necessary optimality condition then gives

$$\frac{\partial \mathcal{H}}{\partial v} = v^T Q + \mu^T = 0 \Rightarrow v = -Q^{-1}\mu,$$

$$\dot{\lambda}(t) = -\left(\frac{\partial \mathcal{H}}{\partial x}\right)^T = -A^T P A x(t) - A^T P B u(t) - A^T \lambda(t), \quad (10.42)$$

$$\dot{\mu}(t) = -\left(\frac{\partial \mathcal{H}}{\partial u}\right)^T = -B^T P A x(t) - B^T P B u(t) - B^T \lambda(t).$$

In other words, by letting $z(t) = [x(t)^T, u(t)^T, \lambda(t)^T, \mu(t)^T]^T$, we obtain the Hamiltonian system

$$\dot{z}(t) = M z(t), \qquad (10.43)$$

---

[10]See Notes and References for pointers to references on optimal control.

where

$$
M = \begin{bmatrix} A & B & 0 & 0 \\ 0 & 0 & 0 & -Q^{-1} \\ -A^T P A & -A^T P B & -A^T & 0 \\ -B^T P A & -B^T P B & -B^T & 0 \end{bmatrix} ;
$$

it now suffices to find the initial conditions on the co-states. For this, we let the initial state be given by

$$
z_0 = [x_0^T, u_0^T, \lambda_0^T, \mu_0^T]^T ,
$$

where $x_0 = -A^{-1} B u_0$, and λ_0 and μ_0 are unknown parameters that should be properly chosen. In fact, the problem is exactly that of selecting λ_0 and μ_0 in such a way that, through this choice, we get

$$
x(T) = -A^{-1} B u_T = x_T .
$$

In order to achieve this, we partition the matrix exponential in the following way

$$
e^{MT} = \begin{bmatrix} \phi_{xx} & \phi_{xu} & \phi_{x\lambda} & \phi_{x\mu} \\ \phi_{ux} & \phi_{uu} & \phi_{u\lambda} & \phi_{u\mu} \\ \phi_{\lambda x} & \phi_{\lambda u} & \phi_{\lambda\lambda} & \phi_{\lambda\mu} \\ \phi_{\mu x} & \phi_{\mu u} & \phi_{\mu\lambda} & \phi_{\mu\mu} \end{bmatrix} . \tag{10.44}
$$

We can find the initial conditions of the co-states by solving

$$
\begin{bmatrix} x_T \\ u_T \end{bmatrix} = \begin{bmatrix} \phi_{xx} & \phi_{xu} & \phi_{x\lambda} & \phi_{x\mu} \\ \phi_{ux} & \phi_{uu} & \phi_{u\lambda} & \phi_{u\mu} \end{bmatrix} \begin{bmatrix} x_0 \\ u_0 \\ \lambda_0 \\ \mu_0 \end{bmatrix} .
$$

Now, let

$$
\Phi_1 = \begin{bmatrix} \phi_{xx} & \phi_{xu} \\ \phi_{ux} & \phi_{uu} \end{bmatrix} \quad \text{and} \quad \Phi_2 = \begin{bmatrix} \phi_{x\lambda} & \phi_{x\mu} \\ \phi_{u\lambda} & \phi_{u\mu} \end{bmatrix} ,
$$

which gives

$$
\begin{bmatrix} \lambda_0 \\ \mu_0 \end{bmatrix} = \Phi_2^{-1} \left(\begin{bmatrix} x_T \\ u_T \end{bmatrix} - \Phi_1 \begin{bmatrix} x_0 \\ u_0 \end{bmatrix} \right) .
$$

Since we are considering a quasi-static process, we have

$$
x_0 = -A^{-1} B u_0 \quad \text{and} \quad x_T = -A^{-1} B u_T ,
$$

and hence the initial conditions of the co-states become

$$\begin{bmatrix} \lambda_0 \\ \mu_0 \end{bmatrix} = -\Phi_2^{-1}\Psi \begin{bmatrix} u_0 \\ u_T \end{bmatrix}, \quad \Psi = \begin{bmatrix} \phi_{xx}A^{-1}B - \phi_{xu} & -A^{-1}B \\ \phi_{ux}A^{-1}B - \phi_{uu} & I \end{bmatrix}.$$

The invertibility of Φ_2 follows directly from the fact that this particular point-to-point transfer problem always has a unique solution.

As an example, Figure 10.11 shows a scalar quasi-static process, where the dynamics of the system is given by $\dot{x}(t) = -x(t) - u(t)$, and where P and Q are both set to 1. The system starts from $x_0 = 1, u_0 = -1$ and the desired final position is $x_T = -1, u_T = 1$. The dash-dotted line shows the subspace $\{(x, u) \mid x = -A^{-1}Bu\}$, while the solid line is the actual trajectory of the system under the optimal control law with $T = 2$.

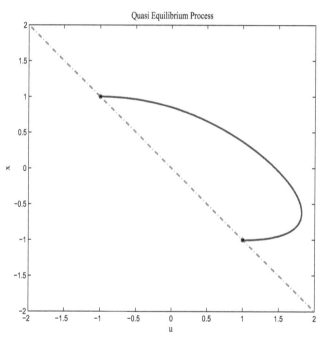

Figure 10.11: A quasi-static process for the system $\dot{x} = -x - u$, with $P = Q = 1, x_0 = 1, u_0 = -1, x_T = -1, u_T = 1$

10.6.1 Leader-Follower Herding

By applying the optimal control laws for quasi-static equilibrium processes, we can now move the leader agents in such a way that the followers are moved, in finite time, between desired positions. In Figure 10.12, snap-shots of a herding process are shown where the leaders (black) move the followers (white) from an initial position to a final position. The leaders' initial and final positions are $x_{l0} = \{(-1, -1), (0, 1), (1, -1)\}$ and $x_{lT} = \{(-1, -1), (0, 1), (1, -1)\}$, respectively. The followers' positions (equilibria) are determined by (10.36), and the time horizon is set to be one second. The matrices P and Q in (10.39) are identity matrices of appropriate dimensions.

SUMMARY

In this chapter, the focus was on what control theoretic properties one can infer from a network by looking solely at the network topology, that is, at the structure of the interaction graph. In particular, we considered the scenario when a few node in the network are allowed to act as input and output nodes to the system, and the remaining nodes are running the agreement protocol. We then investigated the controllablility and observability of the resulting– potentially– steered and observed network in a graph theoretic setting.

For controllability, the question becomes that of determining whether it is possible to "drive" the states of all the floating nodes between arbitrary values by adjusting the value of the input nodes. This would for instance be useful if the nodes are mobile robots that are to be dispatched to a given location, or if they are to execute different control programs based on some internal state that we would wish to control through the input nodes. In particular, in this chapter we showed how the symmetry structure of a network with a single input node, characterized in terms of its automorphism group, directly relates to the controllability of the corresponding input system. Intuitively speaking, what this means is that if some nodes are symmetric with respect to the leader (that is, the leader cannot "tell them apart") then these two nodes constitute an obstruction to controllability, rendering the system not controllable. By duality, we also showed how a similar argument applies to the observability question as well in the case of a single output node co-located with the input node.

These single input/output node results were then extended to the case of multiple input/output nodes. In this case, the notion of symmetry is no longer enough, and instead we introduced network equitable partitions as means by which such controllability and observability characterizations can be extended to networks with multiple inputs and outputs.

One consequence of viewing certain nodes as control inputs, while letting the remaining nodes satisfy the agreement protocol, is that a number of control design tools become available for the resulting controlled LTI systems. We also saw that the unforced system is globally asymptotically stable, and that for a constant input (static leader location), the state of the system con-

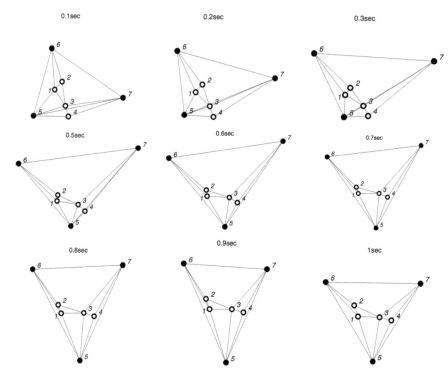

Figure 10.12: A quasi-static process where 3 leaders (black nodes) herd 4 followers (white nodes), where $T = 1$

verges to a point in the convex hull spanned by the static leader agents. This observation was subsequently used to drive the resulting process between quasi-static equilibrium points.

NOTES AND REFERENCES

The explicit study of networks in which certain nodes take on the role of active input or output nodes, while the remaining nodes act as "floaters," for example, by executing the agreement protocol, was introduced in the context of leader-follower robotics by Tanner in [229], where necessary and sufficient conditions for system controllability were given in terms of the eigenvectors of the graph Laplacian. Subsequently, graph theoretic characterizations of controllability for leader-follower multiagent systems were examined by Ji, Muhammad, and Egerstedt [127] and by Rahmani and Mesbahi [199], where graph symmetries (defined through the automorphism group) were introduced as a vehicle for understanding how the network topology impacts the controllability properties. For more information on graphs and their symmetry groups, see for example, Lauri and Scapellato [143]. The matrix-based vantage point, as represented by the Popov-Belevitch-Hautus test is, for example, discussed by Kailath [130].

The idea to view leader-based networks as controlled dynamical systems has appeared repeatedly in the literature. Notably, results along these lines include Swaroop and Hedrick's work on string stability [226], leader-to-follower stability and control, for example, Desai, Ostrowski, and Kumar [66] and Tanner, Pappas, and Kumar [231], virtual leader-based control, as in Egerstedt and Hu [74] and Leonard and Fiorelli [145], and formation control, for example, the works of Eren, Whiteley, Anderson, Morse, and Belhumeur [78] and Beard, Lawton, and Hadaegh [17], just to name a few. The particulars of the optimal control presentation of leader-based multiagent networks draws most of its inspiration from the works of Ji, Muhammad, and Egerstedt [127] and Björkenstam, Ji, Egerstedt, and Martin [31].

SUGGESTED READING

The bulk of the results in this chapter are inspired by the paper by Rahmani, Ji, Mesbahi, and Egerstedt [200], where both the automorphism group and equitable partitions were introduced as tools for connecting graph topologies to controllability. In turn, that work started out from the basic premise established by Tanner in [229]. Together, these two papers provide a rather crisp introduction to the subject. But we also recommend that interested readers examine the related work by Olfati-Saber and Shamma, [184], on consensus filters, in which a similar line of thought has been pursued. For general optimal control, we recommend the book by Bryson and Ho [40].

EXERCISES

Exercise 10.1. Consider a connected, undirected network with input nodes (one or more). Let the floating nodes be running the standard agreement protocol. Show that if the network is not controllable, then the uncontrollable part of the system is asymptotically stable.

Exercise 10.2. Given a controllable leader-follower network, assume that a new agent shows up (that is, the graph gains another node). Construct an algorithm for selecting the minimal number of edges that the new node needs when connecting to the original graph in order to not ruin the controllability properties of the new graph.

Exercise 10.3. Consider a network where a single node acts as an input node and the remaining floating nodes are executing the agreement protocol. In Theorem 10.15, controllability is characterized in terms of network symmetry. Explain how you would have to change this theorem if the network was directed rather than undirected.

Exercise 10.4. Consider again Theorem 10.15. This theorem is directly applicable to observability when the output node and the input node are in fact the same node. Explain how you would have to change this result if the input and output nodes were in fact distinct nodes.

Exercise 10.5. Given an input network and assume that the input nodes' positions can be controlled directly while the floating nodes' dynamics satisfy the agreement protocol. With this setup, consider the networks below, where the input nodes are given in black and the floating nodes in white. Which (if any) of the networks are controllable?

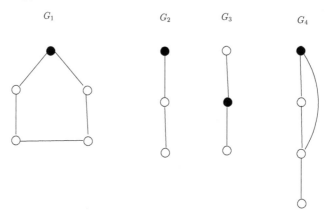

Exercise 10.6. Consider an undirected network with a single anchor node (node v_n, with n the total number of vertices) connected to every one of the floating nodes, that is, the number of floating nodes is $n_f = n - 1$ and $\deg(v_n) = n - 1$.

For this system, find simple expressions for the quantities B_f and $A_f 1$, where the graph Laplacian is partitioned as

$$L = \left[\begin{array}{c|c} A_f & Bf \\ \hline B_f^T & C \end{array}\right].$$

Exercise 10.7. Consider the graph structure in the figure below, which is a quotient graph obtained by grouping vertices together into cells. For example, the notation $\{1, 2, 3, 4\}$ means that vertices $v_1, \ldots, v_4$ are grouped together into that cell. Determine the weights (they should all be nonzero) on the different edges so that the quotient graph is obtained from an equitable partition of an undirected graph with 23 vertices.

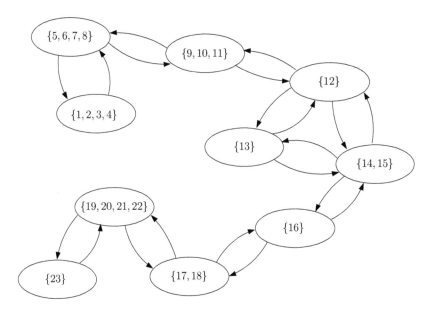

Exercise 10.8. If the network is connected, then the followers will end up (asymptotically) at

$$x_f = -A_f^{-1} B_f x_l,$$

given the static leader positions x_l. Show that each component of x_f above is in fact given by a convex combination of the components of x_l.

Exercise 10.9. Consider the linear-quadratic optimal control problem

$$\min_u \int_0^\infty \left(u(t)^T R u(t) + x(t)^T(t) Q x(t) \right) dt,$$

where the matrices Q and R are, respectively, positive semidefinite and positive definite, and

$$\dot{x}(t) = -A_f x(t) - B_f u(t)$$

corresponds to a controllable network.

Now, just because the followers execute a decentralized control strategy it does not follow that the leaders' optimal control strategy will be decentralized. Determine if this is the case for this infinite horizon optimal control problem.

Exercise 10.10. Sometimes it does not matter which follower ends up at which target position, that is, the follower roles are not assigned. Given a situation in which controllability (or rather the lack thereof) prevents the leader to drive x_i to τ_i and x_j to τ_j (where τ denotes the target position), under what conditions on the pair (A, B), is it possible to drive x_i to τ_j and x_j to τ_i instead?

Exercise 10.11. The previous exercise hints toward a "permutation-based controllability property." In general, how does this notion change the strcuture of the topology-based controllability analysis discussed in this chapter?

Exercise 10.12. Prove Corollary 10.19.

Chapter Eleven

Synthesis of Networks

> "There are only two tragedies in life:
> one is not getting what one wants,
> and the other is getting it."
> — Oscar Wilde

In this chapter, we delve into the problem of network formation (or synthesis), i.e., how networks can be formed in centralized or distributed manners such that the resulting network has certain desirable properties. Our perspective is shaped by viewing the network synthesis problem as a dynamic process by which an initial network is evolved–in lieu of local or global objectives–to reach an equilibrium configuration in the steady state. Reasoning about such a process, as it turns out, critically depends on the information that is available to each node. In the case where the individual agents cannot directly access the entire network structure, it becomes of paramount importance to have a notion whereby each agent come to terms with the particular global or local network structure.

In the first part of this chapter, we examine a candidate notion, where agents in the network can assess what local structures are desirable to them. Not surprisingly, the local nature of decisions and information structure for this part of our discussion assumes a game theoretic flavor. We then turn our attention to cases where the network structure is driven by a centralized algorithm that relies on a global information about the entire network structure.

11.1 NETWORK FORMATION

In this chapter, we consider processes by which agents in a network–aided by local or global knowledge of the network structure–can make decisions about how the structure and parameters of the network should evolve. We will refer to the problem of design and reasoning about this aspect of networked system as the *synthesis of networks* or *network formation* problem.

The solution strategy for this class of problems–as it turns out–is critically dependent on the assumptions that one makes about the information that is available to agents for steering the network toward a satisfying configuration. For example, in the case where the agents can be directed by a "global" algorithm with full access to the structure of the network at each instance, the problem generally reduces to an optimization problem with decision variables involving the existence or absence of particular edges.

A few examples of such problems will be treated in § 11.4. However, there are two important variations from the case where an agent or the steering algorithm acts globally and has global information about the network. These include situations where (a) agents act locally but the information that is available for their local decisions is global, and (b) agents act locally and information is also local. Our motivation in this chapter is to introduce certain facets of both variations. We start with the "know-global-act-local" perspective, giving the network formation a game theoretic character. We then proceed by examining two scenarios that are approached via a "know-global-act-global" framework, which can best be viewed in the realm of optimization.

11.2 LOCAL FORMATION GAMES

In this section, we consider the situation where agents in the network locally modify the network structure in order to reach a structure that has certain desirable network-wide properties. These global properties might include network connectivity, or, motivated by the agreement problem and its extensions, the spectrum of the graph Laplacian.

As an example, if it is desired to maximize the connectivity of the resulting graph, it is natural that all agents strive to connect to all other agents–leading to a complete graph. The problem is of course more interesting and relevant when there are costs associated with the edges in the graph and agents make strategic choices about edges that should be paid for.

More formally, consider a global utility or cost function for the entire network $\mathcal{G}$, denoted by $\mathbf{U}(\mathcal{G})$ and $\mathbf{C}(\mathcal{G})$, respectively, representing the desirable properties of the network and the cost structure for having these properties. We refer to $\mathbf{U}(\mathcal{G})$ and $\mathbf{C}(\mathcal{G})$ as the *social utility* or *social cost* of the network $\mathcal{G}$. In order to introduce a local nature to the decision process on the part of each agent, we also introduce a local utility, or local cost, for each agent that is a function of the overall network $\mathcal{G}$. Hence, for each agent i, we have a utility $\mathbf{u}_i(\mathcal{G})$ or a cost $\mathbf{c}_i(\mathcal{G})$. Knowing this utility or cost, agent i adopts a decision $s_i \in S_i$, which in turn collectively leads to the realization of a network $\mathcal{G}(s)$, where $s \in S$ is the vector of strategies

$$s = [s_1, s_2, \ldots, s_n]^T \in S, \quad \text{with} \quad S = S_1 \times S_2 \times \cdots \times S_n.$$

The strategies that are of particular interest in this section include the scenario where each agent decides on the existence of an undirected edge between itself and another agent in the network.

A moments reflection on the local nature of the decision making process that we would like to capture points to the necessity of defining an equilibrium notion, where the network formation process reaches a steady state. In the case of a dynamic system governed by a differential equation $\dot{x} = f(x)$, the equilibrium is naturally defined as the set of vectors x^* such that $f(x^*) = 0$; hence, once x assumes the value x^*, the dynamics induced by the function f does not steer the state away from it.

A similar notion of equilibrium, this time for a network structure, proves to be a bit more problematic, particularly if it is required to have a local character. One way to define such an equilibrium, which will be adopted in this section, is that of a Nash equilibrium, one of the central constructs in game theory. In order to define this notion in the context of network formation, let the agents in the network adopt strategies $s = (s_1, s_2, \ldots, s_n)$, that is, each s_i captures the set of edges that agent i likes to establish, and pay for, at a given time. Then s is a Nash equilibrium of the network if, for each agent i, whenever $s_i' \neq s_i$, one has

$$\mathbf{u}_i(s_{-i}, s_i') \leq \mathbf{u}_i(s_{-i}, s_i) \text{ for all } i.$$

In other words, s_i is the best response of agent i with respect to any unilateral change in its strategy; s_{-i} is the vector s with a missing ith component. We proceed to examine some of the particular aspects and ramifications of using the Nash equilibrium as a solution concept for the network formation.

11.2.1 The Local Connection Game, Nash Equilibria, and the Price of Anarchy

Equipped with a notion of an equilibrium for network formation processes, at least two important questions need to be addressed: (1) how such an equilibrium can be reached, and (2) how the social cost or utility obtained from this equilibrium relates to the utility that would have been obtained if agents could implement a global algorithm. We refer to the ratio between the cost obtained at a Nash equilibrium and the social optimal cost as the *price of anarchy.* Thus, the closer the price of anarchy is to 1, the more desirable is the corresponding Nash equilibrium.[1] Although these questions turn out to be rather problematic for a general problem setup, we proceed to address the second aspect via a concrete example. In this venue, consider the scenario where n agents, initialized within an empty interconnection network, have a local cost function of the form

$$\alpha d(i) + \sum_{j \neq i} \mathbf{dist}(i, j), \tag{11.1}$$

where $d(i)$ is the degree of vertex i. The interpretation here is that agent i attempts to minimize the sum of its distance to other agents in the network while being considerate of how much each edge, or link, costs. Since $\mathbf{dist}(i, j) = \infty$ when there is no path from node i and j, the local cost structure (11.1) guarantees that, at the Nash equilibrium, the overall network will be connected.[2]

We now let the social cost of the network be

$$\mathbf{C}(\mathcal{G}) = \sum_{i \neq j} \mathbf{dist}(i, j) + \alpha \, \mathrm{card}(E), \tag{11.2}$$

reflecting the desire to minimize the geodesic distances across the network while being frugal on how many edges are required to achieve this objective. We call the network that minimizes (11.2), an efficient or socially optimal network. We proceed to examine the price of anarchy for this so-called local connection game.

Proposition 11.1. *If $\alpha \geq 2$ then any star graph is socially optimal. However, if $\alpha < 2$, the complete graph is socially optimal.*

Proof. Suppose that the socially optimal network has m edges–hence at least $2m$ pairs of vertices are directly connected, contributing αm to the social cost. The rest of the pairs, specifically, $n(n - 1) - 2m$ of them, contribute at least $2(n(n - 1) - 2m)$ to the social cost since their respective distances will be at least 2. Hence, the lower bound for the social cost is $\alpha m + 2m + 2n(n - 1) - 4m = (\alpha - 2)m + 2n(n - 1)$. When $\alpha \geq 2$, the social cost is minimized when the graph is a tree, and in particular a star. When $\alpha < 2$, the complete graph is socially optimal. ∎

---

[1] The lowest ratio is sometimes referred to as the price of stability.
[2] In this setting, when agent i establishes and pays for edge $\{i, j\}$, this edge is also available for agent j.

Proposition 11.2. *If $\alpha \geq 1$ then any star graph is a Nash equilibrium. However, if $\alpha \leq 1$ then the complete graph is a Nash equilibrium.*

Proof. Let $\alpha \geq 1$ and consider the star graph with the center node connected to all the other nodes. If the center node disconnects one of the edges, then it incurs an infinite cost–thus it does not have an incentive to unilaterally change its $n - 1$ connections with other nodes. The leaf nodes,[3] on the other hand, can deviate by adding edges as they cannot overrule the center node's decision and disconnect with the center by deleting an edge. Adding k edges has a savings of k in distances at a price of αk; since $\alpha \geq 1$, this is not a locally profitable move and thus the star is a Nash equilibrium.

Now, let $\alpha \leq 1$, and consider a complete graph. An agent that stops paying for a set of k edges saves αk in edge price, while increasing its total distances by k; thus the complete graph is a Nash equilibrium in this case.∎

Using the above two results, it follows that *for $\alpha \geq 2$ or $\alpha \leq 1$, the price of anarchy is 1.* Below, we show that for $1 < \alpha < 2$, the price of anarchy is at most $4/3$.

First, note that since every pair of vertices that is not connected by an edge is at least a distance 2 apart, one has the lower bound for the social cost, as

$$\mathcal{C}(\mathcal{G}) \geq \alpha m + 2m + 2(n(n-1) - 2m) = 2n(n-1) + m(\alpha - 2), (11.3)$$

where $m = \operatorname{card}(E)$. It thus follows from (11.3) that when $1 < \alpha < 2$, the social optimum is achieved when $\operatorname{card}(E)$ is maximum, that is, with the complete graph.

Theorem 11.3. *For $1 < \alpha < 2$, the price of anarchy is at most $4/3$.*

Proof. Let $m = \operatorname{card}(E)$. The price of anarchy is bounded by

$$\frac{\mathbf{C}(\text{star})}{\mathbf{C}(\text{complete})} = \frac{(n-1)(\alpha - 2 + 2n)}{n(n-1)((\alpha - 2)/2 + 2)}$$

$$= \frac{4}{2+\alpha} - \frac{4 - 2\alpha}{n(2+\alpha)} < \frac{4}{2+\alpha} < \frac{4}{3}.$$

∎

---

[3] That is, the other nodes.

We now proceed to show that, more generally, the price of anarchy, defined by the local cost (11.1) and social cost (11.2), is characterized by the edge cost α. As this is accomplished by monitoring the diameter of the network that corresponds to the Nash equilibrium, we first state the following result.

Proposition 11.4. *If the diameter of a Nash equilibrium in the local connection game is d, then its social cost is at most $\mathcal{O}(d)$ times the minimum possible cost.*

Proof. In order to achieve the minimum cost, we note that, since at the social optimum the graph has to be connected, we need to have at least $n-1$ edges, costing $\alpha(n-1)$. In the meantime, there are $n(n-1)/2$ distances, each of which is at least 1. Thus the cost of the optimal solution is at least $\Omega(\alpha n + n^2)$. On the other hand, at the Nash equilibrium, the distance cost is at most $n^2 d$ so it is at most d times the minimum.

In order to bound the edge costs, consider two classes of edges: (1) the cut edges, which are at most $n-1$, with their associated cost of at most $\alpha(n-1)$, and (2) the non-cut edges.[4] We now proceed to show that the cost of non-cut edges at the Nash equilibrium is at most $\mathcal{O}(n^2 d)$.

In order to show this, consider node v and pick an edge $e = (u, v)$, which has been paid for by node u. Define the set V_e as the set of nodes w such that the shortest path from u to w passes through e. We now evaluate the cost and benefit of having edge e: absence of edge e would save a cost of α; however, the distance to and from each node in V_e would have increased by at most $2d$. Thus, having edge e brings a total distance savings of at most $2d\,\mathrm{card}(V_e)$. Since edge e is present at the Nash equilibrium, it must be that

$$\alpha \leq 2d\,\mathrm{card}(V_e),$$

and therefore $\mathrm{card}V_e \geq \alpha/2d$. Thus, there is a natural correspondence between the number of non-cut edges e present at the Nash equilibrium and the set V_e constructed above. Therefore, there are $\mathcal{O}(n/(\alpha/2d)) = \mathcal{O}(dn/\alpha)$ such edges, with the total cost of $\mathcal{O}(dn)$ for each node v. Hence, the total cost at the Nash equilibrium is $\mathcal{O}(n^2 d)$. ∎

Proposition 11.5. *The diameter of a Nash equilibrium in the local connection game is at most $2\sqrt{\alpha}$; hence the price of anarchy is at most $\mathcal{O}(\sqrt{\alpha})$.*

Proof. Suppose that $\mathbf{dist}(u, v) \geq 2k$ for some k. Then u could pay for an edge in order to improve the sum of its distances to the nodes in the second

---

[4]Cut edges are those that, when removed from the graph, cause a node to become isolated.

half of the shortest path between u and v by

$$(2k - 1) + (2k - 3) + \cdots + 1 = k^2.$$

Thereby, if $\mathbf{dist}(u, v) \geq 2\sqrt{\alpha}$, node u would benefit from adding the edge uv at the price of α, which leads to a contradiction. ∎

11.3 POTENTIAL GAMES AND BEST RESPONSE DYNAMICS

We now turn our attention to the algorithmic facet of the notion of Nash equilibrium for a network formation process. This is particularly important as it is not clear how each agent can end up in a Nash equilibrium by following a particular, local algorithm.

In this section, we consider a subclass of games where much more can be said about the algorithmic aspects of the game. We refer to this class of games as *potential games*. For any finite game, an exact potential function Φ is a function that maps every strategy S to some real value and satisfies the following condition. If $S = (s_1, s_2, \ldots, s_n)$ and $s_i' \neq s_i$ is an alternative strategy for some agent i, and $S' = (s_{-i}, s_i')$, then

$$\Phi(S) - \Phi(S') = \mathbf{u}_i(S') - \mathbf{u}_i(S).$$

In other words, the amount that an agent can benefit by a unilateral change in its strategy should exactly be reflected by how much the potential function will be reduced. Although, in general, a Nash equilibrium does not have to exist, as shown below, the structure of a potential game always guarantees one.

Theorem 11.6. *Every potential game has at least one Nash equilibrium, namely, the strategy S that minimizes $\Phi(S)$.*[5]

Proof. Let Φ be the potential function for the game and let S be a strategy vector minimizing $\Phi(S)$. Consider any move by agent i that results in a new strategy vector S'. By assumption, $\Phi(S') \geq \Phi(S)$, and by definition $\mathbf{u}_i(S') - \mathbf{u}_i(S) = \Phi(S) - \Phi(S')$. Thus, i's utility cannot increase from this move, and hence S is a Nash equilibrium. ∎

---

[5] Such an equilibrium is referred to as a *pure Nash equilibrium* as opposed to a *mixed Nash equilibrium*, employed when probabilistic strategies are considered.

One of the immediate ramification of the above observation is an "algorithm" for reaching Nash equilibrium, namely, the best response dynamics–also known as the "greedy algorithm." In such a setting, each agent chooses its (local) strategy in order to minimize the potential for the game.

Theorem 11.7. *In any finite potential game, the best response dynamics always converges to the Nash equilibrium.*

We are now ready to state a general result on the price of anarchy for a potential game.

Theorem 11.8. *Suppose that we have a potential game with potential function* Φ*, and assume that for any outcome S, we have*

$$\frac{\mathbf{C}(S)}{\gamma} \le \Phi(S) \le \rho \mathbf{C}(S) \tag{11.4}$$

for some constants $\gamma, \rho > 0$*. Then the price of anarchy is at most* $\gamma\rho$*.*

Proof. Let $\bar{S}$ be the strategy vector that minimizes $\Phi(S)$ (it is a Nash equilibrium) and let S^* be the cost that minimizes the social cost $\mathbf{C}(S)$. By (11.4),

$$\frac{\mathbf{C}(\bar{S})}{\gamma} \le \Phi(\bar{S}) \le \Phi(S^*) \le \rho \mathbf{C}(S^*),$$

and thus $\mathbf{C}(\bar{S}) \le \gamma\rho\mathbf{C}(S^*)$. Thereby, the price of anarchy is bounded by $\gamma\rho$. ∎

As an example, consider the case when the local utility for each agent is denoted by

$$\mathbf{u}_i = -\alpha d(i) + \lambda_2(\mathcal{G}), \tag{11.5}$$

for some $\alpha > 0$, and

$$\Phi(S) = \alpha \operatorname{card}(E) - \lambda_2(\mathcal{G}). \tag{11.6}$$

Thus, each agent attempts to maximize the value of $\lambda_2(\mathcal{G})$ by including edges in the network while being considerate of the total cost for these edge choices.

We show that Φ (11.6) is in fact a potential function for this game. In this venue, note that

$$\Phi(S) - \Phi(S') = \alpha(\operatorname{card}(E) - \operatorname{card}(E')) - \lambda_2(\mathcal{G}) + \lambda_2(\mathcal{G}')$$

and

$$\mathbf{u}_i(S') - \mathbf{u}_i(S) = -\alpha d(i)' + \lambda_2(\mathcal{G}') - (-\alpha d(i) + \lambda_2(\mathcal{G}))$$
$$= \alpha(d(i) - d(i)') + \lambda_2(\mathcal{G}') - \lambda_2(\mathcal{G}),$$

where graphs $\mathcal{G}$ and $\mathcal{G}'$ are, respectively, associated with strategies S and S', and $d(i)$ and $d(i)'$ are the degree of vertex i in graphs $\mathcal{G}$ and $\mathcal{G}'$. As a direct consequence of Theorems 11.6 and 11.7, it follows that when each agent is supplied with a local utility function of the form (11.5), then the best response dynamics leads the network, in a decentralized way, to a Nash equilibrium. Moreover, when the social cost satisfies the inequality (11.4), the best response dynamics is guaranteed to lead to a configuration, whose cost is within a constant multiple of the social optimum.

11.3.1 Growing Nash Networks

Motivated by reaching a Nash equilibrium by following the best response dynamics, we consider the problem faced by each agent in its quest for increasing $\lambda_2(\mathcal{G})$ by adding an edge to the existing network.

Assume that, at a given time instance, the network is represented by $\mathcal{G}$ and an agent is faced with choosing an edge among a set of candidate edges $e_1, \ldots, e_p$ that increases $\lambda_2(\mathcal{G})$ the most. Note that the resulting graph Laplacian, after adding any number of edges from the candidate set, assumes the form

$$L(\mathcal{G}, x) = L(\mathcal{G}) + \sum_i x_i b_i b_i^T, \tag{11.7}$$

where each x_i is either zero or one and b_i represents the column in the incidence matrix that corresponds to edge e_i.

Let us for a moment pretend that the x_i in (11.7) can assume real values on the unit interval, allowing us to consider the directional derivative of $\lambda_2(\mathcal{G}, x)$ along each direction x_i. In this venue, the directional derivative of $\lambda_2(\mathcal{G}, x)$, assuming that it is an isolated eigenvalue of $L(\mathcal{G}, x)$, is

$$\frac{\partial \lambda_2(\mathcal{G}, x)}{\partial x_i} = b_i^T q q^T b_i^T = (q_u - q_v)^2, \tag{11.8}$$

where $e_i = uv$ and q is the normalized eigenvector of $L(\mathcal{G})$ that corresponds to its second smallest eigenvalue. In other words, when $\lambda_2(\mathcal{G})$ is an isolated eigenvalue of $L(\mathcal{G})$, the difference $(q_u - q_v)^2$ gives the first order approximation of the increase to $\lambda_2(\mathcal{G})$ if edge uv is added to the graph.

We now provide lower and upper bounds on the second smallest eigenvalue of the augmented graph in terms of the spectrum of $L(\mathcal{G})$ and the eigenvector corresponding to $\lambda_2(\mathcal{G})$. These bounds can then be used to improve the current network $\mathcal{G}$ by adding an edge, among a candidate set of edges, that results in the largest increase in $\lambda_2(\mathcal{G})$. We note that although these estimates can be used for the local decision process faced by each node to improve $\lambda_2(\mathcal{G})$, they do require global knowledge of the network structure and its spectra. We first need a useful lemma.

Lemma 11.9. *Let $\mathcal{G}$ be a graph whose Laplacian has distinct eigenvalues (and hence is connected) with spectral factorization $Q\overline{\Lambda}(\mathcal{G})Q^T$. Assume that a new edge $e = uv$ is added to $\mathcal{G}$ such that for every normalized eigenvector q associated with nonzero eigenvalues of $L(\mathcal{G})$, $q_u - q_v$ is nonzero. Then the nonzero Laplacian eigenvalues of $\mathcal{G} + e$, $\zeta_2, \ldots, \zeta_n$, are distinct and satisfy the so-called secular equation*

$$f(\zeta) = 1 + \sum_{i=2}^{n} \frac{z_i^2}{\lambda_i(\mathcal{G}) - \zeta} = 0,$$

where $z = Q^T b$ and b is the incidence vector for e. Moreover, the eigenvalues of $L(\mathcal{G}) + bb^T$ satisfy the interlacing inequalities

$$\lambda_2(\mathcal{G}) < \zeta_2 < \lambda_3(\mathcal{G}) < \zeta_3 < \cdots < \lambda_n(\mathcal{G}) < \zeta_n.$$

Proof. Let $L(\mathcal{G}) = Q\overline{\Lambda}(\mathcal{G})Q^T$, where Q consists of an orthogonal set of eigenvectors of $L(\mathcal{G})$. Let b be the incidence vector for edge $e = uv$. Then the Laplacian of $\mathcal{G} + e$ is $L(\mathcal{G}) + bb^T$ and

$$Q(L(\mathcal{G}) + bb^T)Q^T = \overline{\Lambda}(\mathcal{G}) + \bar{z}\bar{z}^T,$$

where $\bar{z} = Q^T b$. We note that $\bar{z}_1 = 0$ and $\bar{z}_i \neq 0$ for $i = 2, \ldots, n$ (by the statement of the Lemma). The nonzero eigenvalues of $L(\mathcal{G}) + bb^T$, therefore, are the eigenvalues of $\Lambda(\mathcal{G}) + zz^T$, where

$$\overline{\Lambda}(\mathcal{G}) = \begin{bmatrix} 0 & 0 \\ 0 & \Lambda(\mathcal{G}) \end{bmatrix} \quad \text{and} \quad \bar{z} = \begin{bmatrix} 0 \\ z \end{bmatrix}.$$

Now, let ζ be a nonzero eigenvalue of $L(\mathcal{G}) + bb^T$ with the corresponding eigenvector v,

$$(\Lambda(\mathcal{G}) + zz^T)v = \zeta v,$$

and thereby

$$(\Lambda(\mathcal{G}) - \zeta I)v + (z^T v)z = 0. \tag{11.9}$$

Our next observation toward the final proof of the lemma hinges on showing that the matrix $\Lambda(\mathcal{G}) - \zeta I$ in (11.9) is invertible and that z is not orthogonal to any eigenvector of $\Lambda(\mathcal{G}) + zz^T$. This is shown as follows. If a nonzero eigenvalue ζ for $\Lambda(\mathcal{G}) + zz^T$ is equal to one of the nonzero eigenvalues of $\mathcal{G}$, say $\lambda_i(\mathcal{G})$, then from (11.9) it follows that

$$1_i^T((\Lambda(\mathcal{G}) - \zeta I)v + (z^T v)z) = (z^T v)z_i = 0,$$

where 1_i is the vector with a 1 at the ith entry and zeros at all other entries. Since $z_i \neq 0$, $z^T v = 0$ and $\Lambda(\mathcal{G})v = \zeta v$. However, since $\lambda_i(\mathcal{G})$ ($i = 2, 3, \ldots, n$), are distinct, it follows that the vector v is a multiple of 1_i and hence $0 = z^T v = z_i$, which is a contradiction.

Returning to the main theme of the proof, we now apply $z^T(\Lambda(\mathcal{G}) - \zeta I)^{-1}$ to both sides of (11.9), leading to

$$(z^T v)(1 + z^T(\Lambda(\mathcal{G}) - \zeta I)^{-1}z) = 0,$$

and since $z^T v \neq 0$, it follows that the nonzero eigenvalues of $L(\mathcal{G}) + bb^T$ satisfy the equation

$$f(\zeta) = 1 + z^T(\Lambda(\mathcal{G}) - \zeta I)^{-1}z = 0. \tag{11.10}$$

We note that $f(\zeta)$ (11.10) is monotonic between its two poles. Hence it has $n - 2$ roots, $\zeta_2, \zeta_3, \ldots, \zeta_{n-1}$, on the intervals

$$(\lambda_i(\mathcal{G}), \lambda_{i+1}(\mathcal{G})) \quad \text{for} \quad i = 2, 3, \ldots, n - 1,$$

and $\zeta_n > \lambda_n(\mathcal{G})$. ∎

Proposition 11.10. *Let b be the incidence vector for the edge $e = uv$ in the graph $\mathcal{G}$ with distinct Laplacian eigenvalues such that for every eigenvector q associated with nonzero eigenvalues of $L(\mathcal{G})$, $q_u - q_v$ is nonzero. Moreover, let q be the normalized eigenvector corresponding to $\lambda_2(\mathcal{G})$ and $\rho_i = \lambda_i(\mathcal{G}) - \lambda_2(\mathcal{G})$, for $i = 3, \ldots, n$. Then*

$$\lambda_2(\mathcal{G} + e) \geq \lambda_2(\mathcal{G}) + \frac{(q_u - q_v)^2}{(3/2) + (6/\rho_3)}$$

and

$$\lambda_2(\mathcal{G} + e) \leq \lambda_2(\mathcal{G}) + \frac{(q_u - q_v)^2}{1 + (2 - (q_u - q_v)^2)/\rho_n}.$$

Proof. To prove the lower bound, consider $Q\Lambda Q^T$ as the orthogonal decomposition of $L(\mathcal{G})$; we have assumed that the entries in Λ are distinct. Note that matrices $L(\mathcal{G})$ and $L(\mathcal{G}) + bb^T$ both have zero eigenvalues with the corresponding eigenvector 1. As shown in Lemma 11.9, the remaining

$n-1$ eigenvalues of $L(\mathcal{G}) + bb^T$ are the $n-1$ roots of the so-called secular equation

$$g(\zeta) = 1, \qquad (11.11)$$

where

$$g(\zeta) = \sum_{i=2}^{n} \frac{z_i^2}{\zeta - \lambda_i(\mathcal{G})}$$

and $z = Q^T b$; we note that $z_2 = q_u - q_v$ and $z_1 = 0$. Denote by $\zeta_2, \ldots, \zeta_n$ the nonzero ordered eigenvalues of $L(\mathcal{G}) + bb^T$. By the interlacing property, these eigenvalues satisfy

$$\lambda_i(\mathcal{G}) < \zeta_i < \lambda_{i+1}(\mathcal{G}), \quad i = 2, \ldots, n-1.$$

Since $g(\zeta)$ is a locally decreasing function of $\zeta \in (\lambda_2, \lambda_3)$, in this interval

$$\zeta \le \zeta_2 \quad \text{if and only if} \quad g(\zeta) \ge g(\zeta_2).$$

Since $g(\zeta_2) = 1$, if then follows that $\zeta \le \zeta_2$ if

$$\frac{z_2^2}{\zeta - \lambda_2(\mathcal{G})} \ge 1 + \sum_{i=3}^{n} \frac{z_i^2}{\lambda_i(\mathcal{G}) - \zeta}.$$

Since $\|z\| = 2$ and $\sum_{i=3}^{n} z_i^2 \le 2$, it follows that

$$\frac{2}{\lambda_3(\mathcal{G}) - \zeta} \ge \sum_{i=3}^{n} \frac{z_i^2}{\lambda_i(\mathcal{G}) - \zeta}.$$

Hence, in order to show that $\zeta \ge \zeta_2$, it suffices to show that

$$\frac{z_2^2}{\zeta - \lambda_2(\mathcal{G})} \ge 1 + \frac{2}{\lambda_3(\mathcal{G}) - \zeta}.$$

Let $\zeta - \lambda_2(\mathcal{G}) = \epsilon$ and $\lambda_3(\mathcal{G}) - \lambda_2(\mathcal{G}) = \delta$. We note that if

$$\epsilon = \frac{z_2^2}{2(1/\delta + 1/4 + \sqrt{1/4 + 4/\delta^2})}, \qquad (11.12)$$

then

$$\frac{z_2^2}{\epsilon} \ge 1 + \frac{2}{\delta - \epsilon}.$$

In other words, when ϵ is defined by (11.12), we have $\zeta_2 \geq \lambda_2(\mathcal{G}) + \epsilon$. However,

$$\sqrt{\frac{1}{4} + \frac{4}{\delta^2}} \leq \frac{1}{2} + \frac{2}{\delta},$$

and it follows that

$$\zeta_2 \geq \lambda_2(\mathcal{G}) + \frac{z_2^2}{(3/2) + (6/\delta)}.$$

But $z_2^2 = (q_u - q_v)^2$, and the lower bound in the statement of the proposition follows.

For the upper bound, note that $\lambda_2(\mathcal{G}+e)$ is the number $\zeta_2 \in (\lambda_2(\mathcal{G}), \lambda_3(\mathcal{G}))$ satisfying

$$\zeta_2 = \lambda_2(\mathcal{G}) + \frac{z_2^2}{1 + \sum_{i=3}^{n} z_i^2/(\lambda_i(\mathcal{G}) - \zeta_2)}.$$

As $\lambda_2(\mathcal{G}) < \zeta_2$ and $\|z\| = 2$, it follows that

$$\zeta_2 \leq \lambda_2(\mathcal{G}) + \frac{z_2^2}{1 + \sum_{i=3}^{n} z_i^2/(\lambda_i(\mathcal{G}) - \lambda_2(\mathcal{G}))}$$

$$\leq \lambda_2(\mathcal{G}) + \frac{z_2^2}{1 + (2 - z_2^2)/\rho_n}.$$

∎

11.4 NETWORK SYNTHESIS: A GLOBAL PERSPECTIVE

In this section, we consider the set of n mobile elements as vertices of a graph, with the edge set determined by the relative positions of the respective elements. Specifically, parallel to our setup in Chapter 7, we let $\mathcal{G}$ denote the set of graphs of order n with vertex set $V = [n]$ and edge set $E = \{ij \mid i = 1, 2, \ldots, n-1, j = 2, \ldots, n, i < j\}$, with the weighting function

$$w : \mathbf{R}^3 \times \mathbf{R}^3 \to \mathbf{R}_+,$$

assigning to each edge ij a function of the Euclidean distance between the two mobile nodes i and j. Thus we have

$$w_{ij} = w(x_i, x_j) = f(\|x_i - x_j\|) \tag{11.13}$$

for some $f : \mathbf{R}_+ \to \mathbf{R}_+$, with $x_i \in \mathbf{R}^3$ denoting the position of element i. In our setup the function f in (11.13) will be required to exhibit a distinct behavior as it traverses the positive real line. For example, we will require that this function assume a constant value of 1 when the distance between i and j is less than some threshold and then rapidly drop to zero (or some small value) as the distance between these elements increases. Such a requirement parallels the behavior of an information link in a wireless network where the signal power at the receiver side is inversely proportional to some power of the distance between transmitting and receiving elements. Using this framework, we now consider the formation configuration problem

$$\mathbf{\Lambda} : \quad \max_x \quad \lambda_2(\mathcal{G}, x), \tag{11.14}$$

where $x = [x_1, x_2, \dots, x_n]^T \in \mathbf{R}^{3n}$ is the vector of positions for the distributed system, the matrix $L(\mathcal{G}, x)$ is a weighted graph Laplacian defined elementwise as

$$[L(\mathcal{G}, x)]_{ij} = \begin{cases} -w_{ij} & \text{if } i \neq j, \\ \sum_{s \neq i} w_{is} & \text{if } i = j, \end{cases} \tag{11.15}$$

and $\lambda_2(\mathcal{G}, x)$ denotes the second smallest eigenvalue of the "state-dependent" Laplacian matrix $L(\mathcal{G}, x)$. Furthermore, we restrict the feasible set of (11.14) by imposing the proximity constraint

$$d_{ij} = \|x_i - x_j\|^2 \geq \rho_1 \quad \text{for all } i \neq j, \tag{11.16}$$

preventing the elements from getting arbitrarily close to each other in their desire to maximize $\lambda_2(\mathcal{G}, x)$ in (11.14).

11.4.1 Network Formation Using Semidefinite Programming

The general formulation of the problem $\mathbf{\Lambda}$ (11.14) does not readily hint at being tractable in the sense of admitting an efficient algorithm for its solution. Generally, maximizing the second smallest eigenvalue of a symmetric matrix, subject to matrix inequalities, does not yield to a desirable convex optimization approach and, subsequently, a solution procedure that relies solely on interior point methods (see Appendix A.5). The above complication, however, is somewhat alleviated in the case of graph Laplacians, where the smallest eigenvalue $\lambda_1(L(\mathcal{G}))$ is always zero with the associated eigenvector of 1 composed of unit entries.

Nevertheless, due to the nonlinear dependency of entries of $L(\mathcal{G})$ on the relative distance d_{ij} and the presence of constraints (11.16), the problem $\mathbf{\Lambda}$ (11.14) assumes the form of a nonconvex optimization. In light of this fact, we will proceed to propose an iterative semidefinite programming based

approach for this problem.[6] However, before we proceed, we make a few
remarks on some judicious choices for the function f in (11.13).

The choice of f in (11.13) is guided not only by particular applications
but also by numerical considerations. Although there are a host of choices
for f, for our analysis and numerical experimentation, we let f assume the
form

$$f(d_{ij}) = \epsilon^{(\rho_1 - d_{ij})/(\rho_1 - \rho_2)}, \quad \epsilon > 0, \qquad (11.17)$$

given that $d_{ij} \geq \rho_1$. Note that $f(\rho_1) = 1$ and $f(\rho_2) = \epsilon$.

Among the advantages of working with functions of the form (11.17) are
their differentiability properties, as well as their ability to capture situations
that are of practical relevance. In many such situations, the strength of an
information link is inversely proportional to the relative distance and decays
exponentially after a given threshold is passed. Furthermore, and possibly
more importantly, function (11.17) leads to a stable algorithm for our numerical experimentation, as will be seen shortly.

11.4.2 Maximizing $\lambda_2(\mathcal{G}, x)$

We first present a linear algebraic result in conjunction with the general
problem of maximizing the second smallest eigenvalue of graph Laplacians.[7]

Proposition 11.11. *Consider the m-dimensional subspace $\mathbf{P} \subseteq \mathbf{R}^n$ that
is spanned by the vectors $p_i \in \mathbf{R}^n$, $i = 1, \ldots, m$. Let $P = [p_1, \ldots, p_m] \in
\mathbf{R}^{n \times m}$. Then, for a $M \in \mathcal{S}^n$, one has*

$$x^T M x > 0 \quad \textit{for all nonzero } x \in \mathbf{P}$$

if and only if

$$P^T M P > 0. \qquad (11.18)$$

Proof. An arbitrary nonzero element $x \in \mathbf{P}$ can be written as

$$x = \alpha_1 p_1 + \alpha_2 p_2 + \cdots + \alpha_m p_m$$

for some $\alpha_1, \ldots, \alpha_m \in \mathbf{R}$, not all zeros, and thus $x = Py$, where $y =
[\alpha_1, \alpha_2, \ldots, \alpha_m]^T$. Consequently the first inequality in (11.18) is equivalent

---

[6]A semidefinite program (SDP) is a convex optimization problem that aims to minimize
a linear function of a symmetric matrix over a set defined by linear matrix inequalities.

[7]Recall that $M > 0$ for a symmetric matrix refers to its positive definiteness.

to

$$(Py)^T M(Py) = y^T P^T M P y > 0$$

for all nonzero $y \in \mathbf{R}^m$, which when phrased differently, translates to the matrix inequality $P^T M P > 0$. ∎

Corollary 11.12. *For a graph Laplacian $L(\mathcal{G})$ the constraint*

$$\lambda_2(\mathcal{G}) > 0, \tag{11.19}$$

is equivalent to

$$P^T L(\mathcal{G}) P > 0, \tag{11.20}$$

where $P = [p_1, p_2, \ldots, p_{n-1}]$, and the unit vectors $p_i \in \mathbf{R}^n$ are chosen such that

$$p_i^T \mathbf{1} = 0 \quad (i = 1, 2, \ldots, n-1)$$

and

$$p_i^T p_j = 0 \quad (i \neq j).$$

Proof. We know that $L(\mathcal{G}) \geq 0$, $L(\mathcal{G}) \mathbf{1} = 0$, and for a connected graph **rank** $L(\mathcal{G}) \leq n - 1$. This implies that

$$x^T L(\mathcal{G}) x > 0 \quad \text{for all nonzero } x \in \mathbf{1}^\perp, \tag{11.21}$$

where

$$\mathbf{1}^\perp = \{x \in \mathbf{R}^n \,|\, \mathbf{1}^T x = 0\}. \tag{11.22}$$

In view of Proposition 11.11, condition (11.21) is equivalent to having

$$P^T L(\mathcal{G}) P > 0,$$

with P denoting the matrix of vectors spanning the subspace $\mathbf{1}^\perp$. Without loss of generality, this subspace can be identified with the basis unit vectors satisfying (11.21). ∎

Corollary 11.13. *The problem Λ (11.14) is equivalent to*

$$\Lambda : \quad \max_x \quad \gamma \tag{11.23}$$

$$\text{s.t.} \quad d_{ij} = \|x_i - x_j\|^2 \geq \rho_1, \tag{11.24}$$

$$P^T L(\mathcal{G}, x) P \geq \gamma I_{n-1}, \tag{11.25}$$

where $i = 1, 2, \ldots, n-1, j = 2, \ldots, n, i < j$, and the pairwise orthogonal unit vectors p_i forming the columns of P span the subspace $\mathbf{1}^\perp$ (11.22).

Proof. The proof follows from Corollary 11.12.   ∎

One of the consequences of Corollary 11.13 pertains to the following graph synthesis problem: determine graphs satisfying an upper bound on the number of their edges, with maximum smallest second Laplacian eigenvalue. Although this problem will not be considered further, we point out that it can be reformulated as

$$\max_{\mathcal{G}\in\mathcal{G}}\{\gamma \mid \textbf{Trace }L(\mathcal{G}) \le \rho,\ P^T L(\mathcal{G})P \ge \gamma I_{n-1}\},$$

where P is defined as in Corollary 11.13 and ρ is twice the maximum number of edges allowed in the desired graph.

11.5 DISCRETE AND GREEDY

We now proceed to view the problem Λ (11.14) in an iterative setting, where the goal is shifted toward finding an algorithm that attempts to maximize the second smallest eigenvalue of the graph Laplacian at each step. Toward this aim, we first differentiate both sides of (11.24) with respect to time as

$$2\{\dot{x}_i(t) - \dot{x}_j(t)\}^T\{x_i(t) - x_j(t)\} = \dot{d}_{ij}(t), \qquad (11.26)$$

and then employ Euler's first discretization method, with Δt as the sampling time,

$$x(t) \to x(k), \quad x(k+1) - x(k) \approx \dot{x}(t)\Delta t,$$

to rewrite the first part of (11.24) as

$$2\left(x_i(k+1) - x_j(k+1)\right)^T\{x_i(k) - x_j(k)\} = d_{ij}(k+1) + d_{ij}(k).$$

Similarly, the state-dependent Laplacian $L(\mathcal{G}, x)$ in (11.25) is discretized by first differentiating the terms w_{ij} with respect to time, and then having

$$w_{ij}(k+1) = w_{ij}(k) - \epsilon^{(\rho_1 - d_{ij}(k))/(\rho_1 - \rho_2)}\left(d_{ij}(k+1) - d_{ij}(k)\right),$$

recalling that we are employing functions of the form (11.17) in (11.13). The discrete version of the state-dependent Laplacian, $L(\mathcal{G}, x_k) = L(\mathcal{G}_k)$, assumes the form

$$[L(\mathcal{G}_k)]_{ij} = \begin{cases} -w_{ij}(k) & \text{if } i \ne j, \\ \sum_{s\ne i} w_{is}(k) & \text{if } i = j. \end{cases}$$

Putting it all together, we arrive at the iterative step for solving the optimization problem, in fact an SDP,

$$\Lambda_k : \quad \max_{x(k+1)} \quad \gamma \tag{11.27}$$

subject to the constraints

$$2\big(x_i(k+1) - x_j(k+1)\big)^T \big(x_i(k) - x_j(k)\big)$$
$$= d_{ij}(k+1) + d_{ij}(k), \tag{11.28}$$
$$d_{ij}(k+1) \geq \rho_1, \tag{11.29}$$
$$P^T L(\mathcal{G}(k+1))P \geq \gamma I_{n-1} \tag{11.30}$$

for $i = 1, 2, \ldots, n-1, j = 2, \ldots, n, i < j$, and

$$x(k) = [x_1(k), x_2(k), \ldots, x_n(k)]^T \in \mathbf{R}^{3n}.$$

The algorithm is initiated at time $k = 0$ with an initial graph $\mathcal{G}(0) = \mathcal{G}_0$, and then for $k = 0, 1, 2, \ldots$, we proceed to iteratively find a graph that maximizes $\lambda_2(\mathcal{G}_{k+1})$ by moving the mobile robots. This greedy procedure is then iterated upon until the value of $\lambda_2(\mathcal{G}_k)$ cannot be improved further. We note that the proposed greedy algorithm converges, as the sequence generated by it is nondecreasing and bounded from above.[8]

11.5.1 Euclidean Distance Matrices

In previous section, we proposed an algorithm that converges to an equilibrium configuration, in search of maximizing the quantity $\lambda_2(L(\mathcal{G}))$ for a state-dependent graph. However, by replacing the nonconvex constraint (11.24) with its linear approximation (11.28) - (11.29), one introduces a potential inconsistency between the position and the distance vectors. In this section, we provide two remedies to avoid such potential complications. Let us first recall the notion of a *Euclidean distance matrix* (EDM).

Given the position vectors $x_1, x_2, \ldots, x_n \in \mathbf{R}^3$, the EDM $E = [d_{ij}] \in \mathbf{R}^{n \times n}$ is defined entrywise as $[E]_{ij} = d_{ij} = \|x_i - x_j\|^2$ for $i, j = 1, 2, \ldots, n$.

The EDMs are nicely characterized in terms of linear matrix inequalities.

Theorem 11.14. *A matrix $E = [d_{ij}] \in \mathbf{R}^{n \times n}$ is an EDM if and only if*

$$JEJ \leq 0, \tag{11.31}$$
$$d_{ii} = 0 \quad for \ i = 1, 2, \ldots, n, \tag{11.32}$$

where $J = I - \mathbf{1}\mathbf{1}^T/n$.

---

[8]The bound follows from the fact that the second smallest eigenvalue of $L(\mathcal{G})$ for a graph of order n is bounded from above by n.

Theorem 11.14 allows us to guarantee that by adding the two convex constraints (11.31) - (11.32) to problem Λ_k (11.27) - (11.30), we always obtain consistency among the position and distance variables at each iteration step. Moreover, by updating the values of the $d_{ij}(k)$ and $[L(k)]_{ij}$ in (11.28) and (11.30), *after* calculating the values of $x(k)$, we can further reduce the effect of linearization in the proposed procedure.

To further expand on this last point, suppose that $x_1(k), x_2(k), \ldots, x_n(k)$, $d_{ij}(k)$, and $[L(k)]_{ij}$, $i = 1, 2, \ldots, n-1$, $j = 2, \ldots, n$, $i < j$, have been obtained after solving the problem Λ_k (11.27) - (11.30). Our proposed modification to the original algorithm thus amounts to updating the values of $d_{ij}(k)$ and $[L(k)]_{ij}$, based on the computed values of $x_1(k), x_2(k), \ldots$, $x_n(k)$, before initiating the next iteration.

11.5.2 An Example

In order to examine how the proposed semidefinite programs dictate the evolution of state-dependent graphs we consider a few representative scenarios.

Figure 11.1 depicts the behavior of six mobile elements under the guidance of the proposed algorithm, leading to a planar configuration that locally maximizes $\lambda_2(L(\mathcal{G}))$. The constants ϵ, ρ_1, and ρ_2 in (11.17) are chosen to be 0.1, 1, and 1.5, respectively. The algorithm was initialized with a configuration that corresponds to a path graph. In this case, the sequence of configurations converges to the truss-shaped graph with the $\lambda_2(L(\mathcal{G}))$ value of 1.6974. For this set of parameters, the truss-shaped graph suggested by the algorithm is the global maximum over the set of graphs on six vertices that can be configured in the plane.[9]

Using the same simulation scenario, but this time, in search of an optimal positional configuration in $\mathbf{R}^3$, the algorithm leads to the trajectories shown in Figure 11.2. In this case, the graph sequence converges to an octahedron-shape configuration with $\lambda_2(\mathcal{G}) = 4.02$.

Increasing the number of nodes to eight, the algorithm was initialized as the unit cube; the resulting trajectories are shown in Fig 11.3. In this figure, the edges between vertices i and j indicate that $d_{ij} \leq \rho_2 = 1.5$. The solid lines in Figure 11.3 represent the final configuration with $\lambda_2(\mathcal{G}) = 2.7658$. Once again, an exhaustive search procedure indicates that the proposed algorithm does lead to the corresponding global optimal configuration.

---

[9]A global maximum may be found in the following exhaustive procedure. First, define a space large enough to contain the optimal configuration. Then grid this region and search over the set of all n grid points for the configuration that leads to the maximum value for $\lambda_2(\mathcal{G})$.

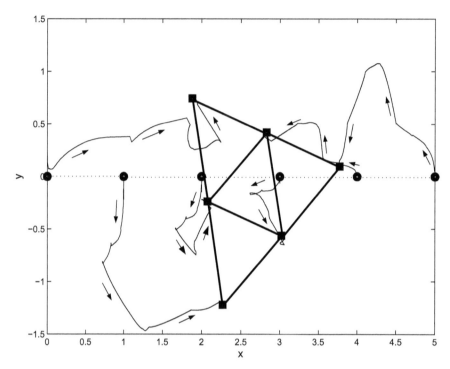

Figure 11.1: The trajectory generated by the proposed algorithm for six nodes in $\mathbf{R}^2$: the configuration evolves from a path graph (circles) to a truss (squares).

11.6 OPTIMIZING THE WEIGHTED AGREEMENT

Next we consider the effect of including weights on the edges of the graph $\mathcal{G} = (V, E)$ with n vertices and m edges. In this case one has

$$L_w(\mathcal{G}) = D(\mathcal{G})WD(\mathcal{G})^T,$$

where $W \in \mathbf{R}^{m \times m}$ is a diagonal matrix. Suppose that the sum of the weights on the edges of the graph is normalized, say, **trace** $W = 1$.

Consider now the selection of the weights $[W]_{ii}$ such that the agreement protocol has a rapid convergence to the agreement subspace. Since $\lambda_2(\mathcal{G})$ provides the slowest mode of convergence for the protocol, it is natural to consider the optimization problem

$$\max_{W} \lambda_2(D(\mathcal{G})WD(\mathcal{G})^T) \tag{11.33}$$

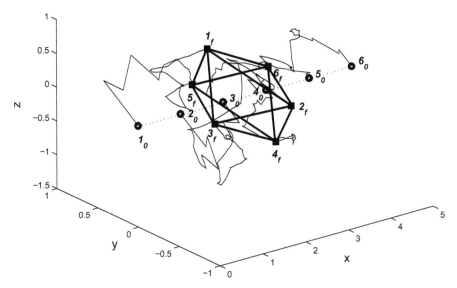

Figure 11.2: The trajectory generated by the proposed algorithm for six nodes in $\mathbf{R}^3$: the configuration evolves from a path graph (circles, $1_0, \ldots, 6_0$) to an octahedron (squares, $1_f, \ldots, 6_f$).

subject to

$$\mathbf{trace}\ W = 1, \qquad (11.34)$$
$$W \text{ is diagonal.} \qquad (11.35)$$

Now, Proposition 11.11 states that if we know the m vectors spanning the null space of the $n \times n$ symmetric matrix M, then we can create an $(n - m) \times (n - m)$ matrix having the same eigenvalues as M, excluding the m zero eigenvalues associated with the underlying null space. As we know that the vector $\mathbf{1}$ belongs to the null space of the weighted graph Laplacian for any $\mathcal{G}$, we can directly access the second smallest eigenvalue of the Laplacian through Proposition 11.11. Thus, the problem (11.33) - (11.35) can be restated as,

$$\gamma^* = \max_{\gamma, W}\ \gamma, \qquad (11.36)$$

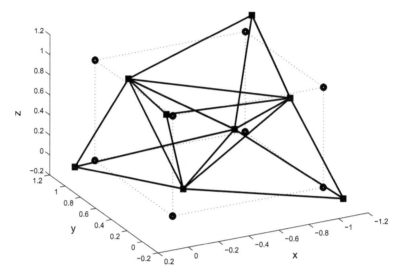

Figure 11.3: Evolution of the proposed algorithm for eight nodes in $\mathbf{R}^3$: The configuration evolves from 3-cube (circles) to octahedron (squares).

subject to

$$P^T D(G) W D(G)^T P \geq \gamma I_{n-1}, \qquad (11.37)$$

$$\textbf{trace } W = 1, \qquad (11.38)$$

$$W \text{ is diagonal}, \qquad (11.39)$$

which is a semidefinite program in the variables γ and W (see Appendix A.5); the columns of the matrix P in (11.37) consists of the normalized basis vectors for the subspace $\mathbf{1}^\perp$.

Example 11.15. *The Laplacian for the weighted graph shown in Figure 11.4 is of the form*

$$L_w(\mathcal{G}) = D(\mathcal{G}) W D(\mathcal{G})^T,$$

where

$$D(\mathcal{G}) = \begin{bmatrix} -1 & 0 & 0 & 0 \\ 1 & -1 & -1 & 0 \\ 0 & 1 & 0 & 0 \\ 0 & 0 & 1 & -1 \\ 0 & 0 & 0 & 1 \end{bmatrix} \quad and \quad W = \begin{bmatrix} w_1 & 0 & 0 & 0 \\ 0 & w_2 & 0 & 0 \\ 0 & 0 & w_3 & 0 \\ 0 & 0 & 0 & w_4 \end{bmatrix}.$$

Moreover, we can let

$$P = \begin{bmatrix} -0.2560 & -0.2422 & 0.7071 & -0.4193 \\ 0.8115 & 0.3175 & 0.0000 & -0.2018 \\ -0.2560 & -0.2422 & -0.7071 & -0.4193 \\ -0.4375 & 0.7031 & -0.0000 & 0.3380 \\ 0.1380 & -0.5362 & 0.0000 & 0.7024 \end{bmatrix}$$

as the matrix P specified in Proposition 11.11 whose columns span the sub-space orthogonal to **span** $\{1\}$. *The optimization problem (11.36) - (11.38) now yields*

$$\gamma^* = 0.1471,$$

and

$$w_1^* = 0.1765, \ w_2^* = 0.1765, \ w_3^* = 0.3824, \ w_4^* = 0.2647.$$

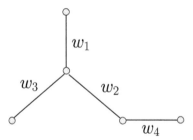

Figure 11.4: Choosing the weights to maximize the second smallest eigenvalue of the weighted Laplacian

SUMMARY

In this chapter, we delved into various aspects of the network formation or synthesis problem using optimization and game theoretic models. The optimization models the we encountered employed matrix variables and where often represented as semidefinite programs. We also used game theoretic and linear algebraic approaches for characterizing means by which the nodes in the network can update their degrees in order to improve the algebraic connectivity of the network. We also examined the notion of price of anarchy and potential games in the context of network formation problems.

NOTES AND REFERENCES

The area of network synthesis has attracted the attention of a number of disciplines, including economics, computer science, and engineering. In economics, for example, it is of interest to characterize how various interdependencies among different entities lead to particular economic indicators. In computer science, on the other hand, network design is often posed and studied in the context of distributed computation and routing over networks—a related construct is that of expander graphs. In various branches of engineering, particularly in the wireless community, these type of problems are referred to as topology design, where the performance of the overall system, that is, battery life and throughput, has a direct relation to the underlying wireless network.

Our point of view is more tinted by the applications in networked dynamic systems where a group of dynamic agents interact over a network and it is desirable to quantify the performance of the system as a function of the network geometry. Our discussion of network formation games closely follows Fabrikant, Luthra, Maneva, Papadimitriou, and Shenker [80] and the excellent review by Tardos and Wexler [232], where other network games as well as relation to the facility location problems are discussed.

We note that Proposition 11.5 can be generalized as follows. The price of anarchy is $\mathcal{O}(1)$ whenever α is $\mathcal{O}(\sqrt{n})$. More generally, the price of anarchy is $\mathcal{O}(1 + \alpha/\sqrt{n})$; see [232]. One common criticism of the notion of Nash equilibrium in the context of social networks is that an explicit account for existence or absence of an edge can potentially depend on the decision of *a pair* of nodes. Although this qualification can be introduced via "coalitions," a more specific notion of equilibrium can be used, namely pairwise stability [122]. In particular, a network $\mathcal{G}$ is pairwise stable if (a) for all $ij \in \mathcal{G}$, $\mathbf{u}_i(\mathcal{G}) \geq \mathbf{u}_i(\mathcal{G} \backslash ij)$ and (b) for all $ij \notin \mathcal{G}$ if $\mathbf{u}_i(\mathcal{G} + ij) > \mathbf{u}_i(\mathcal{G})$ then $\mathbf{u}_j(\mathcal{G} + ij) < \mathbf{u}_j(\mathcal{G})$. Thus a network is pairwise stable if no agent wants to sever a link and no two agents want to add a link. Note that while pairwise stability is a one-link-at-a-time concept, the agent might benefit from severing multiple links at the same time. Using this concept, analogous constructs that parallel the local connection game can be examined. More generally, a theory based on ordinal potential functions, resembling the notion of potential games, can be used to develop algorithms that can lead a group of agents to pairwise stable configurations. Finally, we note that the characterization of EDMs in terms of linear matrix inequalities is shown by Gower [104].

Section 11.3.1 parallels the work of Ghosh and Boyd [97] who proposed algorithms for adding edges, one at a time, to a graph to maximize the second smallest eigenvalue of its graph Laplacian. Section 11.4 is based on the work of Kim and Mesbahi [134]. This work, which is motivated by synthesis of "state-dependent graphs," in turn is related to some other works including those by Fallat and Kirkland [84] where a graph theoretical approach has been proposed to extremize $\lambda_2(L(\mathcal{G}))$ over the set of trees of fixed diameter. Also related to [134] are those by Chung and Oden [49] pertaining to bounding the gap between the first two eigenvalues of graph Laplacians, and Berman and Zhang [21] and Guattery and Miller [108], where, respectively, isoperimetric numbers of weighted graphs and graph embeddings are employed for lower bounding the second smallest Laplacian eigenvalue. The version of this problem where the weights of a given graph are adjusted to lead to fastest convergence was considered by Xiao and Boyd [252]. We note that maximizing the second smallest eigenvalue of state-dependent graph Laplacians over arbitrary graph constraints is a difficult computational problem [177].

Example 11.15 was solved using the CVX package developed by Grant and Boyd [106].

SUGGESTED READING

For much more on the network formation problems as they arise in economics, we suggest Jackson [122] and Goyal [105]. The edited volume [176] has a number of articles devoted to various aspects of network formation with particular emphasis on their algorithmic implications.

EXERCISES

Exercise 11.1. Give an example of a graph $\mathcal{G}$ where adding an edge does not improve $\lambda_2(\mathcal{G})$ and one where it does.

Exercise 11.2. Show that if $\lambda_2(\mathcal{G})$ has an algebraic multiplicity m in the graph $\mathcal{G}$, then adding up to m edges will not improve it.

Exercise 11.3. Show that a graph on n vertices is connected if it contains more than $(n-1)(n-2)/2$ edges.

Exercise 11.4. Show that if the graph $\mathcal{G}$ has two connected components, then the greedy algorithm suggested by (11.8) for adding edges, will result in the addition of an edge that connects these two components.

Exercise 11.5. An ϵ-expander is a k-regular graph $\mathcal{G} = (V, E)$ on n vertices ($k \geq 3$) such that for every $W \subseteq V$ with card$(W) \leq n/2$, the number of vertices in $V \backslash W$ adjacent to some vertex in W is at least ϵ card(W). Show that if $\lambda_2(\mathcal{G}) \geq 2\epsilon k$, then $\mathcal{G}$ is an ϵ-expander.

Exercise 11.6. Show that the Laplacian-based heuristic suggested by (11.8) for adding edges can also be reversed, that is, to remove edges that affect the algebraic connectivity the least.

Exercise 11.7. Show that in any finite potential game, the best response dynamics always converge to the Nash equilibrium; that is, prove Theorem 11.7.

Exercise 11.8. Extend the approach for finding lower and upper bounds on the second smallest eigenvalue of the augmented Laplacian for the case where the edges are weighted.

Exercise 11.9. Consider the cost $\alpha d(i)^2 + \sum_j \mathbf{dist}(i,j)$ in the context of the discussion in §11.2.1. Using this cost, for what values of α is the cycle graph a social optimal for an n-node network?

Exercise 11.10. Use the approach of §11.6 and a semidefinite programming solver (such as the one mentioned in notes and references), to maximize $\lambda_2(\mathcal{G})$ for the weighted versions of K_5, P_5, and S_5, subject to a normalization on the sum of the weights. Comment on any observed patterns for the optimal weight distribution.

Chapter Twelve

Dynamic Graph Processes

> "The best way to have a good idea is
> to have lots of ideas."
> — Linus Pauling

In this chapter, we consider the situation where the geometry of the network is a function of the underlying system's states. Certain aspects of the resulting structure, having a mixture of combinatorial and system theoretic features, are then examined. In this avenue, we will explore the interplay between notions from extremal graph theory and system theory by considering a controllability framework for what we will refer to as state-dependent dynamic graphs. We then explore the ramification of this framework for examining which formations are feasible when the underlying interaction model is specified by a state-dependent graph.

In certain classes of distributed systems, the existence and the quality of the information-exchange mechanism between a pair of dynamic elements is determined–either fully or partially–by their respective states. We refer to the resulting structure, reflecting the dynamic nature of the agents' states on one hand, and the combinatorial character of their interactions on the other, as *state-dependent dynamic graphs*, or *state-dependent graphs*. As we will see in this chapter, state-dependent graphs not only put the study of dynamic networks in the realm of system theory, but also invite us to consider a host of new problems in system theory that are distinctively combinatorial. A problem that highlights both of these facets pertains to the "controllability" of state-dependent graphs. We will also explore the ramification of the state-dependent graph framework when addressing the problem of determining the feasibility of state-dependent graph formations.

12.1 STATE-DEPENDENT GRAPHS

Consider a set of cubes that can rotate, whose sides are color-coded. Let us assume that each color represents one type of sensing or communication

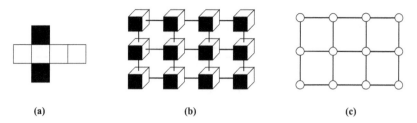

(a)   (b)   (c)

Figure 12.1: (a) The faces of each element are color-coded; (b) the elements can exchange information when the same color coded sides are facing each other; (c) the information graph associated with (b)

device that the element employs to exchange information with the other elements in the system. Moreover, we assume that each pair can exchange information if the correct color sides are facing each other. As an example, when the elements are color coded as in Figure 12.1(a), we may require that they can only exchange information when the white, or the black, sides are facing each other. Hence for the arrangement in Figure 12.1(b), we obtain the information graph of Figure 12.1(c). An analogous scenario also applies to, say, multiple spacecraft systems, not just those consisting of "cubesats," but also systems such as those depicted in Figure 12.2, where each spacecraft is equipped with a directional antenna for accurate relative sensing. In this case, the resulting "relative sensing graph" exhibit a dynamic character that reflects the relative orientation and position of the multiple spacecraft. Evidently, as the translational and rotational states of these cubical elements or robotic systems evolve over time, we obtain a sequence of information graphs; in particular, we realize that the corresponding information-exchange graph is *state-dependent* and, in general, dynamic.

Another example of a state-dependent graph, of particular relevance in the control of multiple unmanned aerial vehicles (UAVs), is the nearest neighbor information exchange paradigm. In this framework, there is an information channel, for example, relative sensing capability, between a pair of UAVs if they are within a given distance of each other. As the positions of the UAVs evolve in time, say during the course of a reconfiguration, the underlying information exchange infrastructure naturally evolves in time as well, resulting in a dynamic proximity graph.

12.1.1 From States to Graphs

A state-dependent graph is a mapping, g_S, from the distributed system state space X to the set of all labeled graphs on n vertices $\mathbf{G}(n)$, that is,

Figure 12.2: Conceptual configuration for the Terrestrial Planet Imager, courtesy of JPL/NASA

$$g_\mathcal{S} : X \to \mathbf{G}(n) \quad \text{and} \quad g_\mathcal{S}(x) = \mathcal{G}, \tag{12.1}$$

as illustrated in Figure 12.3. We will occasionally write $\mathcal{G}_x = g_\mathcal{S}(x)$ to highlight the dependency of the resulting graph on the state x. It is assumed that the order of these graphs, n, is fixed. Their edge set, $E(g_\mathcal{S}(x))$, however, is a function of the state x.

We need to specify further *how* the state of the system dictates the existence of an edge between a pair of vertices in the state-dependent graph. This is achieved by considering the subset $S_{ij} \subseteq X_i \times X_j$, where X_i and X_j are the state spaces of agents i and j, respectively, and requiring that $\{i, j\} \in E(g_\mathcal{S}(x))$ if and only if $(x_i, x_j) \in S_{ij}$; we call S_{ij} the edge states of vertices i and j. It is assumed that the edge sets are such that $(x_i, x_j) \in S_{ij}$ if and only if $(x_j, x_i) \in S_{ji}$ for all i, j.[1]

As an example, for the nearest neighboring scenario obtained from a Δ-disk proximity graph with second-order agents and threshold value of ρ, the

---

[1] Although the existence of an edge between two agents can potentially depend on the states of other agents, we will not consider this case here. Such more general state-dependency schemes would lead us to state-dependent hypergraphs.

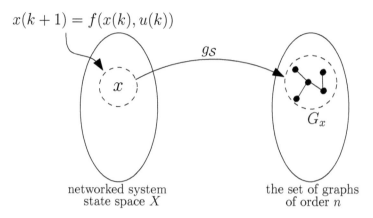

$$x(k+1) = f(x(k), u(k))$$

networked system
state space X

the set of graphs
of order n

Figure 12.3: State-dependent graphs

edge states are of the form

$$S_{ij} = \left\{ \left(\begin{bmatrix} r_i \\ v_i \end{bmatrix}, \begin{bmatrix} r_j \\ v_j \end{bmatrix} \right) \mid \|r_i - r_j\| \leq \rho \right\}, \; i, j \in [n], \; i \neq j.$$

Here r_i and v_i represent, respectively, the position and the velocity of UAV i, and ρ is a given positive number; this set is shown in Figure 12.4. In general, we will denote the collection of the edge states S_{ij} by

$$\mathcal{S} = \{S_{ij}\}_{i,j\in[n],i\neq j} \quad \text{with} \quad S_{ij} \subseteq X_i \times X_j. \tag{12.2}$$

Example 12.1. *Consider two square elements i and j, the four sides of which have alternatively been labeled by "0" and "1." The state of each square, x, is thus represented by one binary state, interpreted as the label that is facing "up." Consider the scenario where there is an edge between the vertices i and j if $x_i + x_j = 0 \pmod 2$. The state space partitions are therefore $x_i(1) = 0$, $x_i(2) = 1$, $x_j(1) = 0$, $x_j(2) = 1$, and the set of edge states is identified as $S_{ij} = \{(0,0),(1,1)\}$.*

Definition 12.2. *Given the set system $\mathcal{S}$ (12.2), we call the map $g_{\mathcal{S}} : X \rightarrow \mathbf{G}(n)$, with an image consisting of graphs of order n, having an edge between vertex i and j if and only if $(x_i, x_j) \in S_{ij}$, a state-dependent graph with respect to $\mathcal{S}$.*[2]

The image of the state-dependent graph $g_{\mathcal{S}}$, is thus

$$\{\mathcal{G} \mid g_{\mathcal{S}}(x) = \mathcal{G}, \text{ for some } x \in X\} = \{\mathcal{G}_x \mid x \in X\},$$

which will be denoted by $g_{\mathcal{S}}(X)$.

---

[2]We could alternatively call $\mathcal{S}$ (12.2) itself the state-dependent graph.

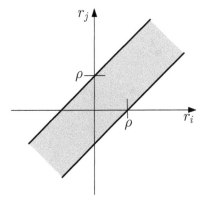

Figure 12.4: The nearest neighbor S_{ij} in $\mathbf{R} \times \mathbf{R}$

12.2 GRAPHICAL EQUATIONS

It seems natural that static state-dependent graphs are examined prior to studying "dynamic" state-dependent graphs. In this section, we consider problems related to checking for the existence of, and possibly solving for, states that have a particular graph realization.

12.2.1 Systems of Inequalities

Given the set S (12.2) and a labeled graph $\mathcal{G}$ of order n, we consider finding solutions to the equation

$$g_S(x) = \mathcal{G}. \tag{12.3}$$

Note that, depending on specific applications, "equality" between a pair of graphs in (12.3) can be considered as a strict equality or up to an isomorphism. In § 12.3, we will consider a scenario where it is more natural to consider a subgraph inclusion relation of the form $\mathcal{G} \subseteq g_S(x)$ rather than the equality in (12.3).

Graphical equation solving can become equivalent to solving systems of equations and inequalities, depending on the characterization of the edge states in S. Let us elaborate on this observation with two examples.

Example 12.3. *Let $\mathcal{G} = (V, E)$ and $V = \{1, 2, 3\}$. The eight possible labeled graphs on three vertices is shown in Figure 12.5, with $\mathcal{G}_8$ denoting K_3. Thus the equation $g_S(x) = \mathcal{G}_8$ has a solution if and only if the set*

$$g^{-1}(\mathcal{G}_8) = \{x \mid (x_1, x_2) \in S_{12}, (x_1, x_3) \in S_{13}, (x_2, x_3) \in S_{23}\}$$

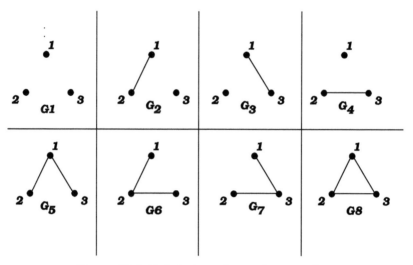

Figure 12.5: Labeled graphs on three vertices

is nonempty. Similarly, the equation $g_S(x) = \mathcal{G}_6$, is solvable if and only if the set

$$g^{-1}(\mathcal{G}_6) = \{x \mid (x_1, x_2) \in S_{12}, (x_1, x_3) \in \overline{S}_{13}, (x_2, x_3) \in S_{23}\}$$

is nonempty.

Example 12.4. *For all $i, j \in \{1, 2, 3\}$, $i \neq j$, assume that*

$$S_{ij} = \{(x_i, x_j) \mid q_{ij}(x_i, x_j) \leq 0\} \subseteq \mathbf{R}^n \times \mathbf{R}^n,$$

where $q_{ij}(x_i, x_j) = x_i^T Q_{ij} x_j$ and $Q_{ij} = Q_{ij}^T \in \mathbf{R}^{n \times n}$. Then the equation $g_S(x) = \mathcal{G}_5$ (Figure 12.5) has a solution if and only if the set

$$\{x \mid q_{12}(x_1, x_2) \leq 0, q_{13}(x_1, x_2) \leq 0, q_{23}(x_2, x_3) > 0\}$$

is nonempty. On the other hand, this set is empty if and only if $q_{12}(x_1, x_2) \leq 0$ and $q_{13}(x_1, x_3) \leq 0$ imply that

$$q_{23}(x_2, x_3) \leq 0.$$

Now extend the q_{ij}, making them functions of $x = [x_1, x_2, x_3]^T$, for example, $\tilde{q}_{ij}(x) = q_{ij}(x_i, x_j) + 0 \times x_k$ where $k \neq i$ and $k \neq j$, and let $\tilde{q}_{ij}(x) = x^T \tilde{Q}_{ij} x$. Using the S-procedure (see Appendix A.5) we conclude that $g_S(x) = G_5$ has no solution if there exist $\tau_1, \tau_2 \geq 0$ such that $\tilde{Q}_{23} \leq \tau_1 \tilde{Q}_{12} + \tau_2 \tilde{Q}_{13}$, where an inequality between two symmetric matrices is interpreted in terms of the ordering induced by positive semidefinite matrices. See also our discussion on graph realizations in §12.4.

12.2.2 Supergraphs and Equation Solving for a Class of Graphs

We now consider the case when the state space of each individual agent is a finite set. This might reflect the discrete nature of the underlying state-space (for example, the cubical agents in Figure 12.1(a) can only assume a finite number of orientations,) or as a result of bounding and creating a mosaic for the underlying continuous state space. In this situation, the state-dependency of the edges among pairs of agents can be represented in a combinatorial way using a bipartite graph. The aim of this section is to further elaborate on this connection.

Let us first construct *supervertices* that represent not only each agent, but also *all the finite states the agent can attain*. Thus, if agent i can assume five distinct states, the supervertex i has five nodes embedded in it–we refer to the vertices in each supervertex that represent a particular state for that agent as a *subvertex*. Each agent thus get inflated in this way. Next, we assign edges between the subvertices of distinct agents using the state-dependency relation. Thus if for agents i and j, the edge set S_{ij} contains (x_i^a, x_j^b) and S_{ji} contains (x_j^b, x_i^a), then there is an edge between the subvertex x_i^a in the ith supervertex and the x_j^b subvertex in the jth supervertex. In this way, each supervertex is connected to another supervertex via a group of edges that connect pairs of subvertices in each. Note that the resulting bipartite graph can be directed; however, as it has been the case in this chapter, we only focus on the symmetric case, that is, we assume that for all pair of agents i and j, if $(a, b) \in S_{ij}$ then $(b, a) \in S_{ji}$. Continuing this procedure for pairs of agents, we obtain the "supergraph" $\mathbf{G}_n(\mathcal{S})$ as the union of bipartite graphs that have been obtained for each pair of agents, representing the state-dependency of the edges between the agents (Figure 12.6).

The supergraph construction is motivated by the graphical equation solving discussed previously in this chapter. In this direction, we first note that the subgraphs of the supergraph $\mathbf{G}_n(\mathcal{S})$, employing "one" subvertex from each supervertex, are exactly those graphs that can be formed when agents assume one of their admissible states. Of course, such constructions have names in graph theory: we call $\mathcal{G}$ a *transversal* subgraph of the supergraph $\mathbf{G}_n(\mathcal{S})$ if

1. its vertices are subvertices of $\mathbf{G}_n(\mathcal{S})$, and

2. its vertices all belong to distinct supervertices of $\mathbf{G}_n(\mathcal{S})$.

If $\mathcal{G}$ is a transversal subgraph of $\mathbf{G}_n(\mathcal{S})$ and contains all of its potential edges (for the same vertex set), it is called an induced transversal. We will denote the subgraph and induced subgraph transversal by $\mathcal{G} \subsetneq_T \mathbf{G}_n(\mathcal{S})$ and $\mathcal{G} \subseteq_I \mathbf{G}_n(\mathcal{S})$, respectively. This is shown in Figure 12.7.

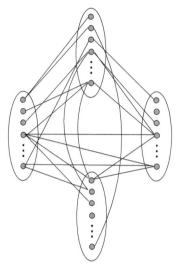

Figure 12.6: Supergraph $\mathbf{G}_n(\mathcal{S})$; the four oval shaped objects denote the supervertices, whereas the embedded nodes denote the subvertices.

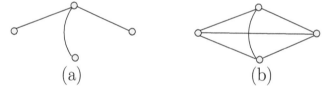

(a)   (b)

Figure 12.7: (a) A transversal subgraph of the supergraph $\mathbf{G}_n(\mathcal{S})$ in Figure 12.6, (b) an induced transversal subgraph of $\mathbf{G}_n(\mathcal{S})$

The supergraph construction has the following immediate ramification.

Proposition 12.5. *Given the collection of edge states $\mathcal{S}$ (12.2), the equation $g_{\mathcal{S}}(x) = G$ has a solution if and only if $G \subseteq_I \mathbf{G}_n(\mathcal{S})$. Furthermore, the inclusion $G \subseteq g_{\mathcal{S}}(x)$ has a solution if and only if $G \subseteq_T \mathbf{G}_n(\mathcal{S})$.*

12.3 DYNAMIC GRAPH CONTROLLABILITY

In this section, we allow the state of each agent to evolve over time, with the resulting graph assuming a dynamic character–the graph is thus elevated to a *graph process*. In fact, we would like to investigate whether one can define a notion of "controllability" for such graph processes. Controllable (state-dependent) graph processes are such that every graph (either labeled

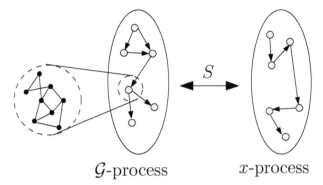

$$\mathcal{G}\text{-process} \qquad\qquad x\text{-process}$$

Figure 12.8: The x-process and the $\mathcal{G}$-process

or up to an isomorphism, depending on the context) is reachable by a judi-cious choice of the control sequence for the agents. Our goal is to explore connections between the controllability of state of the agents, henceforth re-ferred to as the x-process, and the corresponding graph process named as the $\mathcal{G}$-process.[3] For example, it will be of interest for our discussion to exam-ine the conditions under which the controllability of the x- and $\mathcal{G}$-processes become equivalent.

Toward providing a useful notion of graph controllability, we denote by $\mathbf{G}(n, \Delta)$ the set of graphs of order n with maximum vertex degree Δ. Therefore, $\mathbf{G}(n, 0) = \{\overline{K_n}\}$ and $\mathbf{G}(n, n-1) = \mathbf{G}(n)$. Our first notion for graph controllability relies on graph reachability from an arbitrary ini-tial state.

Definition 12.6. *Consider a set of n agents that form a state-dependent graph. Then the resulting $\mathcal{G}$-process is strictly Δ-controllable if, for any $\mathcal{G}_f \in \mathbf{G}(n, \Delta)$ and any initial state $x_0 \in X$, there exist a finite k and an x-process for which $\mathcal{G}_f \subseteq g_{\mathcal{S}}(x(k))$. When $\Delta = n - 1$, we refer to strictly Δ-controllable $\mathcal{G}$-processes as strictly controllable.*

Including the maximum degree Δ qualification in the definition of graph controllability is not purely accidental, although other graph parameters could be employed in this definition as well. The maximum vertex degree, however, does have implications on the relative sensing "overhead" of a net-worked system and the overall graph "complexity," and, as such, is a prudent

---

[3]We will not delve into the various notions of controllability. The x-process is control-lable if for two arbitrary states, there exists a control sequence that steers one to the other.

graph parameter in Definition 12.6.

Our second definition of graph controllability relies more on the properties of the $\mathcal{G}$-process itself rather than the x-process.

Definition 12.7. *Consider a set of n agents that form a state-dependent graph. The $\mathcal{G}$-process is Δ-controllable if, for two graphs $\mathcal{G}_0, \mathcal{G}_f$ belonging to $\mathbf{G}(n, \Delta)$, there exist a finite k and an underlying x-process, for which*

$$\mathcal{G}_0 \subseteq g_{\mathcal{S}}(x(0)) \quad \text{and} \quad \mathcal{G}_f \subseteq g_{\mathcal{S}}(x(k)).$$

When $\Delta = n - 1$, we refer to a Δ-controllable $\mathcal{G}$-process simply as controllable.

The distinction between Definitions 12.6 and 12.7 is the qualification on graph reachability from *an arbitrary state* versus from *an arbitrary graph*, the latter being less stringent.[4]

We note that both the cardinalities of the state space for each agent and the number of elements in the network play an important role in the graph controllability properties. For example, it can be shown that when the cardinality of the state space for each element is p and $2 \log p + 1 < n$, the labeled $\mathcal{G}$-process cannot be controllable.[5]

12.3.1 Calmness

In order to highlight the connection between strict controllability and controllability of the $\mathcal{G}$-process, we introduce the notion of *calmness*. Calmness refers to the situation where a particular graph remains *invariant* as a subgraph of $g_{\mathcal{S}}(x)$ as the state of the agents x evolves during the interval $[t_0, t_f]$.

Definition 12.8. *A graph $\mathcal{G} \in \mathbf{G}(n, \Delta)$ is strictly calm with respect to the controlled x-process, if (1) for any x_0, x_f for which $\mathcal{G} \subseteq g_{\mathcal{S}}(x_0), g_{\mathcal{S}}(x_f)$, there exists a control sequence that steers x_0 to x_f, and (2) for all intermediate states $\mathcal{G} \subseteq g_{\mathcal{S}}(x)$.*

---

[4]Observe that Δ-controllability has a cascading property, in the sense that Δ_0-controllability for some $\Delta_0 > 0$ implies Δ-controllability for all $\Delta \geq \Delta_0$.

[5]This follows from noting that the $\mathcal{G}$-process is certainly not Δ-controllable for some $\Delta > 0$ if the cardinality of the agent's state space is less than $2^{\binom{n}{2}}$, the total number of labeled graphs on n vertices.

Hence the empty graph $\overline{K}_n$ is always strictly calm. When the second qualification in the above definition fails to hold, the graph $\mathcal{G} \in \mathbf{G}(n, \Delta)$ is simply called "calm." Moreover, when every element of a subset of $\mathbf{G}_n^\Delta$ is (strictly) calm with respect to the x-process, the subset itself is referred to as (strictly) calm.

Let us now use the machinery of Δ-controllable graph processes and graph calmness to shed light on the relationship between controllability of the x-process, that is, the controllability of the agents' state and the controllability of the graph process that it generates via the edge state dependency. In this venue, for a group of n agents, let us define the set

$$X_\Delta = \{x \in X \mid \max_{v \in g_S(x)} d(v) \leq \Delta\} = \{x \in X \mid \mathcal{G}_x \in g_S(X) \cap \mathbf{G}(n, \Delta)\},$$

that is, the set of agents' state whose corresponding state-dependent graph on n vertices has a maximum degree less than Δ.

Proposition 12.9. *Consider a group of n agents that form a state-dependent graph. If the $\mathcal{G}$-process is Δ-controllable and $g_S(X)$ is calm with respect to the x-process, then the $\mathcal{G}$-process is strictly Δ-controllable in X_Δ.*

Proof. Let $z \in X_\Delta$ and $\mathcal{G}_f \in \mathbf{G}(n, \Delta)$ be given. It suffices to show that there is a control sequence steering z to x_f where $\mathcal{G}_f \subseteq g_S(x_f)$. The Δ-controllability of the $\mathcal{G}$-process does imply that there is an x-process, taking x_o to x_f, for which $g_S(z) \subseteq g_S(x_o)$ and $\mathcal{G}_f \subseteq g_S(x_f)$. As $g_S(z), g_S(x_o) \subseteq g_S(x_o)$ and $g_S(x_o)$ is Δ-calm, there is a control sequence that steers z to x_o. Joining the two control sequences together now completes the proof. ∎

12.3.2 Regularity and Graph Controllability

We now proceed to explore the *controllability* correspondence between the x- and the $\mathcal{G}$-processes. In this direction, we first note that if there is a natural bijection between the x and the $\mathcal{G}$ processes, then their system theoretic properties will have a more direct interrelationship with each other. Thus a promising direction for allowing us to translate and interpret controllability properties between the agents' state and the resulting graph process is to induce a notion of "pseudo-invertability" that would allow assigning graph topologies to the underlying state of the agents.

Let us thereby digress a bit to introduce a notion that essentially allows us to formalize this notion of pseudo-inverse for the edge state dependency map g_S (12.1).

12.3.3 Szemerédi's Regularity

Given the collection of edge states S (12.2), the *edge state density* between agents i and j is defined as

$$\varrho_S(X_i, X_j) = \frac{\varepsilon(X_i, X_j)}{\operatorname{card}(X_i)\operatorname{card}(X_j)}, \qquad (12.4)$$

where X_i is the state space of agent i, assumed to have finite cardinality, $\operatorname{card}(X_i)$, and $\varepsilon(X_i, X_j)$ is the number of edges between X_i and X_j. In the spirit of §12.2.2, we view X_i, the finite state space of agent i, as a supervertex, that is, it represents not only the vertex i, but all the finite number of states that the agent can attain. The individual supervertices, with the edges that represent the state-dependent edges between each pair of agents, then constitute, as in §12.2.2, the supergraph for the multiagent system. Hence, $\varrho_S(X_i, X_j)$ is the ratio between the states that result in an edge between the two agents i and j and the total number of states that these agents can be in. In order to avoid carrying around the notion $\operatorname{card}(X_i)$, we will assume, without loss of generality, that $\operatorname{card}(X_i) = p$ for all i, and thus (12.4) can be defined as

$$\varrho_S(X_i, X_j) = \frac{\operatorname{card}(S_{ij})}{p^2}, \qquad (12.5)$$

where S_{ij} is the pairs of states for agents i and j, that result in an edge between these agents. Hence for each pair of supervertices, the $\binom{n}{2}$ numbers $\varrho_S(X_i, X_j)$ (12.5), each assuming a value in the unit interval, reflect the ratios of the states that are designated as the edge states.

Now, a moments reflection on how a systematic means of mapping agents' states to graph topologies, and vice versa, can be developed, reveals that we need to require that the edge state assignment between the supervertices for the multi-agent network to be somehow "uniform." However, this uniformity has to be imposed for each subset of subvertices–which leads us to the concept of regularity for the supergraph.

Definition 12.10. *For $\epsilon > 0$, the pair (X_i, X_j) is called ϵ-regular at level ρ if (1) $\varrho_S(X_i, X_j) \geq \rho$, and (2) for every $Y_i \subseteq X_i$ and $Y_j \subseteq X_j$ satisfying*

$$\operatorname{card}(Y_i) > \epsilon \operatorname{card}(X_i) \quad \text{and} \quad \operatorname{card}(Y_j) > \epsilon \operatorname{card}(X_j)$$

one has

$$|\varrho_S(X_i, X_j) - \varrho_S(Y_i, Y_j)| < \epsilon. \qquad (12.6)$$

Let us decipher this definition. We consider two arbitrary supervertices in a supergraph and find the density of the state-dependent edges between them. If the supergraph is ϵ-regular, the density at the supervertex level should be ϵ-close to the density of the state-dependent edges when we zoom in on two arbitrary subsets of subvertices embedded in these respective supervertices. Moreover, we put a bound on how much our lens is required to zoom in inside each pair of supervertices–we only require the regularity in the density of the edge states for a group of subvertices that have cardinality at least ϵ times the cardinality of the supervertex that they belong to. We refer to a supergraph $\mathbf{G}_n(\mathcal{S})$ as ϵ-regular at level ρ if each pair of its supervertices is ϵ-regular at a level of at least ρ. We will also denote an ϵ-regular supergraph at level ρ by $\mathbf{G}_n(\mathcal{S}_{\epsilon,\rho})$ (recall that each supervertex has p subvertices). An important consequence of the notion of regularity is the following observation.

Proposition 12.11. *Consider two ϵ-regular supervertices X_i, X_j with*

$$\varrho(X_i, X_j) = \rho.$$

Let $\Psi \subseteq X_i$ be the set of all subvertices with at least $(\rho - \epsilon)\mathrm{card}(X_j)$ neighbors in X_j. Then

$$\mathrm{card}(\Psi) \geq (1 - \epsilon)\mathrm{card}(X_i).$$

Proof. Suppose that the number of subvertices $x_i \in X_i$, having strictly fewer neighbors in the supervertex X_j than $(\rho - \epsilon)\,\mathrm{card}(X_j)$, is strictly more than $\epsilon\,\mathrm{card}(X_i)$. Denote this set by $\overline{\Psi}$. Then

$$\varepsilon(\overline{\Psi}, X_j) < (\rho - \epsilon)\,\mathrm{card}(\overline{\Psi})\mathrm{card}(X_j),$$

that is, $\varepsilon(\overline{\Psi}, X_j) < (\rho - \epsilon)$ and the pair $(\overline{\Psi}, X_j)$ violates the regularity assumption on the pair (X_i, X_j) (12.6). ∎

We now shift our attention back to the main justification for introducing the notion of supergraph regularity as an effective means of addressing the controllability of state-dependent graph processes. The connection is facilitated by a result that is referred to as the *key lemma* in extremal graph theory, which is discussed next.

12.3.4 Key Lemma

The regularity of the supergraph $\mathbf{G}_n(\mathcal{S})$ provides a level of transparency between the x-process and the $\mathcal{G}$-process, allowing us to make a correspondence between their controllability properties. In fact, the regularity of the supergraph imposes a "pseudo-invertibility" condition on the map $\mathcal{G}$ (12.1) which has a direct controllability interpretation. This connection is made more explicit via the key lemma.

Lemma 12.12. *Consider a set of n agents that form a state-dependent graph, with the corresponding supergraph, representing how edges between each pair of agents depend on their respective states. Let $\rho > \epsilon > 0$ be given. Assume that the supergraph denoted by $\mathbf{G}_n(\mathcal{S}_{\epsilon,\rho})$ is ϵ-regular, and let $\delta = \rho - \epsilon$. Let H be a graph of order n with maximum vertex degree $\Delta(H) > 0$. If*

$$\delta^{\Delta(H)}/(1 + \Delta(H)) \geq \epsilon,$$

then $H \subseteq_T \mathbf{G}_n(\mathcal{S}_{\epsilon,\rho})$, that is, H is a subgraph transversal of the super-graph of the multiagent network. Moreover, the number of such H-subgraph transversals is at least

$$(\delta^{\Delta(H)} - \Delta(H)\,\epsilon)^n\, p^n, \tag{12.7}$$

where p is the cardinality of the state space for all agents.

Let us denote by $\Gamma(x)$ the set of neighboring subvertices of x in the supergraph $\mathcal{G}_n(\mathcal{S})$. Key lemma 12.12 is established through the following constructive algorithm.

Embedding Algorithm

Initialize the sets $C_{0,j} = X_j$ for all $j = 1, \ldots, n$, and set $i = 1$.

1. Pick $x_i \in C_{i-1,i}$, such that for all $j > i$ for which $\{i, j\} \in E(H)$, one has

$$\mathrm{card}(\Gamma(x_i) \cap C_{i-1,j}) > \delta\,\mathrm{card}(C_{i-1,j}). \tag{12.8}$$

Proposition 12.11 guarantees that the set of such states is nonempty; in fact, the cardinality of the set of states that *violate* (12.8) is at most $\Delta(H)\epsilon p$.

2. For each $j > i$, let

$$C_{i,j} = \begin{cases} \Gamma(x_i) \cap C_{i-1,j} & \text{if } \{i, j\} \in E(H), \\ C_{i-1,j} & \text{otherwise.} \end{cases}$$

3. If $i = n$, terminate the algorithm; otherwise, let $i = i + 1$ and go to Step 1.

In Step 2 of the algorithm, denote the cardinality of the set $\{k \in [i] \mid \{k, j\} \in E(H)\}$ by d_{ij}; then one has

$$\text{card}(C_{i,j}) > \delta^{d_{ij}} p \geq \delta^{\Delta(H)} p,$$

when $d_{ij} > 0$, and $\text{card}(C_{ij}) = p$ when $d_{ij} = 0$. In both cases, $\text{card}(C_{i,j}) > \delta^{\Delta(H)} p$ when $j > i$. Moreover, when choosing the exact location of x_i, all but at most $\Delta(H)\epsilon p$ vertices of $C_{i-1,i}$ satisfy (12.8) as needed in Step 1 of the algorithm. Thus when finding the transversal H in $\mathcal{G}(n, \mathcal{S}_{p,\epsilon})$, at least

$$\text{card}(C_{i-1,i}) - \Delta(H)\epsilon p > \delta^{\Delta(H)} p - \Delta(H)p\epsilon \qquad (12.9)$$

free choices exist for each x_i. The estimate (12.7) for the number of embeddings of H in $\mathcal{G}(n, \mathcal{S}_{p,\epsilon})$ now follows from (12.9).

Example 12.13. *In the network of n agents, with agent i having a finite state space X_i of cardinality p, a state $x_i \in X_i$ is called a blind state of i with respect to j if*

$$(x_i, x_j) \notin S_{ij} \quad \text{for all } x_j \in X_j;$$

denote by b_{ij} the number of such states. For this example, let $b_{ij} = b_{ji}$ and furthermore, assume that all other states are edge states,

$$S_{ij} = (X_i \times X_j) \setminus (\{\, ith\ agent's\ blind\ states\ with\ respect\ to\ j\,\}$$
$$\times \{\, jth\ agent's\ blind\ states\ w.r.t.\ i\,\}).$$

Thus $\rho = \varrho_S(X_i, X_j) = 1 - (b_{ij}^2/p^2)$.
   We now proceed to check for the existence of subgraphs with vertex degree of at most 2 for the corresponding state-dependent graph on n agents. Let $\epsilon = m/p$. Lemma 12.12 suggests that we need to ensure the inequality

$$\rho \geq \epsilon + \sqrt{3}\,\epsilon, \qquad (12.10)$$

and that for all $Y_i \subseteq X_i$ and $Y_j \subseteq X_j$ of size strictly greater than m,

$$|\rho - \varrho_S(Y_i, Y_j)| < \frac{m}{p}. \qquad (12.11)$$

The maximum deviation of the quantity $\varrho_S(Y_i, Y_j)$ from the edge state density ρ occurs when

$$\varrho_S(Y_i, Y_j) = 1 - \frac{b_{ij}^2}{(m+1)^2}.$$

Thus for ϵ-regularity it suffices to have

$$\frac{p}{m(m+1)^2} - \frac{1}{mp} < \frac{1}{b_{ij}^2}.\tag{12.12}$$

We now note that for particular values of p and m in (12.11) and (12.12), one can obtain an allowable number of blind states between each pair of agents so that state-dependent subgraphs with vertex degree of at most 2 are still guaranteed to exist for the n-agent network. For example, when $p = 100$ and $m = 20$, having $b = 9$ satisfies both inequalities (12.10) and (12.12). Thereby, almost 10 percent of each agent's states can be blind states with respect to another agent, while still guaranteeing the existence of any state-dependent subgraph with a maximum vertex degree 2. In fact, the bound (12.7) indicates that there are plenty of such subgraph transversals in the corresponding supergraph–in this example, at least $20^!$ of them!

A few remarks are in order at this point. Note that guaranteeing the existence of a transversal embedding H in the supergraph $\mathcal{G}(n, \mathcal{S}_{\epsilon,\rho})$ does not depend on the parameters ρ or ϵ individually. In fact, it is their difference $\rho - \epsilon$ that dictates the number of such embeddings in $\mathcal{G}_n(\mathcal{S}_{\epsilon,\rho})$; that is, it is the *relative* order of density with respect to the "fineness" of regularity that prescribes the number of embeddings. In the meantime, the maximum vertex degree of the desired embedding accounts for the ease by which it can be embedded in the supergraph $\mathcal{G}_n(\mathcal{S}_{\epsilon,\rho})$, that is, to be realized by a judicious choice of the underlying dynamic states. Furthermore, the embedding algorithm suggests a constructive approach through which the desired subgraph can be synthesized.

12.3.5 Graph Controllability

We now reach the main result of this section–stated for a group of dynamic agents with a finite state space. We note however, that the approach can be generalized to other classes of dynamic systems (see Figure 12.9)[6]

Theorem 12.14. *The $\mathcal{G}$-process is Δ-controllable if the x-process is controllable and the supergraph $\mathcal{G}(n, \mathcal{S}_{\epsilon,\rho})$ satisfies*

$$(\rho - \epsilon)^\Delta/(1 + \Delta) \geq \epsilon.$$

On the other hand, the x-process is controllable if the $\mathcal{G}$-process is controllable and $g_S(X)$ is calm with respect to the x-process.

---

[6]The generalization involves partitioning the state space to finitely many regions and employing "measure" in place of "size" to obtain the required extension for the notion of regularity.

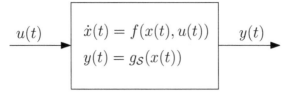

Figure 12.9: Dynamics of state-dependent graphs

Proof. Assuming that the x-process is controllable, consider graphs

$$\mathcal{G}_0, \mathcal{G}_f \in \mathcal{G}(n, \Delta).$$

By regularity of the supergraph $\mathcal{G}(n, \mathcal{S}_{\epsilon,\rho})$, there exist $x_o, x_f \in X$ such that $\mathcal{G}_0 \subseteq g_{\mathcal{S}}(x_o)$ and $\mathcal{G}_f \subseteq g_{\mathcal{S}}(x_f)$. By controllability of the x-process, how-ever, there is a control sequence that steers x_o to x_f; thus the $\mathcal{G}$-process is Δ-controllable. Now assume the controllability of the $\mathcal{G}$-process and con-sider an arbitrary pair $x_o, x_f \in X$, with the corresponding state-dependent graphs $g_{\mathcal{S}}(x_o), g_{\mathcal{S}}(x_f)$. Thereby, there are $\tilde{x}_0, \tilde{x}_f$, and a control sequence such that $g_{\mathcal{S}}(x_o) \subseteq g_{\mathcal{S}}(\tilde{x}_0)$ and $g_{\mathcal{S}}(x_f) \subseteq g_{\mathcal{S}}(\tilde{x}_f)$. As $g_{\mathcal{S}}(\tilde{x}_0) \subseteq g_{\mathcal{S}}(\tilde{x}_0)$ and $g_{\mathcal{S}}(\tilde{x}_f) \subseteq g_{\mathcal{S}}(\tilde{x}_f)$, by the calmness assumption (Definition 12.8), there is a control sequence from x_o to $\tilde{x}_f$, and analogously from $\tilde{x}_f$ to x_f (see Figure 12.10). By joining these three control sequences together, we obtain a control sequence that steers x_o to x_f, and hence the controllability of the x-process. ∎

Hence, when the underlying x-process in Example 12.13 is controllable, the associated $\mathcal{G}$-process is ensured to be 2-controllable.

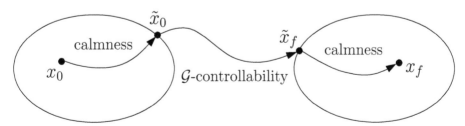

Figure 12.10: Controllability of the x-process in Theorem 12.14

12.4 WHAT GRAPHS CAN BE REALIZED?

In this section, we focus our attention on a particular class of state-dependent graphs, namely, Δ-disk proximity graph encodings of the limited-range adjacency relationships between points in the plane. Suppose we have n such point agents in $\mathbf{R}^2$. Each agent is equipped with a range-limited sensor by which it can sense the position of other agents. All agents have identical sensor ranges Δ. Let the position of each agent be $x_i \in \mathbf{R}^2$, and its dynamics be given by

$$\dot{x}_i = f(x_i, u_i), \qquad (12.13)$$

where $u_i \in \mathbf{R}^m$ is the control for agent i, and $f : \mathbf{R}^2 \times \mathbf{R}^m \to \mathbf{R}^2$ is a smooth vector field. The configuration space of the agent formation is made up of all ordered n-tuples in $\mathbf{R}^2$, with the property that no two points coincide, that is,

$$\mathcal{C}^n(\mathbf{R}^2) = (\mathbf{R}^2 \times \mathbf{R}^2 \times \cdots \times \mathbf{R}^2)\backslash\mathcal{P}, \qquad (12.14)$$

where $\mathcal{P} = \{(x_1, x_2, \ldots, x_n) \mid x_i = x_j \text{ for some } i \neq j\}$. The evolution of the formation can be represented as a trajectory

$$F : \mathbf{R}_+ \to \mathcal{C}^n(\mathbf{R}^2),$$

usually written as $F(t) = [\, x_1(t), x_2(t), \ldots, x_n(t)\,]^T$ to signify time evolution.

Now, let $\mathbf{G}(n)$ denote the space of all possible graphs that can be formed on n. Then we can define the function

$$\Phi_n : \mathcal{C}^n(\mathbf{R}^2) \to \mathbf{G}(n)$$

with $\Phi_n(F(t)) = \mathcal{G}(t)$, where $\mathcal{G}(t) = (V, E(t)) \in \mathbf{G}(n)$ is the Δ-disk proximity graph of the formation $F(t)$. As before, $v_i \in V$ represents agent i at position x_i, and $E(t)$ denotes the edges of the graph, that is, $e_{ij}(t) = e_{ji}(t) \in E(t)$ if and only if $\|x_i(t) - x_j(t)\| \leq \Delta$, $i \neq j$. These graphs are *simple* by construction, that is, there are no loops or parallel edges. The graphs are always undirected because the sensor ranges have been assumed to be identical. The motion of the agents may result in the removal or addition of edges in the graph. Therefore $\mathcal{G}(t)$ is a dynamic structure. Last and most important, every graph in $\mathbf{G}(n)$ is not a valid proximity graph, as we will see shortly.

The last observation is not as obvious as the others, and we say that a *realization* of a graph $\mathcal{G} \in \mathbf{G}(n)$ is a formation $F \in \mathcal{C}^n(\mathbf{R}^2)$ such that $\Phi_n(F) = \mathcal{G}$. An arbitrary graph $\mathcal{G} \in \mathbf{G}(n)$ can therefore be *realized* as a proximity graph in $\mathcal{C}^n(\mathbf{R}^2)$ if $\Phi_n^{-1}(\mathcal{G})$ is nonempty. We denote by the set

$\mathbf{G}_{n,\Delta} \subseteq \mathbf{G}(n)$, the space of all possible graphs on n agents with sensor range Δ, that can be realized in $\mathcal{C}^n(\mathbf{R}^2)$.

For $n = 1$, the configuration space is $\mathcal{C}^1(\mathbf{R}^2) \simeq \mathbf{R}^2$ and the only possible graph on one agent is always realizable, that is, $\mathbf{G}_{1,\Delta} = \mathbf{G}(1)$. For $n = 2$, the situation corresponds to whether the two agents are within Δ distance of each other or not. Therefore all formations in the subset

$$\{(x_1, x_2) \mid \|x_1 - x_2\| \leq \Delta, x_1 \neq x_2\} \subseteq \mathcal{C}^2(\mathbf{R}^2)$$

correspond to the connected graph of two vertices, while the remaining configuration space corresponds to the situation when the graph is disconnected. And so we have $\mathbf{G}_{2,\Delta} = \mathbf{G}(2)$.

Moving on to the case with $n = 3$, there are only four graphs (up to isomorphisms) in $\mathbf{G}(3)$, namely, the ones with edge sets

$$E = \emptyset,\ \{\{1,2\}\}, \{\{1,2\}, \{2,3\}\},\ \text{and}\ \{\{1,2\}, \{2,3\}, \{1,3\}\}.$$

It is clear that these graphs can all be realized in the plane by placing the agents at an appropriate distance from each other, and hence $\mathbf{G}_{3,\Delta} = \mathbf{G}(3)$. A similar enumeration (see Exercise 12.10) of the case when $n = 4$ reveals that also in this case the graph classes are identical, that is, $\mathbf{G}_{4,\Delta} = \mathbf{G}(4)$.

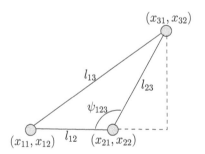

Figure 12.11: Three agents and their interagent distances

An additional observation about the $n = 3$ case will prove useful once we move to $n > 4$. Consider the situation in Figure 12.11, where the three agents are positioned at the points marked by circles. Under the notation in that figure, one can establish (using nothing but basic trigonometry), that whenever we have two edges e_{ij} and e_{ik} in a Δ-disk proximity graph (with agents in the plane) that share a vertex v_i in such a way that there is no edge between vertices v_j and v_k, then

$$\psi_{j,i,k} = \cos^{-1}\left(\frac{\langle x_j - x_i, x_i - x_k\rangle}{\|x_j - x_i\|\|x_i - x_k\|}\right) > \frac{\pi}{3}. \tag{12.15}$$

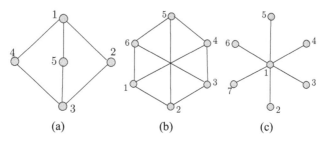

Figure 12.12: Graphs that are not valid proximity graphs

Now, let us put (12.15) to use in the case when $n > 4$. Denote by L_5 and L_6 the graphs in $\mathbf{G}(5)$ and $\mathbf{G}(6)$, respectively, shown in Figures 12.12(a) and 12.12(b). It is straightforward to show that L_5 does not belong to $\mathbf{G}_{5,\Delta}$. This is because, if it is realizable, then the angles ψ_{415}, ψ_{512}, ψ_{123}, ψ_{235}, ψ_{534}, and ψ_{341} are all greater than $\frac{\pi}{3}$, in light of (12.15). As a consequence, if L_5 were indeed realizable, we would have $\psi_{415} + \psi_{512} + \psi_{123} + \psi_{235} + \psi_{534} + \psi_{341} > 6\left(\frac{\pi}{3}\right) = 2\pi$. But since $x_1, x_2, x_3, x_4 \in \mathbf{R}^2$ are vertices of a polygon of 4 sides, we have $\psi_{415} + \psi_{512} + \psi_{123} + \psi_{235} + \psi_{534} + \psi_{341} = 2\pi$, which is a contradiction.

Similarly, one can establish that $L_6 \notin \mathbf{G}_{6,\Delta}$, as well as that the star graph $S_n \in \mathbf{G}(n)$ does not belong to $\mathbf{G}_{n,\Delta}$ for any $n > 6$ (see Figure 12.12(c)). We summarize the findings as a theorem.

Theorem 12.15. *For agents in the plane, $\mathbf{G}_{n,\Delta} = \mathbf{G}(n)$ if and only if $n \leq 4$.*

12.5 PLANNING OVER PROXIMITY GRAPHS

Based on the discussion in the previous section, it would be interesting to know the answer to the following question. Given an arbitrary graph $G \in \mathbf{G}(n)$, can it be realized as a Δ-disk proximity graph for agents in the plane, that is, in $\mathcal{C}^n(\mathbf{R}^2)$? To answer this question, we note that each proximity graph (V, E) for the formation $[x_1, x_2, \ldots, x_n]^T \in \mathcal{C}^n(\mathbf{R}^2)$ can be described by $n(n-1)/2$ relations of the following form:

1. $\|x_i - x_j\| \leq \Delta$ if $e_{ij} \in E$,

2. $\|x_i - x_j\| > \Delta$ if $e_{ij} \notin E$.

Let $x_i = (x_{i,1}, x_{i,2})$ for all $1 \leq i \leq n$. Then each of these relations can be written as inequality constraints, $\{f_k \geq 0\}$, where each

$$f_k \in \mathbf{R}[x_{1,1}, x_{1,2}, \ldots, x_{n,1}, x_{n,2}],$$

is a polynomial in $2n$ variables over the real numbers. Therefore, the realization problem is equivalent to asking if there exist $x_{1,1}, x_{1,2}, \ldots, x_{n,1}, x_{n,2}$ such that the following inequality constraints are satisfied.

$$\Delta^2 - (x_{i,1} - x_{j,1})^2 - (x_{i,2} - x_{j,2})^2 \geq 0 \text{ if } e_{ij} \in E,$$
$$(x_{i,1} - x_{j,2})^2 + (x_{i,2} - x_{j,2})^2 - \Delta^2 > 0 \text{ if } e_{ij} \notin E,$$

where $1 \leq i < j \leq n$. Although these expressions may look messy, they are in fact checkable using tools in algebraic geometry, namely, those pertaining to semialgebraic sets. Without going into the details of these computations, we simply observe that there are plenty of computational tools that will help us solve this and similar feasibility problems. As such, by planning over controllable graph processes, while taking the feasibility of the individual graphs into account, we thus have a method for moving nodes in order to go between target graph topologies, for example, by maintaining connectivity. An example of this is shown in Figure 12.13, in which an initial graph is turned into a path graph by only moving one node at the time to generate the appropriate graph process.

SUMMARY

In this chapter we considered graphs with incidence relations that are dictated by the underlying dynamic states of the agents. We subsequently considered solving graphical equations over such state-dependent graphs, followed by introducing a controllability concept for the corresponding dynamic graphs. The utility of the notion of state-dependent graphs in characterizing feasible formations in the plane concluded this chapter.

NOTES AND REFERENCES

State-dependent graphs as presented in this chapter were introduced by Mesbahi in [156],[157].[7] However, there are a number of earlier works that parallel this framework. First, we mention the work of Aizerman, Gusev, Petrov, Smirnova, and Tenenbaum in [6] where a process of the form

---

[7]The conceptual configuration for the Terrestrial Planet Imager in Figure 12.2 can be found in [3].

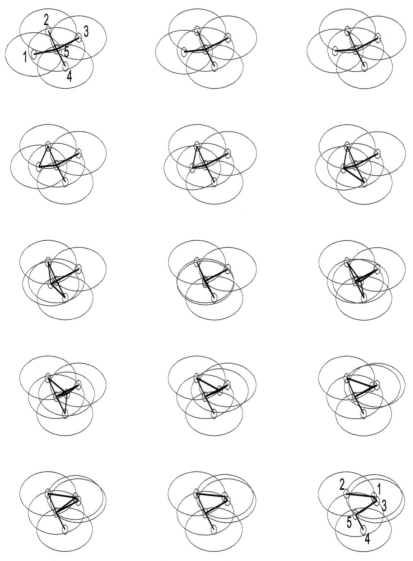

Figure 12.13: A graph process that generates a path graph

$\mathcal{G}(t+1) = F(\mathcal{G}(t))$ is considered, with $\mathcal{G}(t)$ denoting the graph structure at time t, and F a transformation that maps this graph to $\mathcal{G}(t+1)$. As Aizerman and co-authors point out in [6], the motivation for their work had come from mathematical studies on administrative structures, organization of communication and service systems, arrangement for the associative memory of a computer, and so on. In this work, after introducing the notion of "subordinate" functions that operate over trees, the authors consider fixed point and convergence properties of the resulting graph transformation. Related to this work, and in particular to the preceding paper of Petrov [192], the notions of web and graph grammars are also aligned with the state-dependent graphs.

Graph grammars are composed of a finite set of initial graphs, equipped with and closed with respect to a finite set of rules, for their local graph transformation. The general area of graph grammars has flourished as an active area of research in computer science, with many applications in software specification and development, computer graphics, database design, and concurrent systems; see, for example, the handbook on graph grammars edited by Rozenberg [209]. Another area of research related to the present chapter is graph dynamics as described in [196] and references therein. The motivation for this line of work comes from an attempt to generalize a wide array of results in graph theory pertaining to the line graph operator (see Chapter 3). An early result in this area goes back to the early 1930s, where Whitney [244] showed that every finite connected graph, except the triangle, has at most one connected line graph inverse.

The key lemma was originally employed in combinatorial number theory to resolve a famous conjecture of Erdős and Turán. Its application to some open problems in extremal graph theory is more recent; see, for example, Komlós and Simonovits [135]. Finally, the last part of the chapter on the realization problem are based on the works by Muhammad and Egerstedt in [163],[164].

SUGGESTED READING

We refer to the excellent survey of Komlós and Simonovits [135] for more on the key lemma and its various applications in extremal combinatorics. Linear matrix inequalities and S-procedure are discussed in Boyd, El Ghaoui, Feron, and Balakrishnan [33]. For more on theoretical and computational aspects of solving polynomial inequalities we recommend Parrilo [190]; for more applications of positive polynomials in systems and control, see Henrion and Garulli [116].

EXERCISES

Exercise 12.1. Consider four cubes whose faces are colored red (R), blue (B), green (G), and yellow (Y), as shown in the figure. Can one pile up these cubes in such a way that all four colors appear on each side of the pile? Find graphical and algebraic conditions that could be checked for the solvability of this "four cubes problem" for a given choice of the color patterns. *Warning:* this puzzle is also called *Insanity*.

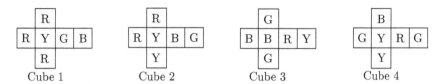

R			
R	Y	G	B
R			

Cube 1

R			
R	Y	B	G
Y			

Cube 2

G			
B	B	R	Y
G			

Cube 3

B			
G	Y	R	G
Y			

Cube 4

Exercise 12.2. The ϵ-regular graphs (in the sense of Szemerédi) are often called "quasi-random," highlighting their resemblance to a random graph in the eyes of a graph theorist. In what sense do such graphs behave like a random graph?

Exercise 12.3. Extend Theorem 12.14 on graph controllability to the case where the state space of each agent is $\mathbf{R}^2$.

Exercise 12.4. Show that Δ_0-controllability for some $\Delta_0 > 0$ implies Δ-controllability for all $\Delta \geq \Delta_0$.

Exercise 12.5. Is a k-regular graph, that is, a graph where every vertex has degree k, ϵ-regular in the sense of Szemerédi for any choice of ϵ?

Exercise 12.6. What is the significance of the inequalities $\mathrm{card}(Y_i) > \epsilon\,\mathrm{card}(X_i)$ and $\mathrm{card}(Y_j) > \epsilon\,\mathrm{card}(X_j)$ in Definition 12.10 in the context of the key lemma 12.12?

Exercise 12.7. Consider a group of ten vehicles in the unit disk; each vehicle can sense another vehicle within the radius of 0.1 units. Is the Peterson graph a feasible sensing network?

Exercise 12.8. Instead of a Δ-disk proximity graph, consider a Δ-square proximity graph. In other words, if $x_i = [x_{i,1}, x_{i,2}]^T$ is the planar position of agent i, the edge $\{i, j\} \in E$ if and only if $|x_{i,1} - x_{j,1}| \leq \Delta$ and $|x_{i,2} - x_{j,2}| \leq \Delta$. For such a proximity sensor, what is the maximum n such that the star graph on n agents is a realization of a feasible formation?

Exercise 12.9. Explain the purpose of introducing the notion of calmness in §12.3.1. Specifically, provide a counterexample for the main statement in Proposition 12.9 if the calmness assumption is violated.

Exercise 12.10. In order to show that $\mathbf{G}_{4,\Delta} = \mathbf{G}(4)$, one needs to enumerate all graphs in $\mathbf{G}(4)$ and show that they can indeed be realized by a planar formation. How many graphs are there in $\mathbf{G}(4)$?

Exercise 12.11. Derive the expression in (12.15).

Chapter Thirteen

Higher-order Networks

> "To be is to be the value of a variable."
> — W. Quine

As network connectivity is one of the key factors determining the performance of coordinated control algorithms, one can take connectivity one step farther and study other types of structures associated with connectivity. For instance, instead of just considering edges, one can consider the areas spanned by the edges, and view these areas as encoding coverage properties. In this chapter, we generalize network graphs and relate them to the so-called higher-dimensional simplicial complexes. In particular, we show how certain proximity graphs generalize to Rips complexes, and we use these complexes to address the coverage problem for sensor networks.

In this chapter, we point out how the concepts developed in this book can be generalized beyond graphs to higher-order structures. However, a disclaimer is already in order at this point; we do not present a particularly mature body of work, as it pertains to networked systems. Rather, we simply point out some possible (and certainly fascinating) extensions.

13.1 SIMPLICIAL COMPLEXES

To take the step from graph models, where the key objects are nodes and edges, to richer structures, one first needs to turn to algebraic topology. In fact, a graph can be generalized to a more expressive combinatorial object known as a *simplicial complex*. Given a set of points V, a k-simplex is an unordered subset $\{v_0, v_1, \ldots, v_k\}$, where $v_i \in V$ and $v_i \neq v_j$ for all $i \neq j$. The *faces* of this k-simplex consist of all $(k-1)$-simplices of the form $\{v_0, \ldots, v_{i-1}, v_{i+1}, \ldots, v_k\}$ for $0 < i < k$. A simplicial complex is a collection of simplices which is closed with respect to the inclusion of faces. Graphs are a concrete example, where the vertices of the graph correspond to V, and the edges correspond to 1-simplices.

The ordering of the vertices corresponds to an *orientation*. A k-simplex $\{v_0, \ldots, v_k\}$ together with an order is an *oriented* k-simplex, denoted by $[v_0, \ldots, v_k]$; see Figure 13.1 for examples of simplices of dimensions zero through three. A change in orientation corresponds to a change in the sign of the coefficient, $[v_0, \ldots, v_j, \ldots, v_i, \ldots, v_k] = -[v_0, \ldots, v_i, \ldots, v_j, \ldots, v_k]$.

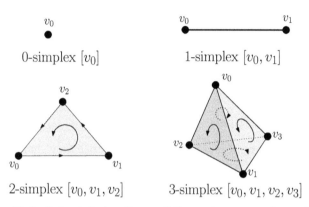

0-simplex $[v_0]$ 1-simplex $[v_0, v_1]$

2-simplex $[v_0, v_1, v_2]$ 3-simplex $[v_0, v_1, v_2, v_3]$

Figure 13.1: Oriented simplices of dimensions zero through three

For easy reference, we here summarize the key notions in dealing with simplicial complexes:

> Given a collection of points, a combinatorial *k-simplex* is an unordered, nonrepeating subset of this collection of size $k + 1$. A combinatorial simplicial complex is a collection of simplices that is closed with respect to *faces*, or sub-simplices, that is, if a simplex belongs to the complex, then so do its faces.

In order to take advantage of this new construction, we also need to generalize the concepts of adjacency and degree from standard graph theory. Two k-simplices σ_i and σ_j of a simplicial complex X are *upper adjacent* if both are faces of some $(k + 1)$-simplex in X. We denote this adjacency by $\sigma_i \sim_U \sigma_j$. The *upper degree* of a k-simplex σ, denoted $\deg_U(\sigma)$, is the number of $(k + 1)$-simplices in X of which σ is a face.

In a similar fashion, we also define *lower adjacency* and *lower degree* of simplices. Two k-simplices σ_i and σ_j of an unordered simplicial complex X are *lower adjacent* if they have a common face. We denote this by $\sigma_i \sim_L \sigma_j$. The *lower degree* of a k-simplex σ, denoted $\deg_L(\sigma)$, is the number of faces in σ.

Examples of these constructions are given in Figure 13.2. In that figure,

$$v_1 \sim_U v_2, \quad v_2 \sim_U v_3, \quad v_1 \sim_U v_3, \quad v_2 \sim_U v_4,$$
$$v_3 \sim_U v_4, \quad v_4 \sim_U v_5, \quad v_4 \sim_U v_6,$$

that is, the normal adjacency relations between vertices become upper adjacencies in this setting. Moreover, the edges also satisfy adjacency relations in that

$$e_1 \sim_U e_2, \quad e_1 \sim_U e_3, \quad e_2 \sim_U e_3, \quad e_3 \sim_U e_4, \quad e_4 \sim_U e_5, \quad e_3 \sim_U e_5$$

since these edges are (pairwise) common faces of the same 2-simplex. Also, the edges that share vertices are lower adjacent, that is,

$$e_1 \sim_L e_2, \quad e_1 \sim_L e_3, \quad e_2 \sim_L e_3, \quad e_1 \sim_L e_4, \quad e_3 \sim_L e_4, \quad e_3 \sim_L e_5,$$
$$e_2 \sim_L e_5, \quad e_4 \sim_L e_5, \quad e_4 \sim_L e_6, \quad e_4 \sim_L e_7, e_5 \sim_L e_6, \quad e_5 \sim_L e_7, \quad e_6 \sim_L e_7.$$

Finally, the two 2-simplices are lower adjacent since they share an edge (e_3), that is, $\xi_1 \sim_L \xi_2$.

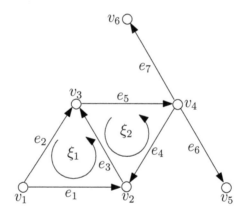

Figure 13.2: Simplicial complex

Now, given an oriented simplicial complex X, suppose that the k-simplex σ is a face to the $(k+1)$-simplex ξ. If the orientation of σ agrees with the one induced by ξ, then σ is said to be *similarly oriented* to ξ. If not, we say that the simplex is *dissimilarly oriented*. Returning to Figure 13.2, e_1 and e_3 are similarly oriented with respect to ξ_1, while e_2 is not. Also, e_3, e_4, and e_5 are all dissimilarly oriented with respect to ξ_2.

Consider next the simplicial complex X. Let, for each $k \geq 0$, the vector space $C_k(X)$ be the vector space whose basis is the set of oriented k-simplices of X. For k larger than the dimension of X, we set $C_k(X) = 0$.

The *boundary map* is defined to be the linear transformation

$$\partial_k : C_k \to C_{k-1},$$

which acts on basis elements $[v_0, \ldots, v_k]$ via

$$\partial_k[v_0, \ldots, v_k] = \sum_{i=0}^{k}(-1)^i[v_0, \ldots, v_{i-1}, v_{i+1}, \ldots, v_k], \qquad (13.1)$$

as illustrated in Figure 13.3.

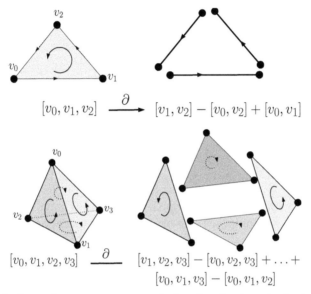

$$[v_0, v_1, v_2] \xrightarrow{\ \partial\ } [v_1, v_2] - [v_0, v_2] + [v_0, v_1]$$

$$[v_0, v_1, v_2, v_3] \xrightarrow{\ \partial\ } [v_1, v_2, v_3] - [v_0, v_2, v_3] + \ldots +$$
$$[v_0, v_1, v_3] - [v_0, v_1, v_2]$$

Figure 13.3: Boundary operator on a 2-simplex (top) and a 3-simplex (bottom)

13.2 COMBINATORIAL LAPLACIANS

One interesting property of the boundary map is that it gives rise to the so-called *chain complex*, a sequence of vector spaces and linear transformations

$$\cdots \xrightarrow{\partial_{k+1}} C_{k+1} \xrightarrow{\partial_k} C_k \xrightarrow{\partial_{k-1}} C_{k-1} \cdots \xrightarrow{\partial_2} C_1 \xrightarrow{\partial_1} C_0.$$

And, since we are dealing with a finite simplicial complex, X, the vector spaces $C_k(X)$ are also of finite dimension. Therefore, we can represent the boundary maps $\partial_k : C_k(X) \to C_{k-1}(X)$ in matrix form.

Furthermore, if the chain complex is defined over a field of characteristic zero[1] (such as the real or complex fields), then one can give an inner

---

[1] A field is of characteristic zero if a repeated sum of the multiplicative identity can never be equal to the additive identity of the field.

product structure to each $C_k(X)$ and, subsequently, get an adjoint operator $\partial_k^* : C_{k-1}(X) \rightarrow C_k(X)$ for each map ∂_k. Each adjoint map ∂_k^* can be expressed as the transpose of the matrix representation of ∂_k. We therefore have another chain complex as

$$\cdots \xleftarrow{\partial_{k+2}^*} C_{k+1} \xleftarrow{\partial_{k+1}^*} C_k \xleftarrow{\partial_k^*} C_{k-1} \cdots \xleftarrow{\partial_2^*} C_1 \xleftarrow{\partial_1^*} C_0.$$

In fact, one can now define the *Laplacian* operator $\Delta_k : C_k(X) \rightarrow C_k(X)$ by

$$\Delta_k = \partial_{k+1}\partial_{k+1}^* + \partial_k^*\partial_k, \qquad (13.2)$$

with the corresponding set of *harmonic k-forms* defined as

$$H_k(X) = \{c \in C_k(X) : \Delta_k c = 0\}. \qquad (13.3)$$

Then, from a branch of algebraic topology known as Hodge theory, we know that each $C_k(X)$ decomposes into orthogonal subspaces as

$$C_k(X) = H_k(X) \oplus \text{Im}(\partial_{k+1}) \oplus \text{Im}(\partial_k^*), \qquad (13.4)$$

where the Laplacian operator $\Delta_k = \partial_{k+1}\partial_{k+1}^* + \partial_k^*\partial_k$ becomes invariant on each of these subspaces, and positive definite on the images of ∂_{k+1} and ∂_k^*.

As mentioned earlier, the boundary operators and their adjoints have matrix representations. In other words, we can also give a matrix representation to the Laplacian. We denote the matrix associated with the k-dimensional Laplacian as L_k. And, through this matrix representation, it can be seen that the familiar *graph Laplacian* is synonymous with L_0 (or $\Delta_0 : C_0(X) \rightarrow C_0(X)$) defined above. Since there are no simplices of negative dimension, $C_{-1}(X)$ is assumed to be 0. Also, the maps ∂_0 and ∂_0^* are assumed to be zero maps, so that

$$\Delta_0 = \partial_1\partial_1^*. \qquad (13.5)$$

But this expression looks suspiciously like the standard graph Laplacian,[2]

$$L = DD^T,$$

where D is the incidence matrix. Moreover, as we can think of the boundary map $\partial_1 : C_1(X) \rightarrow C_0(X)$ as mapping edges to vertices–just like the

---

[2]We will continue to suppress the notational dependency of D on the underlying graph in this section for reasons that become evident shortly.

incidence matrix–we can draw the conclusion that its matrix representation is exactly equal to the standard graph theoretic incidence matrix D.

To make this observation a bit more concrete, let us return to the simplicial complex in Figure 13.2. There the incidence matrix is given by

$$
D_1 =
\begin{bmatrix}
-1 & -1 & 0 & 0 & 0 & 0 & 0 \\
1 & 0 & -1 & 1 & 0 & 0 & 0 \\
0 & 1 & 1 & 0 & -1 & 0 & 0 \\
0 & 0 & 0 & -1 & 1 & -1 & -1 \\
0 & 0 & 0 & 0 & 0 & 1 & 0 \\
0 & 0 & 0 & 0 & 0 & 0 & 1
\end{bmatrix},
$$

where we explicitly include the subscript 1 to highlight the fact that this is the matrix representation of the boundary operator ∂_1, defined over the edges, that is, one-dimensional objects. To see this, note that

$$
D_1
\begin{bmatrix}
1 \\ 0 \\ 0 \\ 0 \\ 0 \\ 0 \\ 0
\end{bmatrix}
=
\begin{bmatrix}
-1 \\ 1 \\ 0 \\ 0 \\ 0 \\ 0
\end{bmatrix},
$$

that is, edge e_1 has v_1 and v_2 as its boundary vertices, and it originates at v_1 and ends at v_2.

When constructing D_2, we need to think of it as an incidence-type matrix that operates on surfaces and returns edges. In that matrix, the signs of the nonzero entries will be determined by whether the orientations of the 1-simplices are similar to the 2-simplex for which they are faces. In other words, for the simplicial complex in Figure 13.2, we get

$$
D_2 =
\begin{bmatrix}
1 & 0 \\
-1 & 0 \\
1 & -1 \\
0 & -1 \\
0 & -1 \\
0 & 0 \\
0 & 0
\end{bmatrix}.
$$

As a sanity check, we note that

$$D_2 \begin{bmatrix} 1 \\ 0 \end{bmatrix} = \begin{bmatrix} 1 \\ -1 \\ 1 \\ 0 \\ 0 \\ 0 \\ 0 \end{bmatrix},$$

that is, the faces to ξ_1 are e_1, e_2, and e_3, with e_1 and e_3 similarly oriented and e_2 dissimilarly oriented to ξ_1,.

With D_1 and D_2, we can now compute the Laplacian operator

$$L_1 = D_2 D_2^T + D_1^T D_1 = \begin{bmatrix} 3 & 0 & 0 & 1 & 0 & 0 & 0 \\ 0 & 3 & 0 & 0 & -1 & 0 & 0 \\ 0 & 0 & 4 & 0 & 0 & 0 & 0 \\ 1 & 0 & 0 & 3 & 0 & 1 & 1 \\ 0 & -1 & 0 & 0 & 3 & -1 & -1 \\ 0 & 0 & 0 & 1 & -1 & 2 & 1 \\ 0 & 0 & 0 & 1 & -1 & 1 & 2 \end{bmatrix}.$$

13.3 TRIANGULATIONS AND THE RIPS COMPLEX

The discussion so far has focused on simplicial complexes in general. However, in communication networks, or for that matter any other network in which the edges correspond to the existence of a communication/sensing link between adjacent nodes, a particular simplicial complex appears naturally.

Triangulated surfaces form concrete (and by now somewhat familiar) examples of simplicial complexes, as already displayed in Figure 13.2. In these triangulations, the vertices correspond to 0-simplices, edges correspond to 1-simplices, and faces correspond to 2-simplices. In fact, when looking at a network graph, for example, one obtained from a Δ-disk proximity graph, drawn in the plane using straight lines between nodes, we see various triangles that overlap and mash together in a complex manner. These triangulations are given by a projection of the so-called *Rips complex* on the plane. A natural question is whether these triangles can be arranged or chosen so as to form a clean triangulation of the planar region "bounded" by the network, as was the case for the coverage problem in Chapter 7.

The Rips complex, which traces its origins back to the work of Vietoris on homology theory in the 1930s, is a way of lifting a graph to a higher-dimensional complex. The briefest definition of the Rips complex, $\mathcal{R}$, associated with a graph is that it is the largest simplicial complex having the graph as a 1-dimensional skeleton. In the context of networked systems, interacting over sensing of communication channels, the nodes in the network correspond to the 0-simplices (or vertices). Likewise, the 1-simplices of $\mathcal{R}$ are precisely the edges of the Δ-disk proximity graph. In general, the k-simplices correspond to sets of $k + 1$ nodes that are pairwise within communication or sensing range of each other.

It is helpful to visualize the Rips complex as drawn in the two bottom drawings of Figure 13.4. Intuitively, the Rips complex lifts the proximity graph to a higher-dimensional space that collates relationships between more than two agents. In contrast, in a proximity graph, the edges allow the study of pairwise relations only. Therefore, the Rips complex is a more powerful way of capturing the spatial and communication relations between agents.

For a multiagent system equipped with radios, the Rips complex can be constructed using synchronous broadcast protocols. What is needed is that each agent becomes aware of those simplices of which it is a member, as well as a knowledge of which other agents share those simplices. To achieve this, suppose that each agent carries a unique identification tag. Also, assume that each agent is capable of communicating its identification tag to its neighboring agents along with some other information of interest. Each agent also maintains an array of lists of identification tags, where each list corresponds to a simplex of which the agent is a part of, as seen in Figure 13.5.

Initially, each agent is aware only of its own identification tag. The first entry in the list is this identification tag, which generates the 0-simplices of the Rips complex $\mathcal{R}$. The agents simultaneously broadcast their identification. The agents within communication range receive this information and add the received tags paired with their own tags to their respective lists. This generates the 1-simplices, or the edges in the simplicial complex. Following this, the agents broadcast their list of edges. After reception, each agent compares the received list of edges to its own list and searches for a cycle, thus generating the 2-simplices, and so on. In this way, all simplices of dimension k or lower are discovered in k broadcasts.

13.3.1 A Triangulation Algorithm

The discovery of the Rips complex, for example, using the algorithm discussed in the previous section, allows us to produce triangulations of the network in a decentralized manner. Loosely speaking, such a triangulation lets us study the *shape* of a network from a planar perspective, for example,

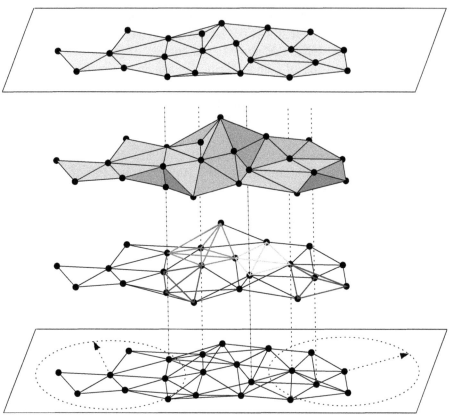

Figure 13.4: A visualization of a Δ-disk proximity graph and its Rips complex. The proximity graph obtained from nodes in the plane is shown in the bottom figure with the communication radii depicted as dotted circles. The Rips complex induced from the proximity graph is given in the center-bottom figure. A surface in the Rips complex corresponding to the triangulation of the proximity graph is given in the center-top figure; and a triangulation of the proximity graph is shown in the top figure.

one can detect the presence of holes in the triangulation, which would create an obstruction to achieving a global objective, such as coverage or routing for sensor and communication networks.

A decentralized triangulation algorithm can be obtained by letting the physical nodes with their communication network *lift* to the 1-dimensional proximity graph. The graph, in turn, *lifts* to the higher-dimensional Rips complex. This geometric complex provides the "superstructure" for extracting a triangulated surface, which projects down to a triangulation adapted to the original network.

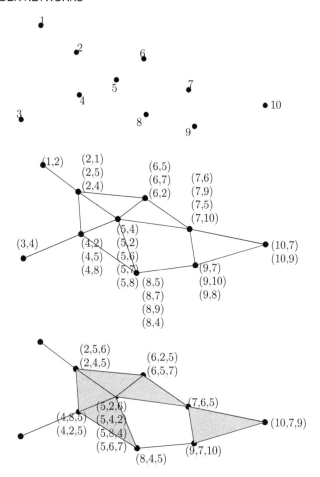

Figure 13.5: A decentralized protocol for obtaining the Rips complex. Using a broadcast protocol, the network discovers higher-dimensional simplices in the Rips complex $\mathcal{R}$. At each broadcast cycle, the information gathered from the previous broadcasts is sorted and passed on to neighboring nodes. The figure shows a progression from 0-simplices (top) to 2-simplices (bottom).

To be more concrete, in order to obtain a triangulation, the first thing to note is that whenever two crossing edges occur in the image of a graph, they span a subgraph in the original graph that is isomorphic to one of the three *crossing generator* graphs shown in Figure 13.6.

These crossing generators can be amalgamated in a certain way to produce more complex graphs with multiple crossing edges in an iterative manner. In fact, through such amalgamations, it is possible to encode any planar

Figure 13.6: Crossing generators

Δ-disk proximity graph. Once we have such an encoding, local rules can
be defined for removing the crossing edges such that the underlying triangulation is preserved and the removal of edges is guaranteed never to create
artificial holes. An example of the application of this algorithm is given in
Figure 13.7, where a Δ-disk proximity graph is shown together with the
triangulation (the simplicial complex) obtained by removing the crossing
edges.

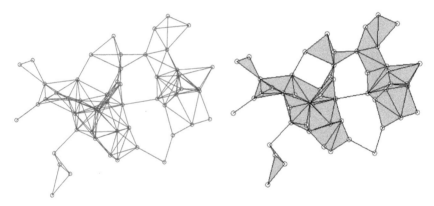

Figure 13.7: The simplicial complex obtained by the removal of crossing
edges, resulting in a triangulation

But, what does the topology of the triangulation associated with a proximity graph characterize? Topologically, one triangulation differs from another
in the number and configuration of *holes*, the presence of which is of significant importance for the study of routing algorithms in sensor networks.
The existence and configuration of holes regulate nonunique optimal routing paths between nodes. On a deeper level, the global topology of the Rips
complex impacts coverage problems in sensor networks.

13.4 THE NERVE COMPLEX

As we have seen repeatedly in this book, the graph Laplacian is a powerful
tool that allows the network topology to be directly incorporated into the

equations of a networked dynamical system. The most direct application of this idea is the agreement protocol, with the simple averaging law

$$\dot{x}_i(t) = - \sum_{v_i \sim v_j} (x_i(t) - x_j(t)),$$

which can be rewritten as

$$\frac{dc(x, j)}{dt} = -L_0 c(x, j),$$

where the component operator is given through $c(x, j) = (x_{1,j}, \ldots, x_{n,j})^T$. We know that for a connected graph the spectral properties of L_0 imply that all states converge toward a common state. It would be interesting to see whether the higher-order combinatorial Laplacians could be used to design distributed algorithms as well. In this section, we study the particular problem of controlling the coverage radii in planar sensor networks.

Let there be n sensor nodes. Each sensor node i is located at position x_i, and has a circular coverage domain of radius r_i. We assume that each sensor node is capable of adjusting its area of coverage by increasing or decreasing r_i. We further assume that each sensor node has the ability to communicate with its neighboring nodes.

Now, given a collection of sets $\mathcal{U} = \{U_\alpha\}_{\alpha \in A}$, where A is an indexing set, the *nerve complex* of $\mathcal{U}$, $\mathcal{N}(\mathcal{U})$, is the abstract simplicial complex whose k-simplices correspond to nonempty intersections of $k + 1$ distinct elements of $\mathcal{U}$. Hence, the vertices of $\mathcal{N}(\mathcal{U})$ correspond to the individual sets $\{U_\alpha\}_{\alpha \in A}$ themselves.

The 0-chain C_0 is therefore a vector space spanned by $\{U_\alpha\}$. An edge in $\mathcal{N}(\mathcal{U})$ exists between two vertices U_{α_i} and U_{α_j} if and only if $U_{\alpha_i} \cap U_{\alpha_j} \neq \emptyset$. Therefore C_1 is a vector space spanned by all nonempty mutual intersections between the elements of $\mathcal{U}$. Similarly, k-dimensional simplices correspond to nonempty intersections $\bigcap_{i=0}^{k} U_{\alpha_i}$ of $k + 1$ of elements of $\mathcal{U}$. We will abbreviate the intersection $\bigcap_{i=0}^{k} U_{\alpha_i}$ by $U_{\alpha_0 \alpha_1 \cdots \alpha_k}$. The boundary of a k-simplex is now defined as,

$$\partial \left(U_{\alpha_0 \alpha_1 \cdots \alpha_k} \right) = \sum_{i=0}^{k} (-1)^i \left(U_{\alpha_0 \cdots \alpha_i \alpha_{i+1} \cdots \alpha_k} \right).$$

By linearity, the boundary operator can be defined for any element of the vector space. Therefore $\partial_k : C_k \to C_{k-1}$ maps a linear combination of k-fold set intersections in C_k to a linear combination of $(k - 1)$-fold set intersections in C_{k-1}. An example of a collection of sets and its nerve complex is depicted in Figure 13.8.[3]

---

[3]It should be noted that the nerve complex is not restricted to circular disks only.

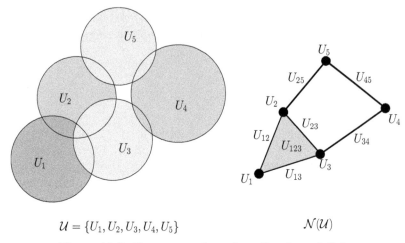

$$\mathcal{U} = \{U_1, U_2, U_3, U_4, U_5\} \qquad\qquad \mathcal{N}(\mathcal{U})$$

Figure 13.8: Nerve complex of a collection of disks

13.4.1 Control of Coverage Radii

To keep the computations simple, we here consider the situation in which we fix the positions of the sensor nodes; the area of intersection is increased or decreased by adjusting the sensing radii, which we denote by r_i, $i = 1, \ldots, n$. Using this simple construction, one can control the area of intersections by implementing the control of individual radii. To take advantage of our new set of tools, we give a coverage control algorithm that is based on the theory of combinatorial Laplacians.

For example, a simple scheme can be based on an averaging rule similar to the agreement protocol. For a given network topology, we can thus let

$$\frac{dr_i(t)}{dt} = - \sum_{\{j \,|\, B_{r_i}(x_i) \cap B_{r_j}(x_j) \neq \emptyset\}} (r_i(t) - r_j(t)),$$

where $B_{r_i}(x_i)$ is the closed ball of radius r_i centered at x_i. = Now, let $\mathbf{r}(t) = [r_1(t), r_2(t), \ldots, r_n(t)]^T$. We can then write the system as

$$\dot{\mathbf{r}}(t) = -L_0(t)\mathbf{r}(t),$$

where $L_0(t)$ is the instantaneous 0-Laplacian of the nerve complex. But, as this is just the agreement protocol, all the radii will converge to some consensus value $\hat{r} = r_1 = r_2 = \cdots = r_n$, if the network is connected for all times.[4]

---

[4]For convergence, the connectivity condition can be maintained by starting from reasonably large values $\mathbf{r}(0) = [r_1(0), \ldots, r_n(0)]^T$.

Although simple and self-configuring, this scheme has several defects. The most severe shortcoming is that $\hat{r}$ is heavily dependent on the initial values of $\mathbf{r}(0)$. The radii may converge to a very small value of $\hat{r}$, thus leaving large gaps in the coverage, or to such a large value that most areas are covered by more than one node, resulting in large and excessive power usage.

In terms of the topological properties of the nerve complex, one would ideally like to design $\mathbf{r}$ such that there are no coverage holes. For any three nodes located at $x_{\alpha_1}, x_{\alpha_2}, x_{\alpha_3}$, by choosing $r_{\alpha_1}, r_{\alpha_2}, r_{\alpha_3}$ large enough, the set intersection $\mathcal{A}_{\alpha_1\alpha_2\alpha_3} = \bigcup_{i=1}^{3} B_{r_{\alpha_i}}(x_{\alpha_i})$ can be made nonempty, thereby inserting a 2-simplex in the nerve complex and annihilating a potential hole. On the other hand, the number of higher-order simplices should be minimized in order to minimize wastage of power. To meet these contradictory goals, a simple control scheme can be based on L_1 instead of L_0. In fact, by letting $\mathcal{A} = (\mathcal{A}_{\alpha_1\alpha_2}, \ldots, \mathcal{A}_{\alpha_m\alpha_n})^T$ represent all 1-simplices of the nerve complex, we can let the system evolve using the 1-Laplacian as

$$\dot{\mathcal{A}}(t) = -L_1(t)\mathcal{A}(t). \tag{13.6}$$

The result is shown in Figure 13.9. However, from this figure it can be seen that even though appropriate sensing radii are selected based on the information encoded in the 1-skeleton, higher-order simplices are most likely needed (in particular, second-order simplices) to eliminate the resulting hole.

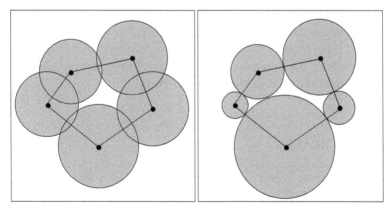

Figure 13.9: Application of the first-order Laplacian to the coverage problem, showing the time progression of the system

SUMMARY

This chapter generalizes the concept of a graph, that is, a combinatorial structure consisting of edges and vertices, by observing that edges are essentially 1-dimensional objects, while vertices are 0-dimensional. In a similar manner, one can form surfaces of various dimensions in order to obtain the so-called simplicial complexes, which leads to the field of algebraic topology. The standard graph is in fact a special case of a simplicial complex, and in this chapter we have discussed two such complexes, namely, the Rips complex and the nerve complex. The Rips complex is useful for describing triangulations (of various orders) over graphs, while the nerve complex can be used to encode the overlapping sensing regions in sensor networks.

Simplicial complexes allow us to define higher-order Laplacian operators as generalizations of the "normal" graph Laplacian. In fact, these higher-order combinatorial Laplacians are defined through

$$\Delta_k = \partial_{k+1}\partial_{k+1}^\star + \partial_k^\star\partial_k,$$

where the ∂_k operator is the boundary operator (thought of as a higher-order incidence matrix) and $\partial_k^\star$ denotes the adjoint operator of ∂_k.

NOTES AND REFERENCES

The idea to define higher-order Laplacian operators for describing dynamic flows in networked systems was outlined by Muhammad and Egerstedt in [166], and extended by Muhammad and Jadbabaie in [167],[168]. However, algebraic topology and networked systems has a much richer history, with key references in the networked area including those by de Silva, Ghrist, and Muhammad [68] and de Silva and Ghrist [69].

SUGGESTED READING

The use of algebraic topology for networked systems is very elegantly used by de Silva and Ghrist in [70], where the persistence of dynamic "holes" in a sensor network implies that it is possible to evade detection indefinitely. A general and fairly accessible discussion of the (much) larger topic of algebraic topology is given in Armstrong [11] and Munkres [169]. The companion topic of discrete exterior calculus is discussed by Desbrun, Hirani, and Marsden in [67].

EXERCISES

Exercise 13.1. As we have seen, one way of thinking about graphs is to view them as 1D simplicial complexes. A 2D simplicial complex thus consists not only of vertices and edges, but also of surfaces. These surfaces can be obtained through triangulations, as seen below.

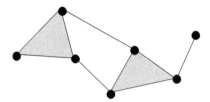

Now, assume that the edges in the graph are directed and let D_1 be the graph incidence matrix, with the graph Laplacian being $L_0 = D_1 D_1^T$.

In a similar way, we can give surfaces an orientation, as seen below.

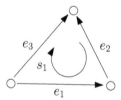

By orienting surfaces, we can also define the matrix $D_2 = \{\iota_{ij}\}$, with

$$\iota_{ij} = \begin{cases} 1 \text{ if } \overset{\circlearrowleft s_j}{\underset{e_i}{\longrightarrow}} \\ -1 \text{ if } \overset{\circlearrowleft s_j}{\underset{e_i}{\longleftarrow}} \\ 0 \text{ o.w.} \end{cases}$$

Through both D_1 and D_2 we can generalize the graph Laplacian to a higher order combinatorial Laplacian

$$L_1 = D_2 D_2^T + D_1^T D_1.$$

Compute L_1 for the following graph.

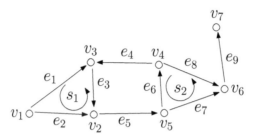

Exercise 13.2. In a (scalar) directed graph, let $e_1, \ldots, e_m$ correspond to the edges obtained by $e_{ij} = x_i - x_j$ if $(v_j, v_i) \in E$, that is, a directed edge has v_j as its tail and v_i as its head. For such a system, we could define an agreement protocol directly over the edges as

$$\dot{e}(t) = -L_1 e(t),$$

where $e = [e_1, \ldots, e_m]^T \in \mathbf{R}^m$. Show that this is indeed the same protocol as the standard (node-based) agreement protocol

$$\dot{x}(t) = -L_0 x(t).$$

Exercise 13.3. Recall that for scalar node-states x_i, $i = 1, \ldots, n$, the agreement protocol $\dot{x}(t) = -L_0 x(t)$ is indeed capturing the averaging interaction rule

$$\dot{x}_i(t) = -\sum_{j \in N(i)} (x_i(t) - x_j(t)), \quad i = 1, \ldots, n.$$

Now, assume that we have $\dot{\xi}(t) = -L_1 \xi(t)$ for some quantity $\xi \in \mathbf{R}^m$, with m being the number of edges in the network. Rewrite this equation as an averaging rule (of some sort) in the same way as was done for x_i above; that is, complete the equation

$$\dot{\xi}_i(t) = ?$$

Exercise 13.4. Same question as 13.3 but with $\dot{\xi}(t) = -L_2 \xi(t)$.

Exercise 13.5. Same question as 13.3 but with $\dot{\xi}(t) = -L_k \xi(t)$ for any arbitrary $k \geq 1$.

Exercise 13.6. One of the main hurdles when using the nerve complex for practical applications is the difficulty of computing intersections for sets of arbitrary shape. Fortunately, it is relatively easy to compute such intersections for circular disks. In fact, we do not even need to compute these intersections since all we need to completely describe the set intersections corresponding to each simplex in the nerve complex is a certificate about whether a nonempty intersection exists. Describe how you would go about producing such certificates.

Exercise 13.7. If the Rips complex is time-varying, it would be interesting to see whether holes in the complex persist over time. Explain what implications the existence (or lack thereof) of such holes would have for an evader trying to avoid getting detected, that is, getting too close to any vertex.

Exercise 13.8. What is the connection between L_1 and the edge Laplacian discussed in Chapter 2?

Appendix A

―――――――――――――――――――――――――――

> "A man is like a fraction whose numerator is what he is
> and whose denominator is what he thinks of himself.
> The larger the denominator, the smaller the fraction."
> — Leo Tolstoy

The appendix gathers a number of concepts and constructs that have been used in the book. These include rudiments of analysis, linear algebra, random sequences, control theory, optimization, and games.

A.1 ANALYSIS

Given a "vector" in $\mathbf{R}$, otherwise known as a real number, there is generally one accepted way to measure its magnitude, namely, its absolute value, which incidentally also captures the notion of its distance from zero, the origin of $\mathbf{R}$. Given a vector in $\mathbf{R}^n$ for $n \geq 1$, or more generally a finite-dimensional vector space, there are a number of ways to measure its magnitude. Let $\mathbf{V}$ be a vector space over $\mathbf{R}$. A function

$$\|\cdot\| : \mathbf{V} \to \mathbf{R}$$

is a norm if it is positive (except when its argument is the zero vector), positively homogeneous (with respect to scalars),

$$\|\alpha\,x\| = |\alpha|\,\|x\|, \quad \alpha \in \mathbf{R},$$

and satisfies the triangular inequality

$$\|x + y\| \leq \|x\| + \|y\|.$$

A norm generalizes the notion of a distance to the origin in $\mathbf{R}$ to an arbitrary vector space. In order to generalize the notion of "angle" to general vector spaces, we consider a function of the form

$$\langle .,. \rangle : \mathbf{V} \times \mathbf{V} \to \mathbf{R},$$

which is called an inner product if it is symmetric, self-positive ($\langle x, x \rangle \geq 0$ for all x and $\langle x, x \rangle = 0$ if and only if $x = 0$), additive (individually, in each of its arguments), and homogeneous (with respect to scalar multiplication). Equipped with the notion of inner product, one has

$$| \langle x, y \rangle |^2 \le \langle x, x \rangle \langle y, y \rangle, \tag{A.1.1}$$

for all $x, y \in \mathbf{V}$ (Cauchy-Schwarz). The inner product induces a (canonical) norm

$$\|x\| = \langle x, x \rangle^{1/2};$$

examples of such norms include the 2-norm in $\mathbf{R}^n$, which is induced by $\langle x, y \rangle = x^T y$, and the Frobenius norm on the space of symmetric matrices, induced by $\langle X, Y \rangle = \mathbf{trace}\,(XY) = \sum_i \sum_j [X]_{ij} [Y]_{ij} = \mathbf{trace}\,(YX)$. The vector 2-norm is the primary norm used in this book.

Norms facilitate the notion of distance and convergence in $\mathbf{R}^n$ and other Euclidean spaces. For example, we say that a sequence of vectors x_i, $i = 1, 2, \ldots$, converges to a vector x^*, denoted as $x_i \to x^*$, if

$$\lim_{i \to \infty} \|x_i - x^*\| = 0.$$

Similarly, a sequence of vectors in $\mathbf{R}^n$ is said to converge to the set $S \subseteq \mathbf{R}^n$ if

$$\lim_{i \to \infty} \inf_{y \in S} \|x_i - y\| = 0.$$

Similarly, the notions of continuity and smoothness for functions can be defined via norms. For example, a function $f : \mathbf{R}^n \to \mathbf{R}^m$ is called Lipschitz continuous if there exists a constant $K > 0$ such that

$$\|f(x_1) - f(x_2)\| \le K \|x_1 - x_2\|. \tag{A.1.2}$$

If (A.1.2) is valid only on a subset of $\mathbf{R}^n$, say S, then we call the function f locally Lipschitz on S.

A.2 MATRIX THEORY

Matrices represent linear operators on finite-dimensional vector spaces. However, it is convenient to work with their representation in terms of two-dimensional arrays of numbers. Thus we write $A \in \mathbf{R}^{n \times m}$ to signify that this array has n rows and m columns. Although such an array representation seems to void profound possibilities for matrices, the contrary seems to have prevailed. For example, let our underlying vector space be $\mathbf{R}^n$ and

$A \in \mathbf{R}^{n \times n}$, and consider the situation where, for some $x \in \mathbf{R}^n$ and $\lambda \in \mathbf{C}$, one has

$$Ax = \lambda x;$$

then the vector x is called the eigenvector of A associated with the eigenvalue λ. A matrix is nonsingular if none of its eigenvectors are zero. The algebraic multiplicity of the eigenvalue of A, λ, is its multiplicity as a root of the characteristic polynomial

$$\det(\lambda I - A) = 0.$$

The geometric multiplicity of the eigenvalue λ, on the other hand, is the number of linearly independent eigenvectors corresponding to it. An eigenvalue is called simple if its algebraic multiplicity is equal to one.

The Kronecker product of two matrices $A \in \mathbf{R}^{n \times m}$ and $B \in \mathbf{R}^{p \times q}$, with $a_{ij} = [A]_{ij}$ and $b_{ij} = [B]_{ij}$, denoted by $A \otimes B$, is defined as the $np \times mq$ matrix

$$\begin{bmatrix} a_{11}B & \cdots & a_{1m}B \\ a_{21}B & \cdots & a_{2n}B \\ a_{31}B & \cdots & a_{3n}B \\ \vdots & \vdots & \\ a_{n1}B & \cdots & a_{nm}B \end{bmatrix}.$$

Among the many algebraic properties of the Kronecker products, we mention the identity

$$(A \otimes B)(C \otimes D) = (AC) \otimes (BD),$$

where the matrices A, B, C, and D have appropriate dimensions.

If we have that the matrix $A \in \mathbf{R}^n$ is such that $[A]_{ij} = [A]_{ji}$, we call A a real symmetric matrix.

Theorem A.1. *A real symmetric matrix A can be factored as $Q \Lambda Q^T$, where Λ is the diagonal matrix with the eigenvalues of A on the diagonal, and the columns of Q are the corresponding orthonormal set of eigenvectors.*

In the case that A is symmetric, the eigenvalues of A are real, and one can order them in such a way that

$$\lambda_1(A) \leq \lambda_2(A) \leq \cdots \leq \lambda_n(A).$$

The variational characterization of eigenvalues of a symmetric matrix states that

$$\lambda_1 = \inf_{x \neq 0} \frac{x^T A x}{x^T x} \quad \text{and} \quad \lambda_n = \sup_{x \neq 0} \frac{x^T A x}{x^T x}.$$

A matrix A is called positive semidefinite if the quadratic form $x^T A x \geq 0$ for all x and positive definite if $x^T A x > 0$ for all $x \neq 0$. The variational characterization of eigenvalues then implies that a matrix is positive semidefinite if and only if all its eigenvalues are nonnegative, and positive definite if they are all positive.

A matrix $A \in \mathbf{R}^{n \times n}$ is called nonnegative if $[A]_{ij} \geq 0$ and positive if $[A]_{ij} > 0$ for $i, j \in \{1, 2, \ldots, n\}$. One of the cornerstones of the theory of nonnegative matrices is the Perron-Frobenius theorem. The main part of the theory goes as follows. If A is a positive matrix, its spectral radius $\rho(A)$ is its simple eigenvalue, which in turn corresponds to an eigenvector with all positive entries. On the other hand, suppose that we associate a nonnegative matrix $A \in \mathbf{R}^{n \times n}$ with a digraph $\mathcal{D} = (V, E)$ as follows: $V = [n]$ and $(j, i) \in E$ if $[A]_{ij} > 0$ for $i, j \in [n]$. If this digraph is strongly connected, then the matrix A has a (unique) eigenvalue equal to its (positive) spectral radius with the associated eigenvector with positive entries. A nonnegative matrix such that its rows sum to one is called a stochastic matrix; if both rows and columns of this matrix sum to one, then it is called doubly stochastic. A matrix with only one 1 in each column and row is called a permutation matrix. The following celebrated result is often referred to as Birkhoff's theorem.

Theorem A.2. *Any doubly stochastic matrix is a convex combination of a set of permutation matrices.*

One of the amazing constructions for matrices is that of determinants, defined as

$$\det(A) = \sum_{\sigma} \prod [A]_{1\sigma(1)} [A]_{2\sigma(2)} \cdots [A]_{n\sigma(n)},$$

where σ varies over all permutations on the finite set $\{1, 2, \ldots, n\}$. A few facts about determinants are as follows.

1. If $A, B \in \mathbf{R}^{n \times n}$, then $\det(AB) = \det(A) \det(B)$.

2. (Cauchy-Binet) Let $A \in \mathbf{R}^{m \times n}$ and $B \in \mathbf{R}^{n \times m}$. Then

$$\det(AB) = \sum_{S} \det(A_S) \det(B_S),$$

where S runs over m-element subsets of $[n]$; $A_S \in \mathbf{R}^{m \times m}$ is the submatrix of A that has as its columns the columns in A indexed by S; and $B_S \in \mathbf{R}^{m \times m}$ is the submatrix of B that has as its rows the rows in B indexed by S. It is assumed that $m \leq n$, since otherwise, the determinant is zero.

3. When $A \in \mathbf{R}^{n \times n}$, $\det(A)$ is the product of eigenvalues of A.

We also mention a useful fact on partitioned semidefinite matrices, often referred to as the Schur complement formula. Consider a partitioned $2n \times 2n$ symmetric matrix

$$X = \begin{bmatrix} X_1 & X_2 \\ X_2^T & X_3 \end{bmatrix} \geq 0.$$

Then given that X_1 is nonsingular, and hence positive definite, one has

$$S := X_3 - X_2^T X_1^{-1} X_3 \geq 0;$$

the matrix S is called the Schur complement of X_1 in X.

A.3 CONTROL THEORY

Control theory is concerned with effective means of influencing the behavior of dynamical systems, for example, to make them more efficient or more stable. A canonical model often used in control theory involves a linear time-invariant model of the form

$$\dot{x}(t) = Ax(t) + Bu(t) \quad \text{and} \quad y(t) = Cx(t), \qquad \text{(A.3.1)}$$

where $x \in \mathbf{R}^n$ is the state of the underlying dynamic system, $A \in \mathbf{R}^{n \times n}$ is the system matrix, $B \in \mathbf{R}^{n \times m}$ is the input matrix, $C \in \mathbf{R}^{p \times n}$ is the output matrix, $y(t) \in \mathbf{R}^p$ is the output of the system, and $u(t) \in \mathbf{R}^m$ is the control signal to be designed. If the signal $u(t)$, which plays the supporting role in steering the main character of this scenario, the state x, in some desired way, can be considered independent of the state x, then the control is called open loop. If $u(t) = f(x(t))$ for some function f, then the control is called closed loop.

A state $z \in \mathbf{R}^n$ is called reachable from a state $x = x(0)$ in time T if there exists an open or closed loop control such that $x(T) = z$ when starting from $x(0) = x$. If an arbitrary state z is reachable from an arbitrary state x in an arbitrary interval $[0, T]$, with $T > 0$, then the system is called controllable. A dual notion of controllability is that of observability: a system is observable if, knowing the control input and the observation y on a given time interval, one can uniquely determine the initial condition. Although controllability by definition involves guaranteeing the existence of a set of infinite-dimensional objects (the control input) that corresponds to another set of infinite-dimensional objects (the set of initial and target states), it can be verified by checking the rank of the controllability matrix

$$[A \,|\, B] = [B \ AB \ \cdots \ A^{n-1}B] \in \mathbf{R}^{n \times nm};$$

in particular, the system (A.3.1) specified by the pair (A, B) is controllable if and only if **rank** $[A \mid B] = n$. Alternatively, the system (A.3.1) is controllable if and only if there does not exist a left eigenvector of A that is orthogonal to all columns of B. System theory duality can then be used to derive a similar matrix theoretic characterization for the observability of the system. In fact, this system is observable if and only if **rank** $[A^T \mid C^T] = n$, or, alternatively, if and only if none of the eigenvectors of A are orthogonal to all rows of the observation matrix C.

Assume now that $A \in \mathbf{R}^n$, $B \in \mathbf{R}^{n \times m}$, and **rank** $[A \mid B] = p < n$. Then there exists a nonsingular matrix P such that

$$ PAP^{-1} = \left[\begin{array}{cc} A_{11} & A_{12} \\ 0 & A_{22} \end{array} \right] \quad \text{and} \quad PB = \left[\begin{array}{c} B_1 \\ 0 \end{array} \right], $$

where $A_{11} \in \mathbf{R}^{p \times p}$, $A_{22} \in \mathbf{R}^{(n-p) \times (n-p)}$, and $B_1 \in \mathbf{R}^{p \times m}$, such that the pair (A_{11}, B_1) is controllable. This a partial Kalman decomposition.

A.3.1 Lyapunov Theory

Consider the dynamics

$$ \dot{x}(t) = f(x(t)), \tag{A.3.2} $$

where $f : \mathbf{R}^n \to \mathbf{R}^n$ is locally Lipschitz continuous and satisfies, without loss of generality, $f(0) = 0$.

Definition A.3. *Various forms of stability at the origin are as follows.*

1. **Stability:** *For all $\epsilon > 0$, there exists $\delta > 0$ such that $\|x(0)\| \leq \delta$ implies that $\|x(t)\| \leq \epsilon$, for all $t \geq 0$.*

2. **Asymptotic Stability:** *The origin is stable and there exists $\delta > 0$ such that $\|x(0)\| \leq \delta$ implies that $x(t) \to 0$ as $t \to \infty$.*

3. **Global Asymptotic Stability:** *The origin is asymptotically stable when (A.3.2) is arbitrary initialized.*

4. **Exponential Stability:** *There exist $\delta > 0, c > 0, \lambda > 0$ such that $\|x(0)\| \leq \delta$ implies that $\|x(t)\| \leq c\|x(0)\|e^{-\lambda t}$, for all $t \geq 0$.*

The main complication in the definition of stability is that it involves quantification over parameters that can assume infinitely many values, for example, all initial conditions, all $t \geq 0$, and so on. However, since stability is an "asymptotic" notion, there is a clever way to establish it by identifying a *certificate* for it. Let $\mathcal{C}^1$ be the class of continuously differentiable functions. Moreover, we call $V : \mathbf{R}^n \to \mathbf{R}$ positive definite if $V(x) > 0$ for all $x \neq 0$ and $V(0) = 0$ and positive semidefinite if $V(x) \geq 0$ for all x. If $V(x) \to \infty$ when $\|x\| \to \infty$, then V is called radially unbounded.

1. A **Lyapunov function** for (A.3.2), with respect to the origin, is a real-valued, positive definite $\mathcal{C}^1$ function $V : \mathbf{R}^n \to \mathbf{R}$, such that $\dot{V}(t) < 0$ for all $x \neq 0$ along the trajectories of (A.3.2).

2. A **weak Lyapunov function** for (A.3.2), with respect to the origin, is a real-valued, positive definite $\mathcal{C}^1$ function $V : \mathbf{R}^n \to \mathbf{R}$ such that $\dot{V}(t) \leq 0$ for all x along the trajectories of (A.3.2).

Lyapunov functions serve as certificates for examining the stability properties of (A.3.2). In particular, one has the following correspondences (all with respect to the origin).

1. If (A.3.2) admits a weak Lyapunov function, it implies the stability of the origin.

2. If (A.3.2) admits a Lyapunov function, it implies the asymptotic stability of the origin.

3. If (A.3.2) admits a radially unbounded Lyapunov function, it implies global asymptotic stability of the origin.

4. If (A.3.2) admits a Lyapunov function and $\dot{V}(t) \leq -\alpha V$ along the trajectory of (A.3.2) for some $\alpha > 0$, it implies the exponential stability of the origin.

Extensions of the notion of stability to the origin (or any other point in $\mathbf{R}^n$) can be realized for sets, which in turn requires an appropriate notion of *certificate* for examining the trajectories of (A.3.2).

Definition A.4. *A set $\mathcal{A}$ is an invariant set of (A.3.2) if, whenever $x(\bar{t}) \in \mathcal{A}$ for some $\bar{t}$, $x(t) \in \mathcal{A}$ for all $t \geq \bar{t}$.*

Example A.5. *Let*

$$\begin{bmatrix} \dot{x}_1 \\ \dot{x}_2 \end{bmatrix} = \begin{bmatrix} x_2 + x_1(1 - x_1^2 - x_2^2) \\ -x_1 + x_2(1 - x_1^2 - x_2^2) \end{bmatrix}.$$

Then the corresponding invariant sets are $(0,0)$ and the unit circle: if $x_1(\bar{t})^2 + x_2(\bar{t})^2 = 1$ for some $\bar{t}$, then $x_1(t)^2 + x_2(t)^2 = 1$ for all $t \geq \bar{t}$.

The following extension of the notion of "certificate" for stability is referred to as LaSalle's invariance principle.

Theorem A.6. *Let* $V : \mathbf{R}^n \to \mathbf{R}$ *be a weak Lyapunov function for (A.3.2). Let* $\mathcal{M}$ *be the largest invariant set (with respect to set inclusion) contained in*

$$\{x \in \mathbf{R}^n \,|\, \dot{V}(x) = 0\}.$$

Then every solution $x(t)$ *of (A.3.2) that remains bounded is such that*

$$\inf_{y \in \mathcal{M}} \|x(t) - y\| \to 0 \quad as \quad t \to \infty.$$

We note that when V is a radially unbounded weak Lyapunov function for (A.3.2), all solutions of (A.3.2) remain bounded.

In discrete time, we have the dynamics

$$x(k+1) = f(x(k)),$$

with equilibrium points as the fixed points of f, that is, $x_e = f(x_e)$. We assume that 0 is such an equilibrium point, and the discussion of Lyapunov stability will (without loss of generality) be confined to the issue of stability to the origin. The discrete time versions of Lyapunov-based certificates are almost identical to the continuous setting. We again let $V(x) > 0$ (possibly radially unbounded) for all $x \neq 0$ and, instead of derivatives, we take differences. In other words, if

$$V(x(k+1)) - V(x(k)) < 0 \quad \text{for all } x(k) \neq 0,$$

along the bounded trajectories of the system, then the origin is (globally, if V is radially unbounded) asymptotically stable.

The discrete time version of LaSalle's invariance principle is analogous to the continuous time case in that, if $V(x(k+1)) - V(x(k)) \leq 0$ along trajectories then x will converge to the largest invariant set $\mathcal{M}$ contained in $\{x \,|\, V(f(x)) - V(x) = 0\}$.

For switched systems, one has to be a bit more careful. In fact, given a collection of discrete modes $S = \{1, \ldots, s\}$, we define a *switched system* as

$$\dot{x}(t) = f_{\sigma(t)} x(t),$$

where $\sigma : [0, \infty) \to S$ is the switch signal that dictates what mode is active at any given time t.

To establish whether the origin is (globally) asymptotically stable for such a system, one first has to specify what class of switch signals one is considering.

Definition A.7. *A switched linear system is* universally *(globally) asymptotically stable if it is (globally) asymptotically stable for every switch signal. It is* existentially *(globally) asymptotically stable if there exists at least one switch signal that renders the system (globally) asymptotically stable.*

The following two facts follow directly from this definition.

1. If mode k is (globally) asymptotically stable, then the switched system is existentially (globally) asymptotically stable. (Just let $\sigma(t) = k$ for all t and asymptotic stability is achieved.)

2. If any of the subsystems (say system k) is unstable, then the switched system is not universally (globally) asymptotically stable. (Again, let $\sigma(t) = k$ for all t and the system goes unstable.)

If the switch signal is not a priori known, one is typically interested in the universal property. The following theorem characterizes this.

Theorem A.8. *The switched system*

$$\dot{x}(t) = f_{\sigma(t)}x(t), \ \sigma(t) \in S$$

is universally (globally) asymptotically stable if there exists a common *(radially unbounded) Lyapunov function V, that is, one such that for all $x \neq 0$, $V(x) > 0$ and*

$$\frac{d}{dt}V(x(t)) = \frac{\partial V(x(t))^T}{\partial x} f_i(x(t)) < 0 \ \text{for all } x \neq 0,$$

for all $i \in S$.

Theorem A.9. *Let V be a common weak Lyapunov function for the different subsystems and let M_i be the largest invariant set (under mode i) contained in*

$$\left\{ x \in \mathbf{R}^n \mid \frac{\partial V(x)^T}{\partial x} f_i(x) = 0 \right\}.$$

If $M_i = M_j$ for all $i, j \in S$, x will asymptotically converge to this set.

A.3.2 Passivity

In Chapter 4 we have employed constructs from passivity theory and nonlinear systems for analyzing nonlinear agreement protocols.

Definition A.10. *The set $\mathcal{L}_2$ consists of measurable vector-valued functions $f : \mathbf{R}_+ \to \mathbf{R}^n$ such that*

$$\|f\|_{\mathcal{L}_2} = \int_0^\infty \|f(t)\|^2 \, dt < \infty.$$

In fact, the function space $\mathcal{L}_2$ can be considered a Hilbert space with the inner product between two functions $f, g \in \mathcal{L}_2$ defined as

$$\langle f, g \rangle = \int_0^\infty f(t)^T g(t) \, dt. \tag{A.3.3}$$

Given the vector-valued function $f(t)$, one can also consider its truncated version,

$$f_T(t) = \begin{cases} f(t) & \text{if } t \leq T, \\ 0 & \text{otherwise.} \end{cases}$$

a construction that has become quite important in stability analysis of nonlinear and uncertain feedback systems. The space $\mathcal{L}_{2e}$ is the space of measurable vector-valued functions $f(t)$ such that for all $T \geq 0$, $f_T(\cdot) \in \mathcal{L}_2$.

Consider now the nonlinear system

$$\dot{x}(t) = f(t, x(t), u(t)), \quad y(t) = g(t, x(t), u(t)) \tag{A.3.4}$$

where it is assumed that $f(t, 0, 0) = 0$ and $g(t, 0, 0) = 0$ for all $t \geq 0$.

Definition A.11. *The system (A.3.4) is passive if, for all $u \in \mathcal{L}_{2e}$, $y \in \mathcal{L}_{2e}$, there exists a constant β such that*

$$\langle u, y \rangle \geq \beta.$$

The system is strictly input passive if, in addition, there exist β and $\delta > 0$, such that, for all u,

$$\langle u, y \rangle \geq \delta\|u\|^2 + \beta,$$

and strictly output passive if there exist β and $\epsilon > 0$, such that, for all u,

$$\langle u, y \rangle \geq \epsilon\|y\|^2 + \beta.$$

Example A.12. *Consider the single integrator with an output, $x(t)$, assumed to be in $\mathcal{L}_{2e}$. Then, for any $T \geq 0$,*

$$\langle \dot{x}, x \rangle = \int_0^T \dot{x}(t)^T x(t) dt = \frac{1}{2} \int_0^T \left[\frac{d}{dt} \|x(t)\|^2 \right] dt$$

$$= \frac{1}{2}(\|x(T)\|^2 - \|x(0)\|^2) \geq -\frac{1}{2}\|x(0)\|^2;$$

hence an integrator is passive.

The following characterization of passivity essentially ties it in with the general class of dissipative systems.

Proposition A.13. *Consider a continuously differentiable function $V(t) \geq 0$ and assume that the function $d(t)$ is such that for all $T \geq 0$,*

$$\int_0^T d(t)dt \geq 0.$$

Then, if

$$\dot{V}(t) \leq u(t)^T y(t) - d(t), \qquad (A.3.5)$$

for all $t \geq 0$ and all inputs u, the system (A.3.4) is passive.

A.4 PROBABILITY

Elements of probability theory were mainly used in Chapter 5, where we examined networks and protocols with random constructs. By a random variable X we mean a variable that can assume values in $\mathbf{R}$ according to some probability density $\mu(x)$, where

$$\mathbf{Pr}\{X \in [a, b]\} = \int_a^b \mu(x)dx.$$

The *expected value* of this random variable is then

$$\mathbf{E}\{X\} = \int_{-\infty}^\infty x\mu(x)dx$$

or, in the case where x only assumes discrete values,

$$\mathbf{E}\{X\} = \sum_x x\mathbf{Pr}\{X = x\}.$$

The variance of the random variable X is

$$\mathbf{var}\{X\} = \mathbf{E}\{(x - \mathbf{E}\{x\})^2\};$$

it captures how much the random variable is expected to deviate from its expected value. Capturing this deviation, or more generally bounding the probability of certain events, is in fact the subject of a number of famous inequalities in probability. These include the Markov inequality

$$\mathbf{Pr}\{X \geq \alpha\} \leq \frac{E\{X\}}{\alpha},$$

where X is assumed to be a nonnegative random variable, and the Chebyshev inequality

$$\mathbf{Pr}\{|X - \mathbf{E}\{X\}| \geq \sigma\alpha\} \leq \frac{1}{\alpha^2},$$

where σ is the finite standard deviation of X, that is, $\sigma^2 = \mathbf{var}\{X\}$.

Another powerful bound used in Chapter 5 is the Chernoff bound, stating that if X is the sum of independent variables, each one with a fixed probability of being a one or not, then

$$\mathbf{Pr}\{X < (1 - \delta)\mathbf{E}\{X\}\} < \left(\frac{e^{-\delta}}{(1-\delta)^{1-\delta}}\right)^{\mathbf{E}\{X\}}$$

and

$$\mathbf{Pr}\{X > (1 + \delta)\mathbf{E}\{X\}\} < \left(\frac{e^{\delta}}{(1+\delta)^{1+\delta}}\right)^{\mathbf{E}\{X\}}.$$

Conditional probabilities and expectations allow one to reason about the relationship between *multiple* random variables. In this direction, consider two random variables X and Y. Then the conditional expectation of X with respect to Y is defined as

$$\mathbf{E}\{X \mid Y = y\} = \int_{-\infty}^{\infty} x\mu(x \mid y)dx,$$

where $\mu(x \mid y)$ is the conditional density function

$$\mu(x \mid y) = \frac{\mu(x, y)}{\mu(y)},$$

with $\mu(x, y)$ being the joint density function of the two random variables X and Y. The latter density function, in turn, parameterizes the probability that two events regarding the two random variables occur simultaneously. For example,

$$\mathbf{Pr}\{X \in [a, b], Y \in [c, d]\} = \int_{a}^{b} \int_{c}^{d} \mu(x, y)dy\, dx.$$

In Chapter 5 we also extensively used the notion of a random sequence for analyzing random networks and noisy protocols over networks. A random sequence $\{V(k)\}_{k\geq 0}$ converges to a random variable V^* with probability one (w.p.1), if, for every $\epsilon > 0$,

$$\mathbf{Pr}\left\{\sup_{k\geq N} \|V(k) - V^*\| \geq \epsilon\right\} \to 0 \quad \text{as} \quad N \to \infty.$$

On the other hand, $\{V(k)\}_{k\geq 0}$ converges in the mean if the deterministic sequence $\{\mathbf{E}\{V(k)\}\}_{k\geq 0}$ converges to (a constant number) V^*. Moreover, a random sequence $\{x(k)\}$ in $\mathbf{R}^n$ converges to x^* in probability if, for each $\epsilon > 0$, $\mathbf{Pr}\{\|x(k) - x^*\| \geq \epsilon\} \to 0$ as $k \to \infty$.

Suppose now that the sequence of nonnegative random variables $\{V(k)\}$ is such that

$$\mathbf{E}\{V(k+1)\,|\,V(0),\ldots,V(k)\} \leq \mathbf{E}\{V(k)\} \text{ and } \mathbf{E}\{V(0)\} < \infty;$$

such a sequence is called a nonnegative supermartingale. The celebrated supermartingale convergence theorem then states that when $\{V(k)\}_{k\geq 0}$ is a nonnegative supermartingale, there exists a random variable $V^* \geq 0$ such that $V(k) \to V^*$ w.p.1, with the following as a consequence.

Lemma A.14. *Consider the sequence of nonnegative random variables* $\{V(k)\}_{k\geq 0}$ *with* $\mathbf{E}\{V(0)\} < \infty$. *Let*

$$\mathbf{E}\{V(k+1)\,|\,V(0),\ldots,V(k)\} \leq [1 - c_1(k)]\,V(k) + c_2(k), \quad \text{(A.4.1)}$$

with $c_1(k)$ *and* $c_2(k)$ *satisfying*

$$0 \leq c_1(k) \leq 1, \; 0 \leq c_2(k),$$
$$\sum_{k=0}^{\infty} c_2(k) < \infty, \; \sum_{k=0}^{\infty} c_1(k) = \infty, \; \lim_{k\to\infty} \frac{c_2(k)}{c_1(k)} = 0. \quad \text{(A.4.2)}$$

Then $V(k) \to 0$ *w.p.1.*

We conclude this section by pointing out that, for discrete-time stochastic systems, a Lyapunov-based framework, analogous to that in §A.3, can be developed for proving convergence properties of random sequences. For example, suppose that a positive definite function V is a supermartingale along the sequence generated by the dynamic system. In other words, for some positive semidefinite matrix C, we have

$$\mathbf{E}\{V(z(k+1)) - V(z(k))\,|\,z(k)\} = -z(k)Cz(k).$$

Then $z(k)$ converges to the set

$$M = \{z\,|\,z^T Cz = 0\}$$

w.p.1. Such a theorem, among its many variants, is then referred to as the stochastic version of LaSalle's invariance principle.

A.5 OPTIMIZATION AND GAMES

Given a set $\Gamma \subseteq \mathbf{R}$, the infimum of Γ ($\inf \Gamma$) is the greatest lower bound on Γ; the least upper bound on Γ is the supremum ($\sup \Gamma$). To make sure that inf and sup always exist, we append $-\infty$ and $+\infty$ to $\mathbf{R}$; we write $\mathbf{R} \cup \{+\infty\}$ if necessary. One has, by convention, $\sup \emptyset = -\infty$ and $\inf \emptyset = +\infty$. If, for a given optimization problem, it is known that inf (respectively, sup) exists, then it is more pleasing to write min (respectively, max). Given a subset S of a vector space $\mathbf{V}$ and a function $f : \mathbf{V} \to \mathbf{R}$, an optimization problem is the problem of the forms

$$\inf_{x \in S} f(x) \quad \text{or} \quad \sup_{x \in S} f(x).$$

Optimization problems are in general difficult. Subclasses that make an optimizer happy include linear programming,

$$\min_{Ax=b,\ x \geq 0} c^T x, \quad \text{where} \quad A \in \mathbf{R}^{m \times n}, b \in \mathbf{R}^m, \text{ and } c \in \mathbf{R}^n,$$

and quadratic programming,

$$\min_{Ax=b,\ x \geq 0} x^T Q x, \quad \text{where} \quad Q \in \mathbf{R}^{n \times n}, b \in \mathbf{R}^m, \text{ and } c \in \mathbf{R}^n,$$

and Q is positive semidefinite. More generally, easily manageable instances of optimization problems include those that are convex: those whose objectives are convex functions and their constraint sets are convex sets. A subset $C \in \mathbf{V}$ is called convex if for all $x, y \in C$ and $\alpha \in (0, 1)$, $\alpha x + (1-\alpha)y \in C$; for a convex set C, the function $f : C \to \mathbf{R}$ is convex if, for all $x, y \in C$ and all $\lambda \in [0, 1]$,

$$f(\lambda x + (1 - \lambda)y) \leq \lambda f(x) + (1 - \lambda) f(y). \tag{A.5.1}$$

In the case when the optimization is unconstrained, for example, when $f : \mathbf{R}^n \to \mathbf{R}$ and the constraints set S is the entire $\mathbf{R}^n$, assuming that f is convex and differentiable, the minimizer can be found by solving $\nabla f(x) = 0$. In general, the set $\{x \mid \nabla f(x) = 0\}$ is called the stationary points of f. This set contains the minimizers and maximizers of f, as well as other points called the saddle points of f.

A recent addition to the list of optimization problems that can be solved with reasonable efficiency is semidefinite programming (SDP). SDPs have found many applications in systems and control theory as well as in combinatorics, probability, and statistics.

Given $A_i \in \mathcal{S}^n$ and $c \in \mathbf{R}^n$, a SDP is defined as

$$\min_x c^T x$$
$$\text{s.t. } x_1 A_1 + \cdots + x_n A_n > 0,$$

where the expression $B > A$ for two symmetric matrices implies that the matrix difference $B - A$ is positive definite; similarly, $B \geq A$ captures the semidefiniteness of $B - A$. An equivalent form for an SDP is

$$\max_{Y} \ \textbf{trace } BY,$$
$$\text{s.t. } \langle A_i, Y \rangle = c_i, \ i = 1, \ldots, m,$$
$$Y \geq 0,$$

where $B \in \mathcal{S}^n$.

Optimization is closely related to solving systems of inequalities. For example, the linear matrix inequality (LMI) is the problem of finding the set of real numbers $x_1, \cdots, x_n$ such that, for a given set of symmetric matrices $A_o, A_1, \ldots, A_n$, one has

$$A_o + \sum_{i=1}^{n} x_i A_i \geq 0,$$

that is, that the linear combination of these symmetric matrices should be positive semidefinite. LMIs have in fact been used for finding sufficient conditions for the feasibility of *nonconvex* sets. For example, suppose that it is desired to find a feasible point in a set defined by the quadratic inequalities

$$x^T Q_1 x \geq 0, \quad x^T Q_2 x < 0,$$

for $x \in \mathbf{R}^n$ and $Q_1, Q_2 \in \mathcal{S}^n$. Then one can check to see if, instead, it is valid that for all vectors x that make $x^T Q_1 x \geq 0$, it is the case that $x^T Q_2 x \geq 0$. This can be accomplished by what is known as the S-procedure, which involves checking for the existence of a nonnegative scalar τ such that

$$Q_2 \geq \tau Q_1,$$

which is an LMI. An extension of this idea is as follows. In order to show whether it is true that, for a given set of symmetric matrices,

$$Q_o, Q_1, Q_2, \ldots, Q_n,$$

the set of inequalities

$$x^T Q_1 x \geq 0, \ x^T Q_2 x \geq 0, \ \ldots, \ x^T Q_n x \geq 0$$

would imply that $x^T Q_o x \geq 0$, it suffices to show the existence of nonnegative scalars $\tau_1, \tau_2, \ldots, \tau_n$, such that

$$Q_o \geq \tau_1 Q_1 + \tau_2 Q_2 + \cdots + \tau_n Q_n,$$

which can be cast as an LMI as well.

A conceptually pleasing generalization of optimization problems is in terms of games. Whereas in an optimization model one is interested in finding the best decision x, subject to constraints, that minimizes or maximizes a given objective f, in a game problem the objective is a function of arguments that can be chosen independently by multiple decision makers, possibly with conflicting individual objectives.

Games come in a variety of flavors: static, dynamic, 2-player, n-player with $n \geq 3$, nonzero sum, zero sum, repeated, and evolutionary, to name a few. The difference is often reflected in the assumptions on the nature of the systems that the players have vested interested in. To motivate some of the key concepts, including the notion of Nash equilibria, we resort to the classical game called the *prisoner's dilemma*.

The setup of the prisoner's dilemma is as follows. Consider two prisoners who under individual interrogation about a crime that they have committed together can either keep quiet or confess. We refer to these prisoners as "agents" or "players." In the case that these players keep quiet, we say that they are collaborating with each other. The cost structure for the game is as follows: (1) if the players both keep quiet, then they each get two years in prison, (2) if the players both confess, then they each get four years in prison, (3) if the first player confesses but the second player keeps quiet, then the first player is free to go and the second player gets five years in prison, and finally, (4) if the second player confesses but the first player keeps quiet, then the second player is free to go and the first player gets five years in prison. We adopt the convention of denoting the result of the decision of agent i chosen from the set Σ_i by π_i, as the prisoners adopt their respective strategies of keeping quite or to confess. In fact, we can summarize this game in a table shown below, depicting the decisions of the prisoners and their "payoff," shown inside each box.

		Prisoner 2	
		Quiet	Confess
Prisoner 1	Quiet	(r, r)	(t, s)
	Confess	(s, t)	(p, p)

For this example, we have $r = -2$, $s = 0$, $t = -5$, $p = -4$, and $\pi_1(\text{confess}, \text{quiet}) = 0$. In this game, there is a notion of utility or payoff and strategy. A Nash equilibrium is a pair of strategies (σ_1^*, σ_2^*) such that

$$\pi_1(\sigma_1^*, \sigma_2^*) \geq \pi_1(\sigma_1, \sigma_2^*) \quad \text{for all } \sigma_1 \in \Sigma_1,$$
$$\pi_2(\sigma_1^*, \sigma_2^*) \geq \pi_2(\sigma_1^*, \sigma_2) \quad \text{for all } \sigma_2 \in \Sigma_2.$$

Thus, at the Nash equilibrium, each player cannot benefit from unilaterally changing its strategy. Hence (confess, confess) in the example above is a Nash equilibrium, and in fact, it is exactly the source of the dilemma.

Bibliography

[1] *http://www.esa.int*

[2] *http://fewcal.kub.nl/sturm/software/sedumi.html*

[3] *http://planetquest.jpl.nasa.gov/TPF/*

[4] R. Agaev and P. Chebotarev. On the spectra of nonsymmetric Laplacian matrices. *Linear Algebra and its Applications*, 399: 157–168, 2005.

[5] R. K. Ahuja, T. L. Magnanti, and J. B. Orlin. *Network Flows: Theory, Algorithms, and Applications*. Prentice-Hall, 1993.

[6] M. A. Aizerman, L. A. Gusev, S. V. Petrov, I. M. Smirnova, and L. A. Tenenbaum. Dynamic approach to analysis of structures described by graphs (foundations of graph-dynamics). In *Topics in the General Theory of Structures*, edited by E. R. Caianiello and M. A. Aizerman. Reidel, 1987.

[7] P. Alriksson and A. Rantzer. Distributed Kalman filtering using weighted averaging, *Proceedings of the 17th International Symposium on Mathematical Theory of Networks and Systems*, July 2006.

[8] B. D. O. Anderson and J. B. Moore. *Optimal Filtering*. Dover, 2005.

[9] H. Ando, Y. Oasa, I. Suzuki, and M. Yamashita. Distributed memoryless point convergence algorithm for mobile robots with limited visibility. *IEEE Transactions on Robotics and Automation*, 15 (5): 818–828, 1999.

[10] M. Arcak. Passivity as a design tool for group coordination. *IEEE Transactions on Automatic Control*, 52 (8): 1380–1390, 2007.

[11] M. Armstrong. *Basic Topology*. Springer, 1997.

[12] L. Arnold. On the asymptotic distribution of the eigenvalues of random matrices. *Journal of Mathematical Analysis and Applications*, 20: 262–268, 1967.

[13] R. J. Aumann. Agreeing to disagree. *Annals of Statistics*, 4 (6): 1236–1239, 1976.

[14] T. Balch and R. C. Arkin. Behavior-based formation control for multi-robot teams. *IEEE Transactions on Robotics and Automation*, 14 (6): 926–939, 1998.

[15] B. Bamieh, F. Paganini, and M. Dahleh. Distributed control of spatially-invariant systems. *IEEE Transactions on Automatic Control*, 47 (7): 1091–1107, 2002.

[16] A-L. Barabási and A. Réka. Emergence of scaling in random networks. *Science*, 286: 509–512, 1999.

[17] R. W. Beard, J. R. Lawton, and F. Y. Hadaegh. A coordination architecture for spacecraft formation control. *IEEE Transactions on Control Systems Technology*, 9 (6): 777–790, 2001.

[18] A. Bensoussan and J. L. Menaldi. Difference equations on weighted graphs. *Journal of Convex Analysis*, 12 (1): 13–44, 2005.

[19] C. Berge. *Hypergraphs*. North-Holland, 1989.

[20] A. Berman and R. Plemmons. *Nonnegative Matrices in the Mathematical Sciences*. SIAM, 1994.

[21] A. Berman and X-D. Zhang. Lower bounds for the eigenvalues of Laplacian matrices. *Linear Algebra and its Applications*, 316: 13–20, 2000.

[22] D. P. Bertsekas and J. N. Tsitsiklis. *Parallel and Distributed Computation*. Prentice-Hall, 1989.

[23] N. L. Biggs, E. K. Lloyd, and R. J. Wilson. *Graph Theory: 1736–1936*. Oxford University Press, 1998.

[24] N. Biggs. *Algebraic Graph Theory*. Cambridge University Press, 1993.

[25] V. Blondel, J. M. Hendrickx, and J. Tsitsiklis. On Krause's consensus formation model with state-dependent connectivity. arXiv:0807.2028v1, 2008.

[26] B. Bollobás. *Combinatorics: Set Systems, Hypergraphs, Families of Vectors, and Combinatorial Probability*. Cambridge University Press, 1986.

[27] B. Bollobás. *Random Graphs*. Cambridge University Press, 2001.

[28] B. Bollobás. *Modern Graph Theory*. Springer, 2002.

[29] B. Bollobás. *Extremal Graph Theory*. Dover, 2004.

[30] J. A. Bondy and U.S.R. Murty. *Graph Theory*. Springer, 2008.

[31] S. Björkenstam, M. Ji, M. Egerstedt, and C. Martin. Leader-Based Multi-Agent Coordination Through Hybrid Optimal Control. *Allerton Conference on Communication, Control, and Computing*, 2006.

[32] M. Born and E. Wolf. *Principles of Optics* (sixth edition). Cambridge University Press, 1997.

[33] S. P. Boyd, L. El Ghaoui, E. Feron, and V. Balakrishnan. *Linear Matrix Inequalities in System and Control Theory*. SIAM, 1994.

[34] S. Boyd, A. Ghosh, B. Prabhakar, and D. Shah. Randomized gossip algorithms. *IEEE Transactions on Information Theory*, 52 (6): 2508–2530, 2006.

[35] B. Brogliato, R. Lozano, B. Maschke, and O. Egeland. *Dissipative Systems Analysis and Control: Theory and Applications*. Springer, 2006.

[36] M. E. Broucke. Disjoint path algorithms for planar reconfiguration of identical vehicles. *Proceedings of the American Control Conference*, June 2003.

[37] M. E. Broucke. Reconfiguration of identical vehicles in 3D. *Proceedings of the IEEE Conference on Decision and Control*, December 2003.

[38] R. Bru, L. Elsner, and M. Neumann. Models of parallel chaotic iteration methods. *Linear Algebra and its Applications*, 103: 175–192, 1988.

[39] L. Bruneau, A. Joye, and M. Merkli, Infinite products of random matrices and repeated interaction dynamics. arXiv:math/0703675v2, Feburary 2008.

[40] A. E. Bryson, Jr. and Y-C. Ho. *Applied Optimal Control*. Taylor and Francis, 1975.

[41] F. Bullo, J. Cortés, and S. Martínez. *Distributed Control of Robotic Networks: A Mathematical Approach to Motion Coordination Algorithms*. Princeton University Press, 2009.

[42] R. E. Burkard. Selected topics in assignment problems. *Discrete Applied Mathematics*, 123: 257–302, 2002.

[43] R. Carli, F. Fagnani, A. Speranzon, and S. Zampieri. Communication constraints in coordinated consensus problem. *Proceedings of the American Control Conference*, June 2006.

[44] S. Chatterjee and E. Seneta. Towards consensus: some convergence theorems on repeated averaging. *Journal of Applied Probability*, 14: 89–97, 1977.

[45] D. Chazan and W. L. Miranker. Chaotic relaxation. *Linear Algebra and its Applications*, 2: 199–222, 1969.

[46] B-D. Chen and S. Lall. Dissipation inequalities for distributed systems on graphs. *Proceedings of the IEEE Conference on Decision and Control*, December 2003.

[47] F. Chung, L. Lu, and V. Vu. The spectra of random graphs with given expected degrees. *Proceedings of the National Academy of Sciences*, 100 (11): 6313–6318, 2003.

[48] F. Chung and L. Lu. *Complex Graphs and Networks*. American Mathematical Society, 2006.

[49] F. R. K. Chung and K. Oden. Weighted graph Laplacians and isoperimetric inequalities. *Pacific Journal of Mathematics*, 192 (2): 257–273, 2000.

[50] F. R. K. Chung. *Spectral Graph Theory*. American Mathematical Society, 1997.

[51] R. Cogburn. On products of random stochastic matrices. *Random Matrices and Their Applications: Proceedings of the AMS-IMS-SIAM Joint Summer Research Conference*. J. E. Cohen, H. Kesten, C. M. Newman (editors). American Mathematical Society, 1984.

[52] R. Conner, M. Heithaus, and L. Barre. Complex social structure, alliance stability and mating access in a bottlenose dolphin "super alliance." *Proceedings of the National Academy of Sciences*: 987–990, 1992.

[53] J. Cortés. Distributed algorithms for reaching consensus on general functions. *Automatica*, 44 (3): 726–737, 2008.

[54] J. Cortés and F. Bullo. Coordination and geometric optimization via distributed dynamical systems. *SIAM Journal of Control and Optimization*, 45 (5): 1543–1574, 2005.

[55] J. Cortés, S. Martínez, and F. Bullo. Robust rendezvous for mobile autonomous agents via proximity graphs in arbitrary dimensions. *IEEE Transactions on Automatic Control*, 51 (8): 1289–1298, 2006.

[56] I. D. Couzin, Behavioral ecology: social organization in fission-fusion societies. *Current Biology*, 16: 169–171, 2006.

[57] I. D. Couzin and N. R. Franks, Self-organized lane formation and optimized traffic flow in army ants. *Proceedings of the Royal Society of London*, 270: 139-146, 2003.

[58] H. S. M. Coxeter and S. L. Greitzer. *Geometry Revisited*. Mathematical Association of America, 1967.

[59] J. L. Crassidis and J. L. Junkins. *Optimal Estimation of Dynamic Systems*. Chapman and Hall/CRC, 2004.

[60] C. H. Caicedo-Nunez and M. Zefran. Rendezvous under noisy measurements. *Proceedings of the IEEE Conference on Decision and Control*, December 2008.

[61] R. D'Andrea and G. E. Dullerud. Distributed control design for spatially interconnected systems. *IEEE Transactions on Automatic Control*, 48 (9): 1478–1495, 2003.

[62] A. Das and M. Mesbahi. Distributed parameter estimation over sensor networks. *IEEE Transactions on Aerospace Systems and Electronics*, 45 (4): 1293–1306, 2009.

[63] P. Dayawansa and C. F. Martin. A converse Lyapunov theorem for a class of dynamical systems which undergo switching, *IEEE Transactions on Automatic Control*, 44 (4): 751–760, 1999.

[64] M. H. DeGroot. Reaching a consensus. *Journal of the American Statistical Association*, 69: 118–121, 1974.

[65] J. Desai, J. Ostrowski, and V. Kumar. Modeling and control of formations of nonholonomic mobile robots. *IEEE Transactions on Robotics and Automation*, (17) 6: 905–908, 2001.

[66] J. Desai, J. Ostrowski, and V. Kumar. Controlling formations of multiple mobile robots. *Proceedings of IEEE International Conference Robotics and Automation*, May 1998.

[67] M. Desbrun, A. Hirani, and J. Marsden. Discrete exterior calculus for variational problems in computer graphics and vision. *Proceedings of the IEEE Conference on Decision and Control*, December 2003.

[68] V. de Silva, R. Ghrist, and A. Muhammad. Blind Swarms for Coverage in 2-D. *Robotics: Science and Systems*, MIT Press, 2005.

[69] V. de Silva and R. Ghrist. Coordinate-free coverage in sensor networks with controlled boundaries via homology. *International Journal of Robotics Research*, 25 (12): 1205–1222, 2006.

[70] V. de Silva and R. Ghrist. Coverage in sensor networks via persistent homology. *Algebraic and Geometric Topology*, 7: 339-358, 2007.

[71] R. Diestel. *Graph Theory*. Springer, 2000.

[72] W. B. Dunbar and R. M. Murray. Receding horizon control of multivehicle formations: a distributed implementation. *Proceedings of the IEEE Conference on Decision and Control*, December 2004.

[73] M. Egerstedt, X. Hu, and A. Stotsky. Control of mobile platforms using a virtual vehicle approach. *IEEE Transactions on Automatic Control*, 46 (11): 1777–1782, 2001.

[74] M. Egerstedt and X. Hu. Formation constrained multi-agent control. *IEEE Transactions on Robotics and Automation*, (17) 6: 947–951, 2001.

[75] L. Elsner, I. Koltracht, and M. Neumann. Convergence of sequential and asynchronous nonlinear paracontractions. *Numerische Mathematik*, 62 (1): 305–319, 1992.

[76] P. Erdős and A. Rényi. On the evolution of random graphs. *Publication of the Mathematical Institute of the Hungarian Academy of Sciences*, 5: 17–61, 1960.

[77] T. Eren, P.N. Belhumeur, B. D. O. Anderson, and A.S. Morse. A framework for maintaining formations based on rigidity. *Proceedings of the 15th IFAC World Congress*, 2002.

[78] T. Eren, W. Whiteley, B. D. O. Anderson, A. S. Morse, and P. N. Belhumeur. Information structures to secure control of rigid formations with leader-follower architecture. *Proceedings of the American Control Conference*, June 2005.

[79] L. C. Evans. *Applied Optimal Control*. American Mathematical Society, 1998.

[80] A. Fabrikant, A. Luthra, E. Maneva, C. H. Papadimitriou, and S. Shenker. On a network creation game. *ACM Symposium on Principles of Distributed Computing*, 2003.

[81] F. Fagnani and S. Zampieri. Randomized consensus algorithms over large scale networks. *IEEE Journal on Selected Areas in Communications*, 26 (4): 634–649, 2008.

[82] A. Fagiolini, E. M. Visibelli, and A. Bicchi. Logical consensus for distributed network agreement. *Proceedings of the IEEE Conference on Decision and Control*, December 2008.

[83] C. Fall, E. Marland, J. Wagner, and J. Tyson (Editors). *Computational Cell Biology*. Springer, 2005.

[84] S. Fallat and S. Kirkland. Extremizing algebraic connectivity subject to graph theoretic constraints. *The Electronic Journal of Linear Algebra*, 3 (1): 48–74, 1998.

[85] A. Fax and R. M. Murray. Graph Laplacian and stabilization of vehicle formations. *Proceedings of the 15th IFAC World Congress*, 2002.

[86] A. Fax and R. M. Murray. Information flow and cooperative control of vehicle formations. *IEEE Transactions on Automatic Control*, 49 (9): 1465–1476, 2004.

[87] M. Feinberg. *Lectures on Chemical Reaction Networks*. Notes of lectures given at the Mathematics Research Centre. University of Wisconsin, 1979.

[88] G. Ferrari-Trecate, A. Buffa, and M. Gati. Analysis of coordination in multi-agent systems through partial difference equations. Part I: The Laplacian control. *Proceedings of the 16th IFAC World Congress*, 2005.

[89] G. Ferrari-Trecate, M. Egerstedt, A. Buffa, and M. Ji. Laplacian sheep: a hybrid, stop-go policy for leader-based containment control. *Hybrid Systems: Computation and Control*, Springer, 2006.

[90] M. Fiedler. Algebraic connectivity of graphs. *Czechoslovak Mathematical Journal*, 23 (98): 298–305, 1973.

[91] M. Franceschelli, M. Egerstedt, A. Giua, and C. Mahulea. Constrained invariant motions for networked multi-agent systems. *Proceedings of the American Control Conference*, June 2009.

[92] D. Freedman. *Markov Chains*. Springer, 1983.

[93] R. A. Freeman, P. Yang, and K. M. Lynch. Stability and convergence properties of dynamic average consensus estimators, *Proceedings of the IEEE Conference on Decision and Control*, December 2006.

[94] Z. Füredi and J. Komlós. The eigenvalues of random symmetric matrices. Combinatorica, 1 (3): 233–241, 1981.

[95] V. Gazi and K. M. Passino. A class of attraction/repulsion functions for stable swarm aggregations. *International Journal of Control*, 77: 1567–1579, 2004.

[96] A. Gelb (Editor). *Applied Optimal Estimation*. MIT Press, 1974.

[97] A. Ghosh and S. Boyd. Growing well-connected graphs. *Proceedings of the IEEE Conference on Decision and Control*, December 2006.

[98] A. M. Gibbons. *Topological Graph Theory*. Cambridge University Press, 1985.

[99] E. N. Gilbert. Random graphs. *Annals of Mathematical Statistics*, 30: 1141–1144, 1959.

[100] H. Gluck. Almost all simply connected closed surfaces are rigid. *Geometric Topology*. Lecture Notes in Mathematics, 438: 225–239, Springer, 1975.

[101] C. Godsil and G. Royle. *Algebraic Graph Theory*. Springer, 2001.

[102] R. L. Graham, M. Grötschel, and L. Lovász. *Handbook of Combinatorics*. MIT Press/North-Holland, 1995.

[103] J. L. Gross and T. W. Tucker. *Topological Graph Theory*. Dover, 2001.

[104] J. Gower. Properties of Euclidean and non-Euclidean distance matrices. *Linear Algebra and its Applications*, (67) 1: 81–97, 1985.

[105] S. Goyal. *Connections: An Introduction to the Economics of Networks*. Princeton University Press, 2007.

[106] M. Grant and S. Boyd. *CVX: Matlab Software for Disciplined Convex Programming. www.stanford.edu/ boyd/cvx/*

[107] D. Grünbaum, S. Viscido and J. K. Parrish. Extracting interactive control algorithms from group dynamics of schooling fish. *Proceedings of the Block Island Workshop on Cooperative Control*, V. Kumar, N. E. Leonard and A. S. Morse (editors). Springer, 2004.

[108] S. Guattery and G. L. Miller. On the quality of spectral separators. *SIAM Journal on Matrix Analysis and Applications*, 19 (3): 701–719, 1998.

[109] H. Guo, M.Y. Li, and Z. Shuai. A graph-theoretical approach to the method of global Lyapunov functions. *Proceedings of the American Mathematical Society*, 136: 2793–2802, 2008.

[110] T. Gustavi, D.V. Dimarogonas, M. Egerstedt, and X. Hu. On the number of leaders needed to ensure network connectivity in arbitrary dimensions. *Proceedings of the Mediterranean Conference on Control and Automation*, June 2009.

[111] M. Haque and M. Egerstedt. Decentralized formation selection mechanisms inspired by foraging bottlenose dolphins. *Mathematical Theory of Networks and Systems*, July 2008.

[112] H. R. Hashemipour, S. Roy, and A. J. Laub. Decentralized structures for parallel Kalman filtering. *IEEE Transactions on Automatic Control*, 33 (1): 88–94, 1988.

[113] Y. Hatano and M. Mesbahi. Agreement over random networks. *IEEE Transactions on Automatic Control*, 50 (11): 1867–1872, 2005.

[114] Y. Hatano, A. Das, and M. Mesbahi. Agreement in presence of noise: pseudogradients on random geometric networks. *Proceedings of the IEEE Conference on Decision and Control and European Control Conference*, December 2005.

[115] J. M. Hendrickx, B. D. O. Anderson, J-C. Delvenne, and V. D. Blondel. Directed graphs for the analysis of rigidity and persistence in autonomous agent systems. *International Journal of Robust and Nonlinear Control*, 17: 960–981, 2000.

[116] D. Henrion and A. Garulli (Editors). *Positive Polynomials in Control*. Springer, 2005.

[117] J. Hespanha. Uniform stability of switched linear systems: extensions of LaSalle's invariance principle. *IEEE Transactions on Automatic Control*, 49 (4): 470–482, 2004.

[118] D. J. Hoare, I. D. Couzin, J-G. Godin, and J. Krause. Context-dependent group size choice in fish. *Animal Behaviour*, 67: 155–164, 2004.

[119] J. Hofbauer and K. Sigmund. *Evolutionary games and population dynamics*. Cambridge University Press, 1998.

[120] R. A. Horn and C. Johnson. *Matrix Analysis*. Cambridge University Press, 1985.

[121] W. Imrich and S. Klavar. *Product Graphs: Structure and Recognition*. Wiley, 2000.

[122] M. O. Jackson. *Social and Economic Networks*. Princeton University Press, 2008.

[123] D. J. Jacobs and B. Hendrickson. An algorithm for two-dimensional rigidity percolation: the pebble game, *Journal of Computational Physics*, 137 (2): 346–365, 1997.

[124] A. Jadbabaie, J. Lin, and A. S. Morse. Coordination of groups of mobile autonomous agents using nearest neighbor rules. *IEEE Transactions on Automatic Control*, 48 (6): 988–1001, 2003.

[125] S. Janson, T. Luczak, and A. Rucinski. *Random Graphs*. Wiley, 2000.

[126] M. Ji and M. Egerstedt. Distributed coordination control of multi-agent systems while preserving connectedness. *IEEE Transactions on Robotics*, 23 (4): 693–703, 2007.

[127] M. Ji, A. Muhammad, and M. Egerstedt. Leader-based multi-agent coordination: controllability and optimal control. *Proceedings of the American Control Conference*, June 2006.

[128] F. Juhász. On the spectrum of a random graph. Colloquia Mathematica Societatis János Bolyai. *Algebraic Methods in Graph Theory,Szeged (Hungary)*,   V. Sós and L. Lovász (editors). North-Holland, 1978.

[129] M. Juvan and B. Mohar. Laplacian eigenvalues and bandwidth-type invariants of graphs. *Journal of Graph Theory*, (17): 393–407, 1993.

[130] T. Kailath. *Linear Systems*. Prentice-Hall, 1980.

[131] H. K. Khalil. Nonlinear Systems. Prentice-Hall, 2001.

[132] U. A. Khan and J. M. F. Moura. Distributing the Kalman filter for large-scale systems. *IEEE Transactions on Signal Processing*, 56 (10): 4919–4935, 2008.

[133] Y. Kim, M. Mesbahi, and F. Y. Hadaegh. Multiple-spacecraft reconfigurations through collision avoidance, bouncing, and stalemates. *Journal of Optimization Theory and Applications*, 122 (2): 323–343, 2004.

[134] Y. Kim and M. Mesbahi. On maximizing the second smallest eigenvalue of a state-dependent graph Laplacian, *IEEE Transactions on Automatic Control*, (51) 1: 116-120, 2006.

[135] J. Komlós and M. Simonovits. Szemerédi's regularity lemma and its applications in graph theory. *Combinatorics: Paul Erdős is Eighty*, vol. 2. Bolyai Society Mathematical Studies 2, Budapest, 1996.

[136] S. Kirti and A. Scaglione. Scalable distributed Kalman filtering through consensus. *Proceedings of the IEEE International Conference on Acoustics, Speech and Signal Processing,* April 2008.

[137] H. W. Kuhn. The Hungarian method for the assignment problem. *Naval Research Logistics Quarterly*, 2: 83–97, 1955.

[138] H. J. Kushner. *Stochastic Stability and Control.* Academic Press, 1967.

[139] H. J. Kushner. *Introduction to Stochastic Control.* Holt, Reinhart, and Winston, 1971.

[140] G. Lafferriere, A. Williams, J. Caughman, and J. J. P. Veerman. Decentralized control of vehicle formations. *System and Control Letters*, 54: 899–910, 2005.

[141] G. Laman. On graphs and rigidity of plane skeletal structures. *Journal of Engineering Mathematics*, 4 (4): 331–340, 1970.

[142] A. N. Langville and C. D. Meyer. *Google's PageRank and Beyond: The Science of Search Engine Rankings*. Princeton University Press, 2006.

[143] J. Lauri and R. Scapellato. *Topics in Graph Automorphism and Reconstruction.* Cambridge University Press, 2003.

[144] J. Lawton, R. Beard, and B. Young. A decentralized approach to formation maneuvers. *IEEE Transactions on Robotics and Automation,* 19 (6): 933–941, 2003.

[145] N. E. Leonard and E. Fiorelli. Virtual leaders, artificial potentials and coordinated control of groups. *Proceedings of the IEEE Conference on Decision Control,* December 2001.

[146] P. Liljeroth, J. Repp, and G. Meyer. Current-induced hydorgen tautomerization and conductance switching of naphthalocyanine molecules. *Science,* 317: 1203–1206.

[147] Z. Lin, M. Broucke, and B. Francis. Local control strategies for groups of mobile autonomous agents. *IEEE Transactions on Automatic Control,* 49 (4): 622–629, 2004.

[148] L. Lovász. *Combinatorial Problems and Exercises.* American Mathematical Society, 2007.

[149] D. Luenberger. *Optimization by Vector Space Methods.* Wiley, 1969.

[150] J. Mann, R. Connor, P. Tyack, and H. Whitehead. *Cetacean Societies.* University of Chicago Press, 2000.

[151] S. Martínez, J. Cortés, and F. Bullo. Motion coordination with distributed information. *IEEE Control Systems Magazine,* 27 (4): 75–88, 2007.

[152] J. M. McNew and E. Klavins. Locally interacting hybrid systems with embedded graph grammars. *Proceedings of the IEEE Conference on Decision and Control,* December 2006.

[153] J. M. McNew, E. Klavins, and M. Egerstedt. Solving coverage problems with embedded graph grammars. *Hybrid Systems: Computation and Control.* Springer, 2007.

[154] R. Merris. Laplacian matrices of graphs: a survey. *Linear Algebra and its Applications,* 197,198: 143–176, 1994.

[155] M. Mesbahi. On a dynamic extension of the theory of graphs. *Proceedings of the American Control Conference,* May 2002.

[156] M. Mesbahi. State-dependent graphs. *Proceedings of the IEEE Conference on Decision and Control*, December 2003.

[157] M. Mesbahi. State-dependent graphs and their controllability properties. *IEEE Transactions on Automatic Control*, 50 (3): 387–392, 2005.

[158] M. Mesbahi and F. Y. Hadaegh. Formation flying control of multiple spacecraft via graphs, matrix inequalities, and switching. *AIAA Journal of Guidance, Control, and Dynamics*, 24 (2): 369–377, 2001.

[159] C. D. Meyer. *Matrix Analysis and Applied Linear Algebra*. SIAM, 2001.

[160] R. E. Mirollo and S. H. Strogatz. Synchronization of pulse-coupled biological oscillators. *SIAM Journal on Applied Mathematics*, 50: 1645–1662, 1990.

[161] B. Mohar. Eigenvalues in combinatorial optimization. In *Combinatorial and Graph-Theoretical Problems in Linear Algebra*, R. A. Brualdi, S. Friedland, and V. Klee (editors). Springer-Verlag, 1993.

[162] L. Moreau. Stability of multiagent systems with time-dependent communication links. *IEEE Transactions on Automatic Control*, 50 (2): 169–182, 2005.

[163] A. Muhammad and M. Egerstedt. Connectivity graphs as models of local interactions. *Journal of Applied Mathematics and Computation*, 168 (1): 243-269, 2005.

[164] A. Muhammad and M. Egerstedt. Positivstellensatz certificates for non-feasibility of connectivity graphs in multi-agent coordination. *Proceedings of the IFAC World Congress*, July 2005.

[165] A. Muhammad and M. Egerstedt. On the structural complexity of multi-agent robot formations. *Proceedings of the American Control Conference*, June 2004.

[166] A. Muhammad and M. Egerstedt. Control using higher order Laplacians in network topologies. *Proceedings of the Mathematical Theory of Networks and Systems*, July, 2006.

[167] A. Muhammad and A. Jadbabaie. Dynamic coverage verification in mobile sensor networks via switched higher order Laplacians. In *Robotics: Science and Systems*, O. Broch (editor). MIT Press, 2007.

[168] A. Muhammad and A. Jadbabaie. Asymptotic stability of switched higher order Laplacians and dynamic coverage. In *Hybrid Systems: Computation and Control*, A. Bemporad, A. Bicchi, and G. Buttazzo (editors). Lecture Notes in Computer Science, Springer, 2007.

[169] J. Munkres. *Elements of Algebraic Topology*. Addison-Wesley, 1993.

[170] J. D. Murray. *Mathematical Biology* (vols. 1 and 2). Springer, 2008.

[171] M. Nabi, M. Mesbahi, N. Fathpour, and F. Y. Hadaegh. Local estimators for multiple spacecraft formation flying. *Proceedings of the AIAA Guidance, Navigation, and Control Conference*, August 2008.

[172] H. S. Nalwa. *Encyclopedia of Nanoscience and Nanotechnology*. American Scientific Publishers, 2004.

[173] A. Nedić, A. Ozdaglar, and P. A. Parrilo. Constrained consensus and optimization in multi-agent network. *IEEE Transactions on Automatic Control*, 55 (4): 922–938, 2010.

[174] M. E. J. Newman. Spread of epidemic disease on networks. Physical Review E, 66 (1): 1–11, 2002.

[175] M. Newman, A. Barabási, and D. J. Watts. *The Structure and Dynamics of Networks*. Princeton, University Press, 2006.

[176] N. Nisan, T. Roughgarden, É. Tardos, and V. Vazirani. *Algorithmic Game Theory*. Cambridge University Press, 2007.

[177] H. Q. Ngo and D.-Z. Du. Notes on the complexity of switching networks. In *Advances in Switching Networks*, H. Q. Ngo and D.-Z. Du (editors). Kluwer Academic, 307–357, 2000.

[178] M. A. Nowak. *Evolutionary Dynamics*. Harvard University Press, 2006.

[179] A. Nguyen and M. Mesbahi. A factorization lemma for the agreement dynamics. *Proceedings of the IEEE Conference on Decision and Control*, December 2007.

[180] P. Ögren, M. Egerstedt, and X. Hu. A control Lyapunov function approach to multi-agent coordination. *IEEE Transactions on Robotics and Automation*, 18 (5): 847–851, 2002.

[181] R. Olfati-Saber and R.M. Murray. Agreement problems in networks with directed graphs and switching topology, *Proceedings of the IEEE Conference on Decision and Control*, December 2003.

[182] R. Olfati-Saber and R. M. Murray. Consensus problems in networks of agents with switching topology and time-delays. *IEEE Transactions on Automatic Control*, 49 (9): 1520–1533, 2004.

[183] R. Olfati-Saber. Ultrafast consensus in small-world networks. *Proceedings of the American Control Conference*, June 2005.

[184] R. Olfati-Saber and J. S. Shamma. Consensus filters for sensor networks and distributed sensor fusion. *Proceedings of the IEEE Conference on Decision and Control*, December 2005.

[185] R. Olfati-Saber. Flocking for multi-agent dynamic systems: algorithms and theory. *IEEE Transactions on Automatic Control*, 51 (3): 401–420, 2006.

[186] R. Olfati-Saber and N. F. Sandell. Distributed tracking in sensor networks with limited sensing range. *Proceedings of the American Control Conference*, June 2008.

[187] A. Olshevsky and J. N. Tsitsiklis. Convergence rates in distributed consensus and averaging. *Proceedings of the IEEE Conference on Decision and Control,* December 2006.

[188] J. G. Oxley. *Matroid Theory*. Oxford University Press, 2006.

[189] D. A. Paley, N. E. Leonard, R. Sepulchre, D. Grünbaum, and J. K. Parrish. Oscillator models and collective motion: spatial patterns in the dynamics of engineered and biological networks. *IEEE Control System Magazine*, 27 (4): 89–105, 2007.

[190] P. A. Parrilo. Semidefinite programming relaxations for semialgebraic problems. *Mathematical Programming, Ser. B*, 96 (2): 293–320, 2003.

[191] M. Penrose. *Random Geometric Graphs*. Oxford University Press, 2003.

[192] S. V. Petrov. Graph grammars and graphodynamics problem. *Automation and Remote Control,* 10: 133–138, 1977.

[193] B. T. Polyak. *Introduction to Optimization*, Optimization Software, 1987.

[194] M. Porfiri and D. J. Stilwell. Consensus seeking over random weighted directed graphs. *IEEE Transactions on Automatic Control*, 52 (9): 1767–1773, 2007.

[195] V. V. Prasolov. *Elements of Combinatorial And Differential Topology*. American Mathematical Society, 2006.

[196] E. Prisner. *Graph Dynamics*. Longman House, 1995.

[197] K. Pryor and K. Norris. *Dolphin Societies*. University of California Press, 1998.

[198] R. Radner. Team decision problems. *Annals of Mathematical Statistics*, 33 (3): 857–881, 1962.

[199] A. Rahmani and M. Mesbahi. On the controlled agreement problem, *Proceedings of the American Control Conference*, June 2006.

[200] A. Rahmani, M. Ji, M. Mesbahi, and M. Egerstedt. Controllability of multi-agent systems from a graph-theoretic perspective. *SIAM Journal on Control and Optimization*, 48 (1): 162–186, 2009.

[201] B. S. Rao and H. F. Durrant-Whyte. Fully decentralized algorithms for multisensor Kalman filtering. *IEE Proceedings-D*, 138 (5): 413–420, 1991.

[202] W. Ren and R. Beard, Consensus of information under dynamically changing interaction topologies. *Proceedings of the American Control Conference*, June 2004.

[203] W. Ren and R. Beard. Consensus seeking in multiagent systems under dynamically changing interaction topologies. *IEEE Transactions on Automatic Control*, 50 (5): 655–661, 2005.

[204] W. Ren and R. Beard. *Distributed Consensus in Multi-vehicle Cooperative Control*. Springer, 2008.

[205] C. Reynolds. Flocks, herds and schools: a distributed behavioral model. *Proceedings of ACM SIGGRAPH Conference*, 1987.

[206] M. Rieth and W. Schommers. *Handbook of Theoretical and Computational Nanotechnology*. American Scientific, 2006.

[207] S. Rosenberg. *The Laplacian on a Riemannian Manifold*. London Mathematical Society Student Texts, 31. Cambridge University Press, 1997.

[208] B. Roth. Rigid and flexible frameworks. *The American Mathematical Monthly*, 88 (1): 6–21, 1981.

[209] G. Rozenberg (Editor). *Handbook of Graph Grammars and Computing by Graph Transformation*. World Scientific, 1997.

[210] S. Salsa. *Partial Differential Equations in Action: From Modelling to Theory*. Springer, 2010.

[211] J. Sandhu, M. Mesbahi, and T. Tsukamaki. Cuts and flows in relative sensing and control of spatially distributed systems. *Proceedings of the American Control Conference*, June 2005.

[212] J. Sandhu, M. Mesbahi, and T. Tsukamaki. Relative sensing networks: observability, estimation, and the control structure. *Proceedings of the IEEE Conference on Decision and Control*, December 2005.

[213] E. Seneta. *Non-negative Matrices and Markov Chains*. Springer, 2006.

[214] R. Sepulchre, D. A. Paley, and N. E. Leonard. Stabilization of planar collective motion with limited communication. *IEEE Transactions on Automatic Control*, 53 (3): 706–719, 2008.

[215] L. Shi, K. H. Johansson, and R. M. Murray. Estimation over wireless sensor networks: tradeoff between communication, computation, and estimation qualities. *Proceedings of the 17th IFAC World Congress*, July 2008.

[216] A. N. Shiryaev. *Probability*. Springer, 1995.

[217] B. Shucker, T. Murphey, and J. Bennett. A method of cooperative control using occasional non-local interactions. *Proceedings of the American Control Conference*, June 2006.

[218] E. Slijper. *Whales and Dolphins*. University of Michigan Press, 1976.

[219] J. J. E. Slotine and W. Wang. A study of synchronization and group cooperation using partial contraction theory. In *Cooperative Control: The 2003 Block Island Workshop on Cooperative Control*. V. Kumar, N. Leonard, and A. S. More (editors). Springer, 2004.

[220] B. Smith, M. Egerstedt, and A. Howard. Automatic generation of persistent formations for multi-agent networks under range constraints. *ACM/Springer Mobile Networks and Applications Journal (MONET)*, 14 (3): 322–335, June 2009.

[221] D. P. Spanos, R. Olfati-Saber, and R. M. Murray. Distributed Kalman filtering in sensor networks with quantifiable performance. *International Symposium on Information Processing in Sensor Networks*, 2005.

[222] D. Spielman. *Spectral Graph Theory and its Applications*. *http://www.cs.yale.edu/homes/spielman/eigs/*

[223] J. L. Speyer. Computation and transmission requirements for decentralized linear-quadratic-Gaussian control problem. *IEEE Transactions on Automatic Control*, 24 (2): 266–269, 1979.

[224] R. Stachnik, K. Ashlin, and K. Hamilton. Space-Station-SAMSI: A spacecraft array for Michelson spatial interferometry. *Bulletin of the American Astronomical Society*, 16: 818–827, 1984.

[225] S. S. Stanković, M. S. Stanković, and D. M. Stipanović. A consensus-based overlapping decentralized estimator in lossy networks: stability and denoising effects. *Proceedings of the American Control Conference*, June 2008.

[226] D. Swaroop and J. K. Hedrick. String stability of interconnected systems. *IEEE Transactions on Automatic Control*, 41 (33): 349–357, 1996.

[227] A. Tahbaz-Salehi and A. Jadbabaie. A necessary and sufficient condition for consensus over random networks. *IEEE Transactions on Automatic Control*, 53 (3): 791–795, 2008.

[228] H. G. Tanner, A. Jadbabaie, and G. J. Pappas. *Flocking in fixed and switching networks*. *IEEE Transactions on Automatic Control*, 52 (5): 863–868, 2007.

[229] H. G. Tanner. On the controllability of nearest neighbor interconnections. *Proceedings of the IEEE Conference on Decision and Control*, December 2004.

[230] H. G. Tanner, A. Jadbabaie, and G. Pappas. Stable flocking of mobile agents. Part II : Dynamic topology. *Proceedings of the IEEE Conference on Decision and Control*, December 2003.

[231] H. G. Tanner, G. J. Pappas and V. Kumar. Leader-to-formation stability. *IEEE Transactions on Robotics and Automation*, 20 (3): 433–455, 2004.

[232] É. Tardos and T. Wexler. Network formation games and the potential function method. In *Algorithmic Game Theory*. Cambridge University Press, 2007.

[233] R. E. Tarjan. *Data Structures and Network Algorithms*. SIAM, 1987.

[234] T. Tay and W. Whiteley. Generating isostatic frameworks, *Structural Topology*, 11: 21–69, 1985.

[235] O. N. Temkin, A. V. Zeigarnik, and D. G. Bonchev. *Chemical Reaction Networks: A Graph-Theoretical Approach*. CRC Press, 1996.

[236] W. T. Tutte. *Graph Theory*. Cambridge University Press, 2001.

[237] S. Utete and H. F. Durrant-Whyte. Routing for reliability in decentralized sensing networks. *Proceedings of the American Control Conference*, June 1994.

[238] T. Vicsek, A. Czirók, E. Ben-Jacob, I. Cohen, and O. Shochet. Novel type of phase transition in a system of self-driven particles. *Physical Review Letters*, 75 (6): 1226–1229, 1995.

[239] G. Walsh, H. Ye, and L. Bushnell. Stability analysis of networked control systems. *Proceedings of the American Control Conference*, June 1999.

[240] P. K. C. Wang and F. Y. Hadaegh. Coordination and control of multiple microspacecraft moving in formation. *Journal of the Astronautical Sciences*, 44: 315–355, 1996.

[241] S. Wasserman and K. Faust. *Social Networks Analysis: Methods and Applications*. Cambridge University Press, 2004.

[242] D. J. Watts and S. H. Strogatz. Collective dynamics of "small-world" networks, *Nature*, 393: 440–442, June 1998.

[243] D. B. West. *Introduction to Graph Theory*. Prentice-Hall, 2001.

[244] H. Whitney. Congruent graphs and the connectivity of graphs. *American Journal of Mathematics*, 54: 150–168, 1932.

[245] E. P. Wigner. On the distribution of the roots of certain symmetric matrices. *Annals of Mathematics*, 67: 325–327, 1958.

[246] A. S. Willsky, M. G. Bello, D. A. Castanon, B. C. Levy, and G. C. Verghese. Combing and updating of local estimates and regional maps along sets of one-dimensional tracks. *IEEE Transactions on Automatic Control*, 27 (4): 799–813, 1982.

[247] R. J. Wilson. *Introduction to Graph Theory*. Prentice-Hall, 1996.

[248] W. Woess. *Random Walks on Infinite Graphs and Groups*. Cambridge University Press, 2000.

[249] J. Wolfowitz. Products of indecomposable, aperiodic, stochastic matrices. *Proceedings of the American Mathematical Society*, 4: 733–737, 1963.

[250] C. W. Wu. Synchronization and convergence of linear dynamics in random directed networks. *IEEE Transactions on Automatic Control*, 51 (7): 1207–1210, 2006.

[251] C. W. Wu. *Synchronization in Complex Networks of Nonlinear Dynamical Systems*. World Scientific, 2007.

[252] L. Xiao and S. Boyd. Fast linear iterations for distributed averaging. *Systems and Control Letters*, 53: 65–78, 2004.

[253] L. Xiao, S. Boyd, and S. Lall. A space-time diffusion scheme for peer-to-peer least squares estimation. *Proceedings of the International Conference on Information Processing in Sensor Networks*, 2006.

[254] F. Xue and P. R. Kumar. The number of neighbors needed for connectivity of wireless networks. *Wireless Networks*, 10: 169–181, 2004.

[255] D. Zelazo, A. Rahmani, and M. Mesbahi. Agreement via the Edge Laplacian. *Proceedings of the IEEE Conference on Decision and Control*, December 2007.

Index

www.ingramcontent.com/pod-product-compliance
Ingram Content Group UK Ltd.
Pitfield, Milton Keynes, MK11 3LW, UK
UKHW030853020325
455718UK00004B/41